Introduction to Heat Transfer

WIT*PRESS*

WIT Press publishes leading books in Science and Technology.
Visit our website for the current list of titles.
www.witpress.com

WIT*eLibrary*

Home of the Transactions of the Wessex Institute, the WIT electronic-library
provides the international scientific community with immediate and permanent
access to individual papers presented at WIT conferences.
Visit the WIT eLibrary athttp://library.witpress.com

Introduction to Heat Transfer

Bengt Sundén
Lund University, Sweden

WITPRESS Southampton, Boston

Bengt Sundén
Lund University, Sweden

Published by

WIT Press

Ashurst Lodge, Ashurst, Southampton, SO40 7AA, UK
Tel: 44 (0) 238 029 3223; Fax: 44 (0) 238 029 2853
E-Mail: witpress@witpress.com
http://www.witpress.com

For USA, Canada and Mexico

WIT Press

25 Bridge Street, Billerica, MA 01821, USA
Tel: 978 667 5841; Fax: 978 667 7582
E-Mail: infousa@witpress.com
http://www.witpress.com

British Library Cataloguing-in-Publication Data

A Catalogue record for this book is available
from the British Library

ISBN (hardback): 978-1-84564-656-1
ISBN (paperback): 978-1-84564-660-8
eISBN: 978-1-84564-657-8

Library of Congress Catalog Card Number: 2011939411

*The texts of the papers in this volume were set
individually by the authors or under their supervision.*

Contents

Nomenclature

A area [m^2]

a thermal diffusivity [m^2/s], (eq. (4.5))

a_λ monochromatic absorption coefficient [–], (eq. (12.48))

Bi   Biot number [–], (eq. (4.5))

Bo   boiling number [–], (eq. (14.69))

b length, thickness [m]

C heat capacity flow rate [W/K], (eq. (15.4))

C_D drag coefficient [–], (eq. (11.1))

C_F shear stress coefficient [–], (eq. (7.30))

C_p pressure coefficient [–], Fig. 11.3

c specific heat [J/(kg K)]

c propagation velocity for electromagnetic wave motion [m/s]

c_p specific heat at constant pressure [J/(kg K)]

D diameter [m]

D_h hydraulic diameter [m], (eq. (8.8))

d_{ij} deviatric stress tensor [N/m^2], (eq. (6.12))

E energy [J]

$\dot{E}$ energy per time unit [W]

E emitted radiation energy, emissive power [W/m^2]

E voltage [V]

E_B emitted black body radiation [W/m^2], (eq. (12.10))

E_λ emitted monochromatic radiation emissve power [W/(m^2 m)]

e specific energy [J/kg]

e_{ij} rate of strain tensor [1/s], (eq. (6.15))

F force [N]

F correction factor [–]

F_{ij} angle factor, view or shape factor [–], (eq. (12.26))

Fo   Fourier number [–], (eq. (4.5))

Fr   Froude number [–], (eq. (14.56))

f Darcy friction factor [–], (eq. (8.6))

f_s vortex frequency [s^{-1}], (eq. (11.2))

G incident radiation [W/m^2], (eq. (12.5))

G average mass velocity [kg/(m^2 s)]

Gr   Grashof number [–], (eq. (10.4))

Gr^*	modified Grashof number [–], (eq. (10.51))	
g	gravity constant [m/s^2]	
H	enthalpy [J]	
$\dot{H}$	enthalpy per time unit [W]	
h	specific enthalpy [J/kg]	
h	Planck's constant [J s], (eq. (12.2))	
h_{fg}	latent heat [J/kg]	
I	radiation intensity [W/m^2 sr], (eq. (12.20))	
I	current [A]	
I	momentum [kg m/s]	
i, j, k	unit vectors	
J	radiosity [W/m^2], (eq. (12.5))	
Ja	Jakob number, (eq. (13.26))	
j	Colburn factor [–], $St \cdot Pr^{2/3}$	
K_c	contraction coefficient [–], (eq. (15.37))	
K_e	expansion coefficient [–], (eq. (15.37))	
L	length, thickness [m]	
L_i	entrance length [m], (eq. (8.4))	
$LMTD$	logarithmic mean temperature difference [K], [°C], (eq. (15.9))	
l_m	mixing length [m], (eq. (9.51))	
M	molecular weight [kg/kmol]	
m	mass [kg]	
$\dot{m}$	mass flow rate [kg/s]	
N	number of tube rows	
NTU	number of transfer units [–], (eq. (15.17))	
Nu	Nusselt number [–], (eq. (6.62))	
n	number of molecules per unit volume	
P	efficiency parameter [–], (eq. (15.12))	
P_R	reduced pressure [–]	
Pr	Prandtl number [–], (eq. (6.50))	
Pr_t	turbulent Prandtl number, [–], (eq. (9.33))	
p	pressure [Pa]	
p'	fluctuating pressure [Pa]	
$\bar{p}$	time averaged pressure [Pa]	
Q	heat [J]	
$\dot{Q}$	heat transfer rate [W]	
q	heat flux [W/m^2]	
R	radius [m]	
R	gas constant [J/(kg K)]	
R	heat capacity flow rate ratio [–], (eq. (15.13))	
R_P	surface roughness [μm]	
Re	Reynolds number [–], (eq. (6.60))	
r	radius [m]	
r_s	latent heat melting [J/kg]	
S_L	lateral tube pitch [m]	

S_T	longitudinal tube pitch [m]
Sr	Strouhal number [–], (eq. (11.2))
St	Stanton number [–], (eq. (7.42))
T	absolute temperature [K]
T^+	dimensionless temperature [–], (eq. (9.82))
TR	thermal resistance [K/W], (eq. (15.1))
t	temperature [°C]
t'	fluctuating temperature [°C]
$\bar{t}$	time averaged temperature [°C]
U	internal energy [J]
$\dot{U}$	internal energy per time unit (effekt) [W]
U	(mean) velocity [m/s]
U_∞	freestream velocity [m/s]
U	overall heat transfer coefficient [W/(m² K)], (eq. (15.1))
u, v, w	local velocity [m/s]
u', v', w'	fluctuating velocity [m/s]
$\bar{u}, \bar{v}, \bar{w}$	time averaged velocity [m/s], (eq. (9.4))
u^+	dimensionless velocity [–]
u_{fS}	fictious liquid velocity [m/s], (eq. (14.43))
u_{gS}	fictious gas velocity [m/s], (eq. (14.44))
u_m	mean velocity [m/s]
u_τ, u^*	friction velocity [m/s]
V	volume [m³]
X	coordinate [m]
X	Martinelli parameter [–], (eq. (14.52))
X_F	flowing mass fraction [–], (eq. (14.40))
X_S	static mass fraction [–], (eq. (14.41))
x, y, z	coordinates [m]
y^+	dimensionless coordinate perpendicular to a solid surface [–], (eq. (9.45))
W	work [J]
$\dot{W}$	work per time unit [W]
We	Weber number [–], (eq. (14.23))
Z	width, length [m]
α	heat transfer coefficient [W/(m² K)], (eq. (1.23))
α	absorptance [–], (eq. (12.3))
β	thermal expansion coefficient [1/K]
β	angle [rad]
γ	ratio of specific heats, c_p/c_v [–]
δ	angle [rad]
δ	boundary layer thickness [m]
δ_{ij}	Kronecker's delta [–], (eq. (6.12))
ε	emissivity [–], (eq. (12.11))
ε	void [–], (eq. (14.39))
ε	efficiency [–], (eq. (15.14))

ε_m	turbulent kinematic viscosity [m^2/s], (eq. (9.31))
ε_q	turbulent diffusivity [m^2/s], (eq. (9.32))
η	dimensionless coordinate [–]
η	fin effectiveness [–], (eq. (3.48))
θ	dimensionless temperature [–]
θ	angle [rad]
θ	thermal length [–], (eq. (15.43))
ϑ	temperature [°C]
κ	von Karmans constant [–], (eq. (9.51))
λ	parameter [–]
λ	thermal conductivity [W/(m K)], (eq. (1.1))
λ	wavelength [m]
λ_1	mean free path for molecular motion [m]
μ	dynamic viscosity [kg/(m s)]
ν	kinematic viscosity [m^2/s]
ρ	density [kg/m^3]
ρ	reflectance [–], (eq. (12.3))
ρ	resistivity [Ω m]
σ	Stefan-Boltzmann constant [W/(m^2 K^4)], (eq. (12.10))
σ	shear stress [N/m^2]
σ	surface tension [N/m]
σ	electric conductivity [1/(Ω m)]
σ	area ratio [–], (eq. (15.35))
τ	time [s]
τ	transmittance [–], (eq. (12.3))
ϕ	angle [rad]
φ	fin efficiency [–], (eq. (3.49))
Φ	radiation energy [W/m^2], (eq. (12.21))
ψ	stream function [s^{-1}], (eq. (10.21))
ω	solid angle [sr], (eq. (12.16))

Index

B	bulk, blackbody
C	convective boiling
f	at film temperature, fluid
g	gas
i	inner
KK	nucleate boiling
m	mean
o	outer
TF	two-phase
w	wall
λ	monochromatic
∞	freestream

Preface

This book aims as an introduction of heat transfer at undergraduate and graduate levels. Compared to other similar textbooks it differs significantly as it is much more comprehensive in describing the thermal conductivity of various substances, providing deeper analysis of fin heat transfer, and it includes buried pipes. For convective heat transfer the relation to fluid mechanics is much more highlighted. The turbulent convection, evaporation and heat exchanger chapters are all more comprehensive than other general heat transfer textbooks available. The textbook has been used for exchange students and PhD students for several years. The text has been developed and improved over the years.

At the end of the book a number of problems in heat transfer can be found. These can be solved by the methods presented in the various chapters.

Lund in September 2011

Bengt Sundén
Professor in Heat Transfer
Department of Energy Sciences
Lund University

Introduction

Heat is a form of energy which is always transferred from the hot part to the cold part within a substance or from a body at a high temperature to another body at a lower temperature. The bodies do not need to be in contact but a difference in temperature must exist.

In some cases, the amount of heat transferred can de determined simply by applying basic relations or laws of thermodynamics and fluid mechanics. In other cases, where the mechanisms of the heat transport are not completely known, methods of analogy or empirical methods based on experiments are applied.

Heat can be transferred by three different means, namely heat conduction, convection, and thermal radiation (see Fig. I-l).

Heat conduction is a process where the energy transfer from a region at a high temperature to a low temperature region is governed by the molecular motion as in solid bodies and fluids (gases and liquids) at rest, and by movement of electrons as for metals.

When a fluid is flowing along an exterior surface or inside a duct and as the temperatures of the fluid and the solid surface are different, the amount of heat being exchanged is affected by the macroscopic fluid motion. This type of heat transfer is called convection. Depending on how the macroscopic fluid movement is created, forced convection or free (natural) convection prevails. In some cases, both forced and free convection occur simultaneously. The process is then called mixed convection or combined forced and free convection.

Heat transfer by radiation does not require any medium to propagate. The heat transfer between two surfaces by radiation is in fact maximum when no media is

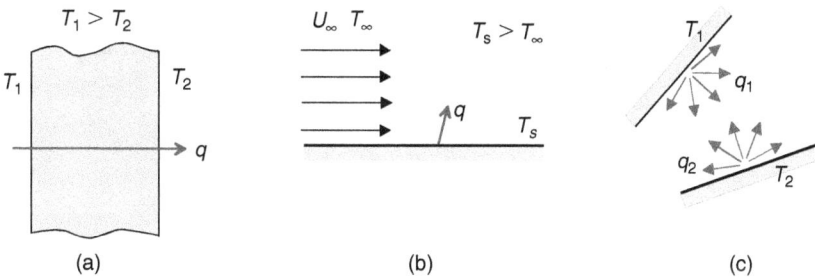

Figure I.1: Heat transfer by (a) heat conduction, (b) convection, and (c) thermal radiation.

present between the surfaces. Radiation may occur between surfaces, between a surface, and a participating medium like a gas. The heat exchange due to radiation is governed by electromagnetic waves according to Maxwell's theory or in the form of discrete photons according to Planck's hypothesis.

In many branches of engineering and technology, it is of great interest to be able to calculate temperature distributions and heat fluxes. In order to design, size, and rate heat exchangers, e.g., condensers, evaporators, radiators etc., analysis of the heat transfer is needed. Huge applications of this type of equipment appear frequently in heat and power generation, process industries, etc. Design and sizing of air conditioning equipment, electronics cooling, and insulation of buildings require knowledge in heat transfer. For vehicles, many heat transfer problems are present.

To enable stress and strain analysis in equipment exposed to high temperature, analysis of the temperature field and heat loads is needed.

In manufacturing, production, and treatment of materials, heat transfer is also important.

Cooling of electronics and other equipment carrying electrical currents is an important application area of heat transfer. Also, in combustion devices heat transfer is of significance by thermal radiation and convection.

Processing and treatment of food require analysis of heat and mass transfer.

1 Heat conduction

1.1 General theory

Heat conduction is a process where heat is diffusing through solid bodies or fluids at rest. The heat flux (W/m^2) is written as

$$q = -\lambda \frac{\partial t}{\partial n} \quad (\text{W/m}^2) \tag{1.1}$$

where λ is defined as the thermal conductivity (W/(m K)). $\partial t / \partial n$ is the temperature gradient in the direction of the surface normal vector. The negative sign means that heat is always transferred from a region at high temperature to another region at a lower temperature.

Consider now a body in which a certain temperature field exists. It is assumed that the material is isotropic, i.e., the thermal conductivity is independent of direction. Consider two isotherms as shown in Fig. 1.1.

Consider a surface element dA on the isotherm $t + dt$ and let n be the normal vector for this surface element. Because the material is isotropic and the isotherms are considered, the heat flux in the direction n in Fig. 1.1 is given by eq. (1.1), i.e.,

$$q_n = \frac{d\dot{Q}}{dA} = -\lambda \frac{\partial t}{\partial n} \tag{1.2}$$

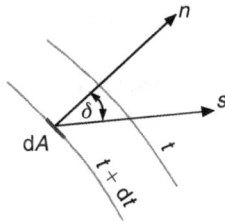

Figure 1.1: Temperature field in a body.

The objective is now to calculate the heat flux in an arbitrary direction s which is related to the n direction by the angle δ. One then has

$$q_s = \frac{d\dot{Q}}{dA} \cos \delta = -\lambda \frac{\partial t}{\partial n} \cos \delta \qquad (1.3)$$

However, the following relation also holds

$$\frac{\partial t}{\partial n} = \frac{\partial t}{\partial s} \cdot \frac{\partial s}{\partial n} \qquad (1.4)$$

and furthermore

$$n = s \cos \delta \qquad (1.5)$$

Equation (1.5) gives $ds/dn = 1/\cos \delta$ which means that eq. (1.3), by using eq. (1.4), can be written as

$$q_s = -\lambda \frac{\partial t}{\partial s} \qquad (1.6)$$

This means that the heat flux in an arbitrary direction is obtained by calculating the temperature gradient in that direction and then multiply it by $-\lambda$.

It is obvious that the heat flux is biggest in a direction perpendicular to the isotherms.

If a Cartesian coordinate system (x, y, z) is used, eq. (1.7) below is valid.

$$q_x = -\lambda \frac{\partial t}{\partial x}, \quad q_y = -\lambda \frac{\partial t}{\partial y}, \quad q_z = -\lambda \frac{\partial t}{\partial z} \qquad (1.7)$$

These fluxes are the components of the heat flux vector $\vec{q}$ according to

$$\vec{q} = \mathbf{i}\, q_x + \mathbf{j}\, q_y + \mathbf{k}\, q_z \qquad (1.8)$$

where $\mathbf{i}$, $\mathbf{j}$, and $\mathbf{k}$ are the unit vectors in the (x, y, z) coordinate system.

Note: Equation (1.1) is commonly referred to as Fourier's heat conduction law after the French mathematician Jean Baptiste Fourier who first formulated this relation in 1822.

1.2   Analogy with electric current

In the theory of electricity, the so-called Ohm's law reads

$$dI = -\sigma\, dA \frac{\partial E}{\partial n} \qquad (1.9)$$

where I is the current, E the electric potential (or voltage), and σ the electric conductivity ($\sigma = 1/\rho$, where ρ is the resistivity). By comparing eqs. (1.1) or (1.2) and (1.9), it is evident that the total heat flow $\dot{Q}$ corresponds to the current I, the temperature t corresponds to the potential E, and finally λ corresponds to σ. Due to this fact, it is possible to study heat conduction problems by models from the theory of electricity. This will be applied in Chapter 3.

1.3   Heat conduction equation for an isotropic material

Consider a body at rest and introduce a Cartesian coordinate system (x, y, z) as in Fig. 1.2. A fixed volume element $dx\,dy\,dz$ within the body is studied. Because the body is at rest, no mass will enter or leave the volume element. The volume element can therefore be treated as a system from a thermodynamic point of view and the first law is applicable, i.e., see Ref. [1]

$$\Delta\dot{E} = \dot{Q} - \dot{W} \tag{1.10}$$

where $\Delta\dot{E}$ is the change of energy within the system, $\dot{Q}$ the heat, and $\dot{W}$ the work.

Because the element is at rest and no forces are acting, one has $\dot{W} = 0$. Consider now the heat flow in the x-direction. Through the left surface one has

$$\dot{Q}_x = -\lambda\frac{\partial t}{\partial x}dy\,dz \tag{1.11}$$

and through the right surface one has

$$\dot{Q}_{x+dx} = -\lambda\frac{\partial t}{\partial x}dy\,dz + \frac{\partial}{\partial x}\left(-\lambda\frac{\partial t}{\partial x}dy\,dz\right)dx \tag{1.12}$$

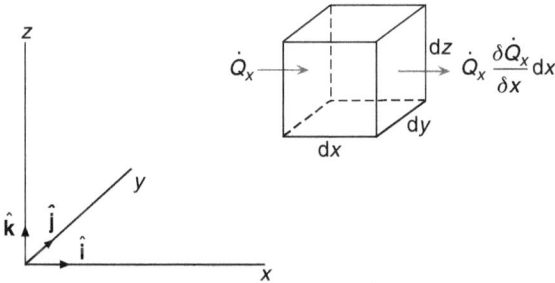

Figure 1.2: Volume element considered in the derivation of the heat conduction equation.

The net diffusing heat out of the volume element in the x-direction is

$$\Delta \dot{Q}_x = -\frac{\partial}{\partial x}\left(\lambda \frac{\partial t}{\partial x}\right) dx\, dy\, dz \tag{1.13}$$

In the y- and z-direction similar approaches give

$$\Delta \dot{Q}_y = -\frac{\partial}{\partial y}\left(\lambda \frac{\partial t}{\partial y}\right) dx\, dy\, dz \tag{1.14}$$

$$\Delta \dot{Q}_z = -\frac{\partial}{\partial z}\left(\lambda \frac{\partial t}{\partial z}\right) dx\, dy\, dz \tag{1.15}$$

Besides diffusion (heat conduction), heat can be generated within the body if, for example, an electric current is conducted through the body (Joule heating) or if some internal reactions occur within the volume element. Such internally generated heat is denoted by Q' and is calculated per unit volume, i.e., W/m^3. The total heat flow $\dot{Q}$ in eq. (1.10) can now be established. However, one has to remember that according to thermodynamics heat is positive if it is supplied to the system. Thus one has

$$\dot{Q} = -(\Delta\dot{Q}_x + \Delta\dot{Q}_y + \Delta\dot{Q}_z) + Q'\, dx\, dy\, dz$$

$$= \left\{\frac{\partial}{\partial x}\left(\lambda\frac{\partial t}{\partial x}\right) + \frac{\partial}{\partial y}\left(\lambda\frac{\partial t}{\partial y}\right) + \frac{\partial}{\partial z}\left(\lambda\frac{\partial t}{\partial z}\right) + Q'\right\} dx\, dy\, dz \tag{1.16}$$

If the change in potential energy is negligible, the energy change $\Delta\dot{E}$ is equal to the change in internal energy $\Delta\dot{U}$ only because the body is at rest. In heat conduction studies, the main focus is on solid bodies and incompressible fluids and then the change of internal energy $\Delta\dot{U}$ can be written as

$$\Delta\dot{U} = \rho c \frac{\partial t}{\partial \tau} dx\, dy\, dz \tag{1.17}$$

where $c =$ specific heat (J/(kg K)) and $\rho =$ density (kg/m^3).

By combining eq. (1.10) with eqs. (1.16) and (1.17), one obtains

$$\rho c \frac{\partial t}{\partial \tau} = \frac{\partial}{\partial x}\left(\lambda\frac{\partial t}{\partial x}\right) + \frac{\partial}{\partial y}\left(\lambda\frac{\partial t}{\partial y}\right) + \frac{\partial}{\partial z}\left(\lambda\frac{\partial t}{\partial z}\right) + Q' \tag{1.18}$$

Equation (1.18) is the general heat conduction equation for isotropic materials. In particular, if the material is homogeneous, i.e., the thermal conductivity is constant,

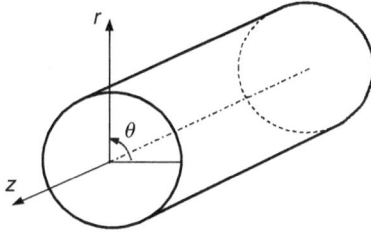

Figure 1.3: Cylindrical coordinate system.

eq. (1.18) can be written as

$$\frac{\partial t}{\partial \tau} = \frac{\lambda}{\rho c}\left\{\frac{\partial^2 t}{\partial x^2} + \frac{\partial^2 t}{\partial y^2} + \frac{\partial^2 t}{\partial z^2}\right\} + \frac{Q'}{\rho c} = \frac{\lambda}{\rho c}\nabla^2 t + \frac{Q'}{\rho c} \qquad (1.19)$$

where $\lambda/\rho c$ is called the thermal diffusivity and is denoted by a, i.e., $a = \lambda/\rho c$.

1.4   Heat conduction equation in cylindrical coordinates

By using coordinate transformation or by applying the first law of thermodynamics directly on a cylindrical volume element, the heat conduction equation in cylindrical coordinates can be derived. From Fig. 1.3, one has $x = r\cos\theta$, $y = r\sin\theta$ and $z = z$.

For a homogeneous material the result is

$$\frac{\partial t}{\partial \tau} = \frac{\lambda}{\rho c}\left\{\frac{\partial^2 t}{\partial r^2} + \frac{1}{r}\frac{\partial t}{\partial r} + \frac{1}{r^2}\frac{\partial^2 t}{\partial \theta^2} + \frac{\partial^2 t}{\partial z^2}\right\} + \frac{Q'}{\rho c} \qquad (1.20)$$

1.5   Heat conduction equation in spherical coordinates

The general heat conduction equation in spherical coordinates can be derived by considering an energy balance (first law of thermodynamics) for a volume element in spherical coordinates. Alternatively, eqs. (1.18) or (1.19) can be transformed to spherical coordinates. With the coordinates in Fig. 1.4, the following equation is valid for a homogeneous material

$$\frac{\partial t}{\partial \tau} = \frac{\lambda}{\rho c}\left\{\frac{1}{r}\frac{\partial^2 (rt)}{\partial r^2} + \frac{1}{r^2\sin\theta}\frac{\partial}{\partial \theta}\left(\sin\theta\frac{\partial t}{\partial \theta}\right) + \frac{1}{r^2\sin^2\theta}\frac{\partial^2 t}{\partial \phi^2}\right\} + \frac{Q'}{\rho c} \qquad (1.21)$$

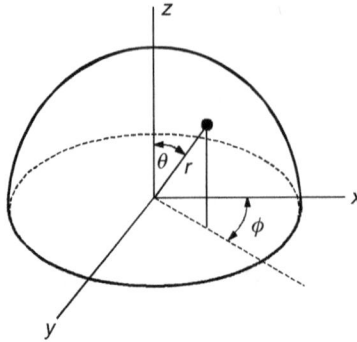

Figure 1.4: Spherical coordinate system.

1.6   Boundary conditions for heat conduction problems

The primary task in studies of heat conduction is to determine the temperature distribution and heat fluxes in the heat conducting body. This means that partial differential equations like (1.18)–(1.21) must be solved. To reach the solution boundary conditions are also needed.

The most simple boundary conditions are:

a) prescribed surface temperature,
b) prescribed heat flux at the surface.

More generally the body is cold or heated by convection (forced or free), and the boundary conditions become more complicated, see Fig. 1.5.

The exact conditions on the surface are (continuity in temperature and heat flux)

$$t_{\text{fluid}_{(x=0^-)}} = t_{\text{solid}_{(x=0^+)}}$$

$$\frac{\dot{Q}}{A} = -\lambda_f \left(\frac{\partial t}{\partial x}\right)_{x=0^-} = -\lambda_s \left(\frac{\partial t}{\partial x}\right)_{x=0^+} \tag{1.22}$$

For solution of such problems, solution of both the convective heat transfer equations (see Chapter 6) in the neighborhood of the surface and the heat conduction equation inside the body is needed. This is a difficult task in general. For convective heat transfer, usually a so-called heat transfer coefficient α is introduced according to

$$\frac{\dot{Q}}{A} = \alpha(t_f - t_w) \tag{1.23}$$

The heat transfer coefficient α is however known only for the cases of uniform wall temperature, $t_w = $ constant, uniform wall heat flux, $q_w = $ constant and for some distributions $t_w(y)$. As heat conduction problems with convective cooling or

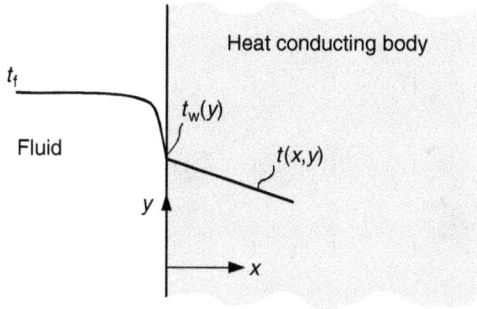

Figure 1.5: Convective heating of a heat conducting body.

heating are studied, the heat transfer coefficient α is assumed known even if this is not strictly correct. However, if a mean value over the surface is used, the accuracy will be sufficient or at least decent.

1.7   Anisotropic material

For some technically important materials, the thermal conductivity depends on the direction. Such materials are said to be anisotropic. Examples are wood, laminated plastics, laminated metal layers, and plywood. The above given heat conduction eqs. (1.18)–(1.21) are not valid. Further studies are advised to the books of Eckert and Drake [2] and Carslaw and Jaeger [3].

References

[1] Y.A. Cengel and M.A. Boles, Thermodynamics: An Engineering Approach, 7th ed., McGraw-Hill, New York (2011).
[2] E.R.G. Eckert and E.M. Drake Jr., Analysis of Heat and Mass Transfer, McGraw-Hill, Tokyo (1972).
[3] H.S. Carslaw and J.C. Jaeger, Conduction of heat in solids, 2nd ed., Oxford University Press, New York (1959).

2 Thermal conductivity

2.1 Introduction

In Chapter 1, the thermal conductivity was introduced in eq. (1.1) and it appeared in the partial differential eqs. (1.18)–(1.21) for heat conduction. To enable meaningful heat transfer calculations or experiments, it is important to know the thermal conductivity and how it depends on the material structure and composition, pressure and temperature. In Table 2.1 below, values of the thermal conductivity for some substances are given at 0°C.

More complete tables for the thermal conductivity can be found in, for example, Holman [1] and Eckert and Drake [2].

By comparing reported values of the thermal conductivity for a huge number of materials, general information about the physical nature of the thermal conductivity has been established.

1. Material in crystalline form, either metallic or nonmetallic, conducts heat better than material in an amorphous form.
2. In crystalline and other materials of oriented structure, e.g., fibrous material like wood, the thermal conductivity has different values relative to the structural axes of the material, such that there are principal axes for the thermal conductivity.
3. Small structural differences in the crystals related to their means of growth influence their thermal conductivity. For this reason, pure and natural crystals have higher thermal conductivities than synthetic varieties.
4. Chemical impurities in crystalline substances result in lower thermal conductivities compared with the pure state. Thus pure metals have much higher thermal conductivities than their respective alloys.
5. Mechanical damage, like cold–working and nuclear irradiation damage, causes changes in the thermal conductivity of a material.
6. Generally, metals are better heat conductors than nonmetals.
7. Solid phases in general have higher thermal conductivity than their respective liquid phases. For example, ice has a higher thermal conductivity than water; the solid phase of lead has a higher thermal conductivity than the liquid phase of lead. However, exceptions exist.
8. The liquid phases have higher thermal conductivities than the gaseous phases.

Table 2.1: Thermal conductivity for some substances at 0°C.

Substance	Thermal conductivity (W/m K)
Metallic solid substances	
Silver (pure)	410
Copper (pure)	385
Aluminum (pure)	202
Nickel (pure)	93
Iron (pure)	73
Carbon steel (1%)	43
Chrome–Nickel–Steel (18% Cr, 8% Ni)	16.3
Nonmetallic solid substances	
Quartz (parallel to axes)	41.6
Marble	1.08–2.94
Ice	2.22
Window glass	0.78
Glass wool	0.038
Liquids	
Mercury	8.21
Sulfur dioxide, SO_2	2.12
Methyl chloride, CH_3Cl	1.79
Water	0.556
Ammonia	0.540
Lubrication oil, SAE 50	0.147
Carbon dioxide, CO_2	0.105
Freon 12, CCl_2F_2	0.073
Gases	
Hydrogen, H_2	0.175
Helium, He	0.141
Air	0.024
Chloroform	0.0066

From Holman [1] and Eckert and Drake [2].

In the following sections, the heat conduction process will be explained. Also how this process is affected by the material structure in the solid, liquid, and gaseous phase, respectively, will be considered. The impact of defects and impurities will be discussed.

2.2   Gases

For gases the kinetic gas theory can be used to calculate the thermal conductivity.

The mechanism for heat conduction in a gas is relatively simple. The kinetic energy of a molecule is related to its temperature. Thus the molecules have a higher velocity at high temperature than at low temperature. The molecules are in a continuous random motion and collide and exchange energy and momentum with each other. If a molecule is moving from a high temperature region to a low temperature region it transports kinetic energy to the low temperature region and delivers this energy by collision with low-energy molecules. The faster the molecules are moving, the faster they transport energy. For this reason, the thermal conductivity will depend on the temperature.

A simplified analysis of the heat conduction process in a gas will now be presented.

Consider the gas flow parallel to a wall according to Fig. 2.1. It is assumed that a velocity gradient as well as a temperature gradient perpendicular to the wall exists.

Molecules are passing an arbitrary plane aa upward and downward. During the time interval $d\tau$ molecules with the velocity v' in the y-direction and situated at an approximative distance $v'd\tau$ will pass the aa.

Let the number of molecules per unit volume be n. The average value of the velocity v' can be expressed as a constant times the mean value of the molecular motion, i.e., $c_1\bar{v}$. All molecules at a distance $c_1\bar{v}\,d\tau$ will pass the plane aa. The number of such molecules is $nc_1\bar{v}\,d\tau$ (per unit area) or per time unit $c_1n\bar{v}$. At an arbitrary position y perpendicular to the wall, the molecules have a velocity $\bar{u}(y)$ in the x-direction. As molecules from the plane 1–1 in Fig. 2.1 are moving upward in the y-direction, they keep their velocity $\bar{u}(y_1)$ until they reach the plane 2–2 where they achieve the velocity $\bar{u}(y_2)$. The distance between the planes 1–1 and 2–2 is of the order of the magnitude of the mean free path of the molecular motion, say $c_1\lambda_1$ (λ_1 is the mean free path length). In a similar way, molecules (as many) are moving downward from plane 2–2 to plane 1–1. Every molecule exchanges the momentum

$$m(\bar{u}(y_2) - \bar{u}(y_1))$$

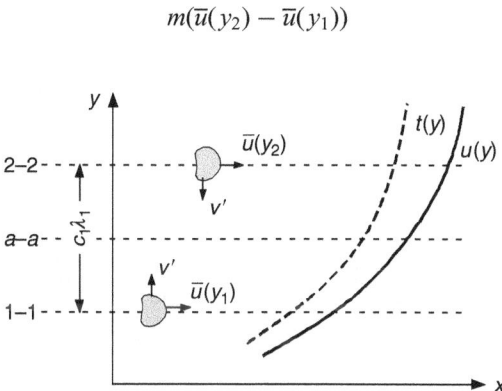

Figure 2.1:   Exchange of momentum between molecules.

and in total per unit area and unit time, the exchanged amount of momentum is

$$I = c_1 n \bar{v} m (\bar{u}(y_2) - \bar{u}(y_1)) \tag{2.1}$$

The velocity difference $\bar{u}(y_2) - \bar{u}(y_1)$ can be written as

$$\bar{u}(y_2) - \bar{u}(y_1) = c_2 \lambda_1 \frac{d\bar{u}}{dy} \tag{2.2}$$

Equation (2.1) can be written as

$$I = c_1 c_2 \, nm\lambda_1 \frac{d\bar{u}}{dy} \tag{2.3}$$

According to Newton's second law, this momentum change corresponds to a force, in this case a shear stress. The shear stress (force per unit area) is for a Newtonian fluid commonly written as

$$\sigma = \mu \frac{d\bar{u}}{dy} \tag{2.4}$$

Equations (2.3) and (2.4) result in

$$\mu = c_1 c_2 nm\bar{v}\lambda_1 \tag{2.5}$$

The product $nm = \rho$ is the density for the considered gas. If the molecules are spherical, Chapman and Enskog (see Ref. [3]) have shown that $c_1 c_2 = 0.499$. The kinetic gas theory has in addition shown that $nm\lambda_1 = \rho\lambda_1$ is independent of pressure and that the mean velocity of the molecular motion can be written as (see Ref. [3])

$$\bar{v} = \sqrt{\frac{8RT}{\pi}} \tag{2.6}$$

where R is the gas constant for the considered gas and T the temperature in degree Kelvin. As is evident from eq. (2.6), $\bar{v}$ is also independent of the pressure and then μ becomes independent of the pressure.

One now finds

$$\mu = 0.499 \, \rho\lambda_1 \sqrt{\frac{8RT}{\pi}} \tag{2.7}$$

If a temperature gradient dt/dy exists in the gas as indicated in Fig. 2.1, heat exchange takes place between various layers similar to the momentum exchange.

Let c_m be the specific heat of a single molecule. Then the heat flux can be written as

$$-q = c_1 c_2 n\bar{v} c_m \lambda_1 \frac{dt}{dy} \tag{2.8}$$

One observes that nc_m can be written as ρc_v, where c_v is the specific heat at constant volume.

In eq. (2.1), the thermal conductivity λ was defined. With the notations above, one finds

$$-q = \lambda \frac{dt}{dy}$$

By comparing this expression with eq. (2.8), one has

$$\lambda = c_1 c_2 \bar{v} c_v \rho \lambda_1 \qquad (2.9)$$

or with eq. (2.5)

$$\lambda = \mu c_v \qquad (2.10)$$

If c_v is constant, it is obvious that λ similar to μ is increasing with temperature as $\sqrt{T}$.

As calculated values of λ from eq. (2.10) are compared with experimental data, only the order of magnitude is correct. Chapman (see Ref. [3]) introduced improvements of the theory, which resulted in an expression of the thermal conductivity with better agreement with experimental data for single atom molecules. The following expression is valid:

$$\lambda = \frac{5}{2}\mu c_v \qquad (2.11)$$

For molecules having several atoms, Eucken [4] developed the following expression in which the number of degrees of freedom has been considered.

$$\lambda = \frac{1}{4}(9\gamma - 5)\mu c_v \qquad (2.12)$$

where $\gamma = c_p/c_v$.

In Fig. 2.2, the thermal conductivity versus temperature is shown for some gases.

2.3   Solid substances

The modern theory for heat conduction in solid substances presents a clear difference between metallic and nonmetallic (dielectric) materials. In metallic conductors or other solid substances conducting electricity, the heat is transported partly by structural waves (so-called phonons) and partly by free electrons. In nonmetallic or dielectric materials, the heat is only transported by phonons or structural waves.

2.3.1   Nonmetallic (dielectrical) solid substances

The thermal conductivity for nonmetallic solid substances is determined by the propagation of the structural waves (thermal vibrations) in the crystal structure.

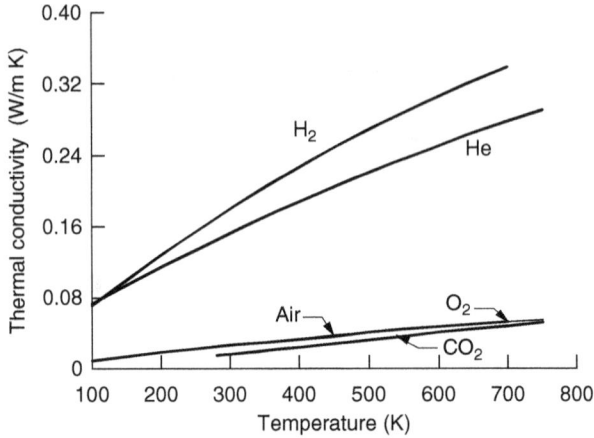

Figure 2.2: Thermal conductivity for some gases. (Based on Ref. [1].)

The waves or vibrations can be regarded as a superposition of elastic and acoustic waves. If a crystal has two surfaces at different temperature, heat will be transferred from the hot to the cold surface by acoustic radiation similar to the energy transmission by electromagnetic waves. The propagation of the structural waves or phonons is affected by a number of scattering processes. These are as follows:

a) Umklapp process
b) Elastic scattering
c) Inelastic scattering
d) Scattering due to impurities
e) Scattering due to mosaic structure.

The Umklapp process can be described in the following way. Consider two phonons that collide. A possible result is that a new wave is created with unchanged total energy and with a direction being determined solely by the colliding waves. Another possibility is that part of the energy is lost and that additional factors influence the direction of the new wave. This latter phenomenon is the so-called Umklapp process.

Elastic and inelastic scattering appears due to natural or induced defects in the material structure. For the elastic process, the energy of the phonons remains constant.

Chemical impurities introduce foreign atoms in the lattice structure of a material. Then the propagation of the phonons is disturbed and as a result the thermal conductivity is reduced. A mosaic structure is created as the crystal is divided into regions with slightly different directions. The non-regular lattice structure may prevent the phonon propagation and as an effect the thermal conductivity is reduced.

The processes described above may to a small or large extent affect the propagation of the thermal structural waves and then the thermal conductivity is reduced (or the thermal resistance is increased). Differences in or absence of any of the mentioned processes may explain the differences in thermal conductivity of nonmetallic substances.

2.3.2 Metallic substances

For the metallic substances, the thermal energy can be transported by structural waves and free electrons. In substances being good electrical conductors, a large number of free electrons are available. These electrons can transport thermal energy from a high temperature region to a low temperature one in a similar way as they can transport an electric charge. Often the free electrons are considered as a gas of electrons.

The thermal conductivity is written as

$$\lambda = \lambda_{fo} + \lambda_e \qquad (2.13)$$

where λ_{fo} is the contribution from the phonon motion and λ_e is the contribution from the free electrons.

In pure metals at room temperature, the contribution by the free electrons to the thermal conductivity is 30 times bigger than the contribution by the phonon motion. As is evident from Table 2.1, the metallic solid substances have one to two orders of magnitude higher thermal conductivity than nonmetallic solid substances.

In alloys on the other hand, the contributions from the phonon motion and the free electrons are of the same order of magnitude.

At room temperature and above, a relation exists between the thermal and electric conductivities for pure metals. This relation is called the Wiedemann–Franz–Lorentz' law and is a result of the so-called electron gas theory. As the contribution from the phonon motion is negligible the relation is written as

$$\lambda = L_0 \sigma T \qquad (2.14)$$

where
λ = thermal conductivity (W/(m K)),
σ = electrical conductivity (1/(m Ω))
T = temperature (K)
L_0 = Lorentz constant = 2.45×10^{-8} (W Ω/K^2)

(1853: G. Wiedemann and R. Franz assumed that λ/σ is constant for all metals. 1872: L. Lorentz correlated λ/σ with temperature and found eq. (2.14), see Ref. [5])
In Fig. 2.3 below the thermal conductivity is shown for some pure metals.

2.3.3 Alloys

The transport of thermal energy in alloys is governed by the phonon motion and the motion of free electrons. Because alloys have various structures and that the phonon

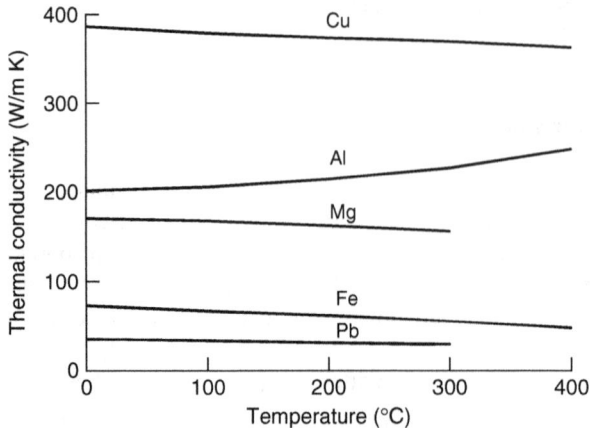

Figure 2.3: Thermal conductivity as function of temperature for some metals.

motion is strongly dependent on the material structure and the electron motion is affected by impurities, different types of alloys need to be addressed separately. In this chapter, four different types of alloys are considered.

a) Alloys of metals which are continuous solid solutions
b) Eutectic alloys
c) Alloys of metals which are limited solid solutions
d) Intermetallic compounds.

In Fig. 2.4 the variation of the thermal conductivity of alloys of type (a) with composition is shown. A and B represent two pure metals. For such alloys, the smallest addition of the alloying metal into the solvent metal acts as an impurity and causes scattering of phonons and electrons. This results in the very sharp reduction of the thermal conductivity. As the amount of the alloying metal is increased, the phonon transport is completely suppressed. Further addition of the alloying metal has a negligible effect on the alloy structure and the thermal conductivity becomes less dependent of the composition (see Fig. 2.4).

For eutectic alloys, which are soluble in the liquid phase but become mechanical mixtures in the solid phase, the change in thermal conductivity with composition will be as indicated in Fig. 2.5. The almost linear relationship indicates that the pure crystal structure of each component is preserved for any composition and heat is conducted independently by each type of crystal.

In alloys which are limited solid solutions, the thermal conductivity varies with composition according to the behavior of the solid phase. This means that λ decreases rapidly with addition of the alloying metal to the solvent metal until the limit of solubility is reached, and two different crystals are formed. Then the thermal conductivity varies approximately linearly as in the case of eutectic alloys. For certain compositions, which may be explained by the existence of an ordered structure

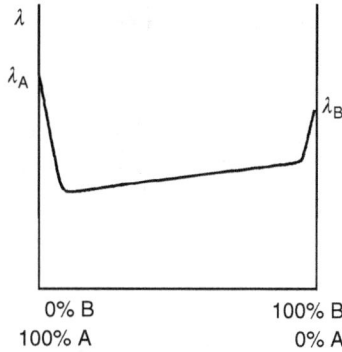

Figure 2.4: Variation of the thermal conductivity with composition for continuous solid solution alloys.

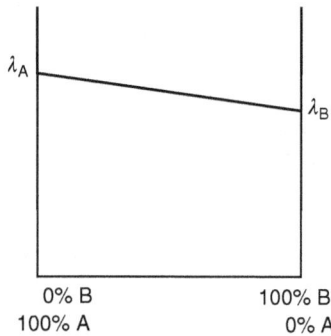

Figure 2.5: Variation of the thermal conductivity with composition for eutectic alloys.

of the alloy (super structure or super lattice), maxima in the thermal conductivity have been observed. Certain heat treatment might be the reason for this.

Some alloys form intermetallic compounds with the solvent metal. These compounds have a definite crystalline structure and offer less resistance to the motion of phonons and free electrons. As an effect, maxima in the thermal conductivity appear at certain compositions.

Table 2.2 shows the thermal conductivity for some alloys.

2.3.4   Phase transitions in solids

Solid substances do not always exist in the same structural form but may have a transition point above which there exists one structural form and below which there exists another. The transition is a result of the rearrangement of the atoms in the lattice, and thus different physical properties occur. In particular, the thermal conductivity is changed. For some substances, e.g., sulfur, the thermal conductivity

Table 2.2: Thermal conductivity of some alloys at 20°C.

Alloy	Thermal conductivity, λ (W/m K)	
Constantan (60% Cu, 40% Ni)	$\lambda = 22.7$	($\lambda_{Cu} = 386$, $\lambda_{Ni} = 90$)
Aluminum bronze (95% Cu, 5% Al)	$\lambda = 83$	($\lambda_{Cu} = 386$, $\lambda_{Al} = 204$)
Bronze (75% Cu, 25% Sn)	$\lambda = 26$	($\lambda_{Cu} = 386$, $\lambda_{Sn} = 64$)
Brass (70% Cu, 30% Zn)	$\lambda = 111$	($\lambda_{Cu} = 386$, $\lambda_{Zn} = 112$)

is changed abruptly at the phase transition. For other substances, the transition is less dramatic, e.g., the transition from α-iron to γ-iron occurs without a discontinuity in the thermal conductivity. However, the temperature gradient of λ, i.e., $d\lambda/dt$, is largely changed.

2.4   Liquids

The mechanisms for the transport of thermal energy in liquids are less explored than for gases and solids. When a solid substance is melting, a transition appears from an ordered structure to a less-ordered structure. The molecular bounds are attenuated and the possibility for free thermal movement increases.

The increased mobility in the liquid phase compared to the solid phase implies that liquids behave closer to gases than to solids. The essential difference between liquids and gases is that the molecules of liquids are more densely packed than for gases and thus the intermolecular forces should be stronger. However, close to the solidification/melting point the liquid structure would be closer to the solid phase.

Several trials have been attempted for to correlate the thermal conductivity to related physical properties which should be easier to measure. These attempts may be diversified as:

1. Relations between the thermal conductivity and the velocity of sound and
2. Relations between the thermal conductivity and other physical properties except the speed of sound.

In general, relations belonging to group (1) are considered to be most accurate.

Without any derivation, a formula for the thermal conductivity of liquids is given. It has been found that this formula agrees with experimental data within about 10%.

$$\lambda_{liquid} = \frac{7}{5}\sqrt{\frac{8}{\pi\gamma}}\left(\frac{\rho}{M}\right)^{2/3} R_M \frac{V_{liquid}}{N_0^{3/2}} \tag{2.15}$$

where
$$\gamma = c_p/c_v$$

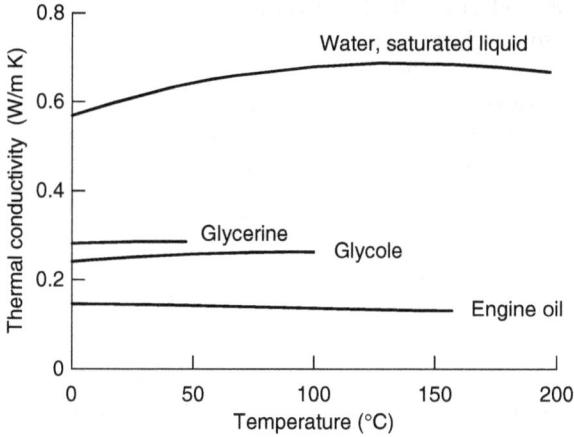

Figure 2.6: Thermal conductivity for some liquids.

$\rho =$ density (kg/m^3)
$M =$ molecular weight (kg/kmol)
$V_{\text{liquid}} =$ sound velocity in the liquid (m/s)
$R_M =$ universal gas constant (N m/kmol K)
$N_0 =$ number of molecules per mol.

In Fig. 2.6 the thermal conductivity for some liquids is presented.

2.4.1   Liquid metals

Liquid metals and electrolytes represent a special class of liquids because the transport of thermal energy is governed by two processes, namely the motion of atoms and free electrons. The contribution from the free electrons is the reason why liquid metals have much higher thermal conductivity than nonmetallic liquids.

2.5   Influence of pressure on the thermal conductivity

For metallic solid substances, it is reasonable to believe that an increase in pressure will distort the crystal structure and induce a resistance against the phonon movement and the motion of free electrons. As an effect, the thermal conductivity will be reduced. Experiments have shown that the thermal conductivity of iron, copper, silver, nickel, platinum, bismuth and antimony is linearly decreasing by increasing pressure, while for lead, tin, cadmium and zinc the thermal conductivity is increasing by increasing pressure.

For amorphous solid substances, an increased pressure would result in improved contact between the molecules and then the thermal conductivity is increased. Experiments have confirmed this behavior.

The thermal conductivity for liquids is affected by pressure in a similar way as for amorphous solid substances.

For gases it has been found that the thermal conductivity is only marginally affected by the pressure at moderate pressure level. At very low pressures, when the mean free path of the molecular motion is of the same order of magnitude as the macroscopic dimensions, the thermal conductivity is significantly affected by the pressure. At high pressures, both μ and c_v are increased by an increased pressure and thus this is true for the thermal conductivity as well.

References

[1] J.P. Holman, Heat Transfer, 10th ed., McGraw-Hill, Tokyo (2009).
[2] E.R.G. Eckert and R.M. Drake, Jr., Analysis of Heat and Mass Transfer, McGraw-Hill, Tokyo (1972).
[3] S. Chapman and T.G. Cowling, The Mathematical Theory of Non-uniform Gases, 2nd ed., Cambridge University Press, London (1958).
[4] A. Eucken, Phys. Z., 14, 32 ff (1913).
[5] S.T. Hsu, Engineering Heat Transfer, Van Nostrand, New York (1963).

Further reading

[1] L.S. Kowalczyk, Thermal conductivity and its variability with temperature and pressure, Trans. ASME, 77, 1021–1035 (1955).
[2] M. Jakob, Heat Transfer, vol. 1, John Wiley and Sons, New York (1949).

3 Steady heat conduction

3.1 Introduction

In this chapter, solutions to the general heat conduction equation, eqs. (1.18)–(1.21), will be studied. The thermal conductivity is assumed to be constant and independent of direction, i.e., the material is homogeneous and isotropic. The heat conduction is steady, which means the temperature is independent of time.

3.2 Heat conduction across a plane wall

Consider a plane wall with thickness b and thermal conductivity λ as shown in Fig. 3.1. The wall boundaries are kept at temperatures t_1 and t_2.

The heat conduction will be one-dimensional, i.e., heat is conducted only in the x-direction. If no internal heat is generated, eq. (1.19) is reduced to

$$\frac{d^2 t}{dx^2} = 0 \tag{3.1}$$

The solution of eq. (3.1) is

$$t = c_1 x + c_2 \tag{3.2}$$

The constants c_1 and c_2 are determined by considering the boundary conditions $x = 0$: $t = t_1$ and $x = b$: $t = t_2$. Application of these conditions gives the temperature

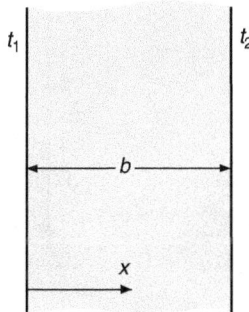

Figure 3.1: Steady heat conduction across a plane wall.

distribution in the wall as

$$t = t_1 + \frac{x}{b}(t_2 - t_1)$$ (3.3)

The heat transfer rate across the wall is given by

$$\dot{Q} = -\lambda A \frac{dt}{dx}$$

By using eq. (3.3), one obtains

$$\dot{Q} = \lambda A \frac{t_1 - t_2}{b}$$ (3.4)

Equation (3.4) can also be written as

$$t_1 - t_2 = \frac{b}{\lambda A}\dot{Q}$$ (3.5)

The analogy with Ohm's law in the theory of electricity is evident. $t_1 - t_2$ corresponds to the difference in electric potential, $b/\lambda A$ is a resistance, and $\dot{Q}$ corresponds to the electric current. Equation (3.5) is preferable to use, as a wall is composed of several layers of various thickness and thermal conductivity.

3.2.1   Multilayered plane wall

Consider a one-dimensional heat conduction across a wall having several layers with different thickness and thermal conductivity as shown in Fig. 3.2a.

The heat (rate $\dot{Q}$) is transferred through all the layers, which are connected in series. By using eq. (3.5), one has

$$t_1 - t_4 = \dot{Q}\left(\frac{b_1}{\lambda_1 A} + \frac{b_2}{\lambda_2 A} + \frac{b_3}{\lambda_3 A}\right)$$

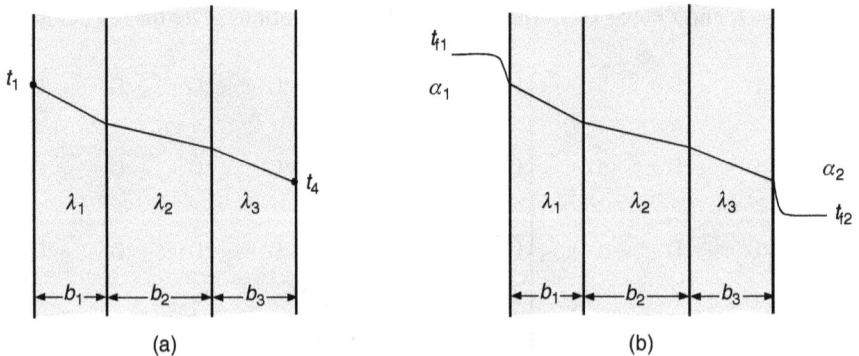

Figure 3.2: Steady heat transfer across a multilayered plane wall: (a) only heat conduction and (b) convective heat transfer and heat conduction.

or

$$\dot{Q} = \frac{t_1 - t_4}{(b_1/\lambda_1 A) + (b_2/\lambda_2 A) + (b_3/\lambda_3 A)} \tag{3.6}$$

In practice, it is common that the wall is heated and cooled by a fluid flow along the boundaries. Convective heat transfer then prevails as shown in Fig. 3.2b. From eq. (1.24), one imagines that the convective thermal resistance can be written as $1/\alpha A$. Equation (3.6) is then modified by introducing the convective resistances and one obtains

$$\dot{Q} = \frac{t_{f_1} - t_{f_2}}{(1/\alpha_1 A) + (b_1/\lambda_1 A) + (b_2/\lambda_2 A) + (b_3/\lambda_3 A) + (1/\alpha_2 A)} \tag{3.7}$$

3.3   Heat conduction across circular tubes and layers

As far as heat conduction across a circular and cylindrical tube or layer is considered, it is more appropriate to use cylindrical coordinates in the analysis, see Fig. 3.3. The heat conduction is assumed to be one-dimensional and without internal heat generation.

Equation (1.20) is reduced to

$$\frac{d^2 t}{dr^2} + \frac{1}{r}\frac{dt}{dr} = 0 \tag{3.8}$$

The solution to eq. (3.8) is given by

$$t = c_1 \ln r + c_2$$

where c_1 and c_2 are determined by the conditions $r = r_i: t = t_i$ and $r = r_o: t = t_o$.

The temperature distribution can be written as

$$\frac{t - t_o}{t_i - t_o} = \frac{\ln(r/r_o)}{\ln(r_i/r_o)} \tag{3.9}$$

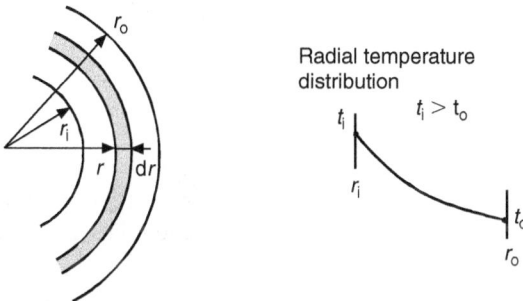

Figure 3.3: Steady heat conduction in a circular tube or layer.

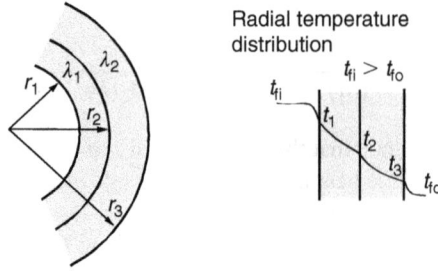

Figure 3.4: Heat transfer across a circular tube composed of different materials.

The heat transfer rate $\dot{Q}$ can be determined by

$$\dot{Q} = -\lambda A \frac{dt}{dr} = -\lambda 2\pi r L \frac{dt}{dr}$$

where L is the axial length of the tube or the layer.

With eq. (3.9), one has

$$\dot{Q} = \frac{t_i - t_o}{(1/2\pi \lambda L)\ln(r_o/r_i)} \tag{3.10}$$

From eq. (3.10), it is evident that the thermal resistance is

$$\frac{1}{2\pi \lambda L} \ln \frac{r_o}{r_i} \tag{3.11}$$

3.3.1   Circular tube or layer composed of different materials

The heat transfer across a circular tube composed of several layers of different thermal conductivity and convective heating and cooling on the inner and outer surfaces, respectively (see Fig. 3.4), can simply be determined as

$$\dot{Q} = \frac{t_{f_i} - t_{f_o}}{(1/2\pi r_1 L\alpha_i) + (1/2\pi \lambda_1 L)\ln(r_2/r_1) + (1/2\pi \lambda_2 L)\ln(r_3/r_2) + (1/2\pi r_3 L\alpha_o)} \tag{3.12}$$

3.4   Heat conduction in a spherical layer

Consider now the one-dimensional heat conduction through a spherical layer. If internal heat is not generated, eq. (1.21) gives

$$\frac{1}{r} \frac{\partial^2 (rt)}{\partial r^2} = 0 \tag{3.13}$$

The solution to eq. (3.13) is

$$t = c_1 + \frac{c_2}{r}$$

where c_1 and c_2 are determined by the boundary conditions $r = r_i$: $t = t_i$ and $r = r_o$: $t = t_o$. The temperature distribution is given by

$$\frac{t - t_o}{t_i - t_o} = \frac{(1/r) - (1/r_o)}{(1/r_i) - (1/r_o)} \tag{3.14}$$

The heat transfer rate is found by applying

$$\dot{Q} = -\lambda A \frac{dt}{dr} = -\lambda 4\pi r^2 \frac{dt}{dr}$$

Equation (3.14) gives

$$\dot{Q} = \frac{t_i - t_o}{1/4\pi\lambda((1/r_i) - (1/r_o))} \tag{3.15}$$

The heat transfer rate for a sphere composed of several spherical layers can be determined, which is similar to eqs. (3.7) and (3.12).

3.5   Critical insulation thickness

In this section it will be shown that under certain conditions insulation of a tube can increase the heat transfer rate or the heat loss from the tube.

Consider a circular tube with insulation as shown in Fig. 3.5.

The fluid inside the tube has the temperature t_i. The convective heat transfer coefficient in the inner surface, α_i, is assumed to be so high that the tube wall achieves the temperature t_i. The thermal resistance in the tube wall is assumed negligible. At the outer surface of the insulation, the convective heat transfer coefficient is α_o. The surrounding fluid has the temperature t_f. Equation (3.12) gives

$$\dot{Q} = \frac{t_i - t_f}{(1/2\pi\lambda L)\ln(r_o/r_i) + (1/2\pi r_o L\alpha_o)}$$

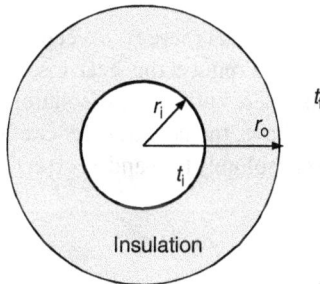

Figure 3.5: Circular tube with insulation.

or

$$\dot{Q} = \frac{2\pi L(t_i - t_f)}{(1/\lambda)\ln(r_o/r_i) + 1/(r_o\alpha_o)} \tag{3.16}$$

As the denominator in eq. (3.16) is minimum, the heat transfer rate will be maximum. If one assumes α_o, λ, and r_i to be constant and investigates what happens as r_o is increased, i.e., the insulation thickness is increased. The denominator in eq. (3.16) is denoted by N.

$$N = \frac{1}{\lambda}\ln\frac{r_o}{r_i} + \frac{1}{\alpha_o r_o}$$

One now obtains

$$\frac{dN}{dr_o} = \frac{1}{\lambda r_o} - \frac{1}{\alpha_o r_o^2}$$

The condition for minimum, $dN/dr_o = 0$, gives

$$r_{o,crit} = \frac{\lambda}{\alpha_o} \tag{3.17}$$

(The second-order derivative shows that eq. (3.17) gives a true minimum of N.)

How should eq. (3.17) be interpreted? In eq. (3.16), the term $(1/\lambda)\ln(r_o/r_i)$ is the thermal resistance due to the insulation material. This is increased by increasing r_o. The term $1/(\alpha_o r_o)$ in eq. (3.16) is the thermal resistance due to convective cooling. This is decreasing by increasing r_o. At $r_o = r_{o,crit}$, the increase in the thermal resistance due to the insulation is equal to the decrease in the convective thermal resistance and then the total resistance is minimum. The conclusion is that tubes having an outer radius (here r_i) less than the critical one, $r_{o,crit}$, achieve an increased heat loss as insulation is supplied. As the insulation thickness exceeds $r_{o,crit}$, the heat loss is decreased, see Fig. 3.6.

In Fig. 3.6, $\dot{Q}_0$ is a dimensionless heat transfer rate:

$$\dot{Q}_0 = \frac{\dot{Q}}{2\pi L(t_i - t_f)\alpha_o r_i} = \frac{1}{(r_i\alpha_o/\lambda)\ln(r_o/r_i) + 1/(r_o/r_i)}$$

For cases where the tube outer radius (here r_i) exceeds $r_{o,crit}$, insulation will always have the desired effect, i.e., to reduce the heat loss. A practical application is isolation of electric conduits where both electric isolation and maximum cooling are requested. In the derivation above, the heat transfer coefficient α_o was independent of r_o and $(t_i - t_f)$. This is not absolutely true and a correction will now be introduced. For forced convection, one has

$$\alpha_o \sim r_o^{-m} \tag{3.18}$$

and for natural convection

$$\alpha_o \sim r_o^{-m}(t_i - t_f)^n \tag{3.19}$$

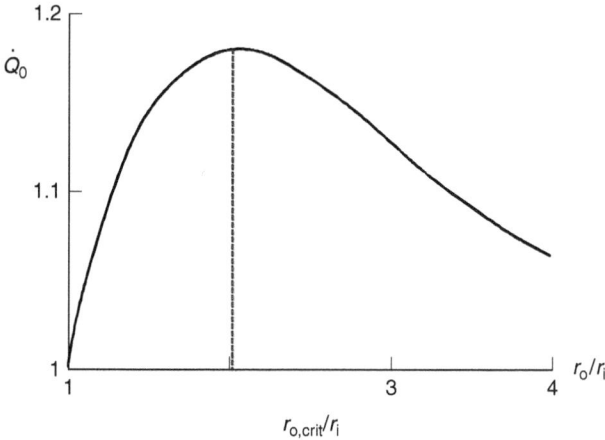

Figure 3.6: Illustration of critical insulation.

(Equations (3.18) and (3.19) will be evident in later chapters.)

For a circular tube as in Fig. 3.5, the following expression for the critical radius r_o is obtained, see Sparrow [1].

$$r_{o,crit} = \frac{\lambda}{\alpha_o} \frac{1 - m}{1 + n} \qquad (3.20)$$

For a circular cylinder in forced convection, one has $m = 0.382$, $n = 0$ for Reynolds numbers in the interval 4000–40,000, while for natural convection $m = n = 0.25$ (if $GrPr < 10^9$, see Chapter 10).

3.6   Plane wall with heat sources (internal heat generation)

In the preceding sections, no internal heat generation occurred. In electric machines, other electric and electronic equipment so-called Joule heating occurs as a current is present. The generated heat is given by the square of the electric current times the resistance. This heat can cause severe temperature increases in the material and generally cooling is needed. Other cases where heat sources are present can be found in nuclear applications and where chemical reactions occur.

To illustrate the phenomenon of heat sources, a simple case as shown in Fig. 3.7 is considered. The wall is cooled by a fluid on both sides and the fluid temperature at far distances is t_f. The heat transfer coefficient between the wall and fluid is α (W/m^2 K). The heat source is uniformly distributed and its strength is Q' (W/m^3). Only one-dimensional and steady heat conduction is considered.

If one introduces $\vartheta = t - t_f$, eq. (1.19) can be written as

$$\frac{d^2 \vartheta}{dx^2} + \frac{Q'}{\lambda} = 0 \qquad (3.21)$$

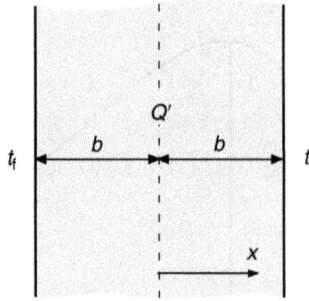

Figure 3.7: Plane wall with heat sources.

The boundary conditions of eq. (3.21) are

$$x = 0: \frac{d\vartheta}{dx} = 0 \qquad (3.22)$$

$$x = b: -\lambda \frac{dt}{dx} = \alpha(t - t_f) \qquad (3.23)$$

or

$$-\lambda \frac{d\vartheta}{dx} = \alpha\vartheta$$

The solution of eq. (3.21) is

$$\vartheta = -\frac{Q'}{2\lambda}x^2 + C_1 x + C_2$$

The constants C_1 and C_2 are determined by the conditions (3.22) and (3.23). The solution becomes

$$\vartheta = \frac{Q'}{2\lambda}(b^2 - x^2) + \frac{Q'b}{\alpha} \qquad (3.24)$$

The maximum temperature is reached at $x = 0$.

$$t_{max} = t_f + \frac{Q'}{2\lambda}b^2 + \frac{Q'b}{\alpha} \qquad (3.25)$$

3.7   Circular rod (or wire) with internal heat generation

Consider a circular rod or wire as shown in Fig. 3.8. Through the rod an electric current is flowing. Heat is then generated and the resulting temperature distribution in the rod is of interest. Equation (1.20) gives

$$\frac{d^2t}{dr^2} + \frac{1}{r}\frac{dt}{dr} = -\frac{Q'}{\lambda} \qquad (3.26)$$

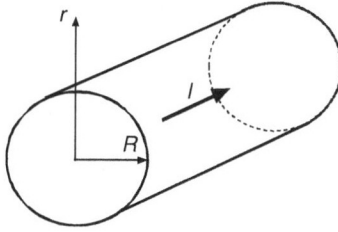

Figure 3.8: Circular rod (wire).

The solution of eq. (3.26) is

$$t = -\frac{Q'}{\lambda}\frac{r^2}{4} + C_1 \ln r + C_2 \qquad (3.27)$$

The boundary conditions are

$$r = 0: \frac{dt}{dr} = 0 \qquad (3.28)$$

$$r = R: -\lambda\frac{dt}{dr} = \alpha(t - t_f) \qquad (3.29)$$

From eqs. (3.27)–(3.29), the constants C_1 and C_2 are found and the temperature distribution can be written as

$$t - t_f = \frac{Q'}{4\lambda}(R^2 - r^2) + \frac{Q'R}{2\alpha} \qquad (3.30)$$

3.8   Finned heat transfer surfaces

A method to increase the heat transfer rate between a solid surface and a fluid is to equip the surface with the so-called fins (extended surface elements). These will increase the heat transferring area between the solid and fluid. Finned surfaces are commonly used in many applications, e.g., heat and power equipment, air-conditioning units, electric motors and transformers, and cooling of electronics. The fin geometry can be quite different. An arbitrary form is shown in Fig. 3.9.

Inside the fin the heat transfer is due to conduction, while between the fin surface and the fluid convection takes place.

3.8.1   Rectangular fins

Consider a rectangular fin as shown in Fig. 3.10.

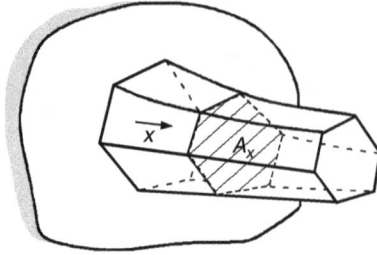

Figure 3.9: Arbitrary fin.

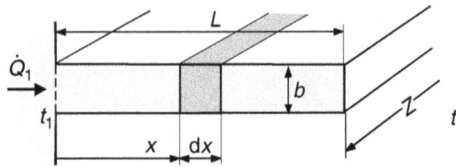

Figure 3.10: Rectangular cooling fin.

The heat conduction in the fin is assumed to be one-dimensional (x-direction) and the temperature at the base ($x = 0$) is t_1. The fluid temperature is t_f. The convective heat transfer coefficient, α, between the fin surface and the fluid is assumed to be known and uniform. Consider now a heat balance for the element dx in Fig. 3.10.

Through the left surface, the amount of heat conducted into the element is

$$\dot{Q}_x = -\lambda A \frac{dt}{dx}$$

Through the right surface, the amount of heat conducted out from the element is

$$\dot{Q}_{x+dx} = -\lambda A \frac{dt}{dx} + \frac{d}{dx}\left(-\lambda A \frac{dt}{dx}\right) dx$$

The heat exchange due to convection is

$$\dot{Q}_{conv} = \alpha C \, dx(t - t_f)$$

where C is the convective perimeter for the element dx. At steady state, the supplied heat must be equal to the delivered heat. Thus one has

$$\lambda A \frac{d^2 t}{dx^2} dx - \alpha C \, dx(t - t_f) = 0$$

or if $\vartheta = t - t_f$ is introduced

$$\frac{d^2 \vartheta}{dx^2} - \frac{\alpha C}{\lambda A}\vartheta = 0 \tag{3.31}$$

The boundary conditions of eq. (3.31) are

$$x = 0: t = t_1 \quad \text{and} \quad \vartheta = \vartheta_1 = t_1 - t_f \tag{3.32}$$

$$x = L: -\lambda \frac{dt}{dx} = \alpha'(t - t_f) \tag{3.33}$$

or

$$-\lambda \frac{d\vartheta}{dx} = \alpha'\vartheta$$

In eq. (3.33), α' is the heat transfer coefficient at the rear edge of the fin (fin tip). The solution of eq. (3.31) is (see, for example, Ref. [2])

$$\vartheta = C_1 e^{mx} + C_2 e^{-mx} \tag{3.34}$$

where

$$m = \sqrt{\frac{\alpha C}{\lambda A}} \tag{3.35}$$

If the fin is thin and very long, the heat transferred at its rear edge (fin tip) is negligible and the boundary condition (3.33) is changed to

$$\left(\frac{d\vartheta}{dx}\right)_{x=L} = 0 \tag{3.36}$$

Equations (3.32) and (3.36) determine the constants C_1 and C_2 according to

$$\left.\begin{array}{l} \vartheta_1 = C_1 + C_2 \\ 0 = C_1 m e^{mL} - C_2 m e^{-mL} \end{array}\right\} \tag{3.37}$$

From eqs. (3.37) and (3.34), one finds

$$\frac{\vartheta}{\vartheta_1} = \frac{e^{m(L-x)} + e^{-m(L-x)}}{e^{mL} + e^{-mL}} = \frac{\cosh m(L-x)}{\cosh mL} \tag{3.38}$$

Especially at $x = L$, one has

$$\vartheta_2 = \frac{\vartheta_1}{\cosh mL} \tag{3.39}$$

From an engineering point of view, it is of great interest to know the amount of heat being transferred from the fin. All this heat must have entered the fin at its base by heat conduction. Thus the heat transfer rate is calculated as

$$\dot{Q}_1 = -\lambda A \left(\frac{dt}{dx}\right)_{x=0} = -\lambda A \left\{-m\vartheta_1 \frac{\sinh m(L-x)}{\cosh mL}\right\}_{x=0}$$

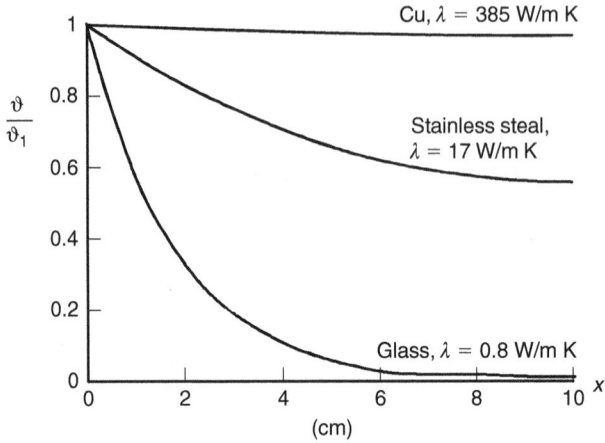

Figure 3.11: Temperature distribution in a rectangular fin, eq. (3.38) ($b = 2$ cm, $L = 10$ cm and $\alpha = 25$ W/m^2 K).

which can be written as

$$\dot{Q}_1 = m\lambda A \vartheta_1 \tanh mL = \sqrt{\alpha C \lambda A} \vartheta_1 \tanh mL \qquad (3.40)$$

In Fig. 3.11, the temperature distribution for a rectangular fin, eq. (3.38), is shown. The influence of the fin thermal conductivity is obvious.

If the fin cannot be assumed to be thin and long, the more accurate boundary condition (3.33) must be used as the constants C_1 and C_2 in eq. (3.34) are determined. Here only the final results are given.

$$\frac{\vartheta}{\vartheta_1} = \frac{\cosh m(L-x) + (\alpha'/m\lambda)\sinh m(L-x)}{\cosh mL + (\alpha'/m\lambda)\sinh mL} \qquad (3.41)$$

$$\frac{\vartheta_2}{\vartheta_1} = \frac{1}{\cosh mL + (\alpha'/m\lambda)\sinh mL} \qquad (3.42)$$

$$\dot{Q}_1 = m\lambda A \vartheta_1 \frac{(\alpha'/m\lambda) + \tanh mL}{1 + (\alpha'/m\lambda)\tanh mL} \qquad (3.43)$$

The presented theory is also applicable for the so-called thermometer pockets as illustrated in Fig. 3.12. In the pocket, a thermocouple or a thermometer is placed. The purpose is that the sensor should detect the fluid temperature. If the fluid temperature is higher and also much different from the temperature of the surroundings, the tube wall will have a lower temperature than the fluid and thus heat is conducted through the pocket material to the tube wall. The sensor records

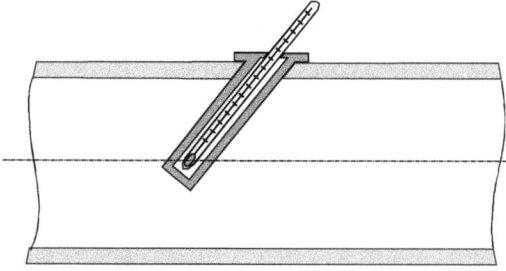

Figure 3.12: Thermometer and thermometer pocket for measuring the fluid temperature in a tube flow.

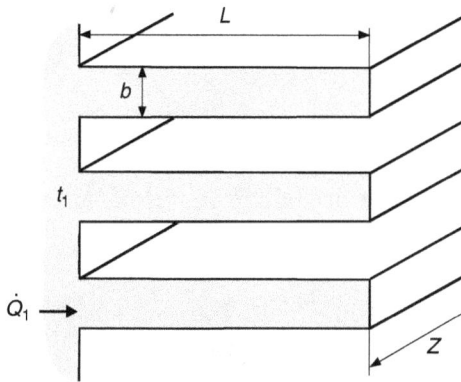

Figure 3.13: Arrangement of rectangular fins.

the temperature at the bottom of the pocket (fin tip). Equation (3.39) or (3.42) determines this temperature. It is then possible to judge the error in the measured fluid temperature or determine the proper length of the pocket to maintain a certain accuracy.

3.8.2 Conditions for fins to be beneficial

In Fig. 3.13, an arrangement of rectangular fins is depicted. Of primary interest is to know at what conditions it is useful or beneficial to apply fins on a surface.

First one needs to find out if fins at all will increase the heat transfer rate. If the heat transferred increases by increasing the fin length, it is obvious that fins are beneficial. The condition to be satisfied can be formulated as

$$\frac{d\dot{Q}_1}{dL} > 0 \qquad\qquad (3.44)$$

If eq. (3.44) is applied on eq. (3.43), one obtains

$$\frac{d\dot{Q}_1}{dL} = \frac{m\lambda A\vartheta_1}{\left(1 + \dfrac{\alpha'}{m\lambda}\tanh mL\right)^2}$$

$$\times \left\{ m\frac{1}{\cosh^2 mL}\left(1 + \frac{\alpha'}{m\lambda}\tanh mL\right) \right.$$

$$\left. - \frac{m\alpha'}{m\lambda}\frac{1}{\cosh^2 mL}\left(\frac{\alpha'}{m\lambda} + \tanh mL\right) \right\} > 0$$

$$\Rightarrow 1 - \left(\frac{\alpha'}{m\lambda}\right)^2 > 0 \tag{3.45}$$

According to eq. (3.35), m is defined as

$$m = \sqrt{\frac{\alpha C}{\lambda A}}$$

By considering Fig. 3.13, one finds $C = 2(Z + b) \approx 2Z$ and $A = bZ$, and then m can be written as

$$m = \sqrt{\frac{2\alpha}{\lambda b}}$$

If the assumption $\alpha' = \alpha$ is applied, eq. (3.45) gives

$$1 - \frac{\alpha^2}{(2\alpha/\lambda b)\lambda^2} > 0$$

or

$$\frac{2\lambda}{\alpha b} > 1 \tag{3.46}$$

In practice, it is not always true that the heat conduction is one-dimensional within the fin. To consider this, commonly a rule of thumb is introduced and eq. (3.46) is instead written as

$$\frac{2\lambda}{\alpha b} > 5 \tag{3.47}$$

3.8.3   Fin performance

To assess the performance of extended surfaces or fins, some measure of the fin thermal performance is needed. Basically two such measures are introduced. The first one is called fin effectiveness and is defined as the ratio of the fin heat transfer rate to the heat transfer rate that would exist without the fin. This effectiveness is denoted by η. The second one is called fin efficiency and is here denoted by φ.

It is defined as the ratio of the fin heat transfer rate to the heat transfer rate of a similar fin but with infinite thermal conductivity. A fin with infinite thermal conductivity will have a uniform temperature equal to the base temperature.

The definitions for η and φ can be expressed as

$$\eta = \frac{\text{heat transfer rate of the fin}}{\text{heat transfer rate from the base without the fin}} \qquad (3.48)$$

$$\varphi = \frac{\text{heat transfer rate of the fin}}{\text{heat transfer rate of a similar fin with } \lambda = \infty} \qquad (3.49)$$

From these definitions it follows that $\eta > 1$ and $\varphi < 1$.

3.8.4 Optimal rectangular fin concerning maximum heat transfer at fixed fin mass (weight)

In the engineering design of finned heat transfer surfaces, it is desirable to use fins being optimized in one way or the other. In this section it will be shown how the fin length should be selected in relation to the fin thickness to maximize the heat transfer rate at a given fin mass (weight).

Consider a rectangular fin as shown in Fig. 3.14.

The mass of the fin can be written as

$$M = \rho b L Z = \rho A_1 Z \qquad (3.50)$$

where A_1 is the fin area in a plane perpendicular to the Z-direction. The length Z is assumed to be fixed and the density ρ of the fin material is assumed to be known. Now the combination of b and L gives maximum heat transfer rate as M and A_1 are constant. The fin is assumed to be thin and long. The heat transfer rate is given by eq. (3.40)

$$\dot{Q}_1 = \sqrt{\alpha C \lambda A}\, \vartheta_1 \tanh mL$$

with $m = \sqrt{\alpha C / \lambda A}$.

Here $C = 2Z$ and $A = bZ$ and then $\dot{Q}_1$ can be written as

$$\dot{Q}_1 = Z \vartheta_1 \sqrt{2\alpha \lambda b} \tanh \sqrt{\frac{2\alpha}{\lambda b}} L \qquad (3.51)$$

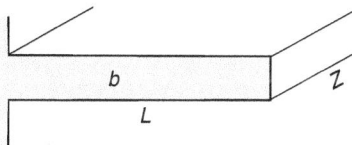

Figure 3.14: Rectangular fin.

Maximum of eq. (3.51) is now being looked for with the condition $A_1 = bL$ is constant. It is more appropriate to write eq. (3.51) in the form

$$\dot{Q}_1 = Z\vartheta_1\sqrt{2\alpha\lambda b}\tanh\left(\sqrt{\frac{2\alpha}{\lambda b}\frac{A_1}{b}}\right) \qquad (3.52)$$

$d\dot{Q}_1/db = 0$ gives the desired result.

$$\vartheta_1\sqrt{2\alpha\lambda}\left\{\frac{\frac{1}{2\sqrt{b}}\tanh\left(\sqrt{\frac{2\alpha}{\lambda b}\frac{A_1}{b}}\right)+}{\cosh^2\left((\sqrt{(2\alpha/\lambda b)}(A_1/b)\right)}\sqrt{\frac{2\alpha}{\lambda}}A_1\left(-\frac{3}{2}\right)b^{-5/2}\right\} = 0 \qquad (3.53)$$

If $u = \sqrt{(2\alpha/\lambda b)}(A_1/b)$ is introduced, eq. (3.53) can be written as

$$\tanh u = \frac{3u}{\cosh^2 u} \qquad (3.54)$$

The solution to eq. (3.54) must be found numerically or graphically. It is found that $u = 1.419$ is the solution to eq. (3.54). One then has

$$\sqrt{\frac{2\alpha}{\lambda b}\frac{A_1}{b}} = 1.419$$

which can be written as

$$\frac{L}{b/2} = 1.419\sqrt{\frac{2\lambda}{\alpha b}} \qquad (3.55)$$

If the fin length L is chosen in relation to the fin thickness b according to eq. (3.55), the heat transfer rate will be maximum at a given fin mass.

What about the fin performance measures for this optimal fin?

For η in eq. (3.48), one obtains with eq. (3.52)

$$\eta = \frac{Z\vartheta_1\sqrt{2\alpha\lambda b}\tanh u}{Zb\alpha\vartheta_1} = \sqrt{\frac{2\lambda}{\alpha b}}\tanh u = 0.889\sqrt{\frac{2\lambda}{\alpha b}} \qquad (3.56)$$

For φ in eq. (3.49), one obtains with eq. (3.52)

$$\varphi = \frac{Z\vartheta_1\sqrt{2\alpha\lambda b}\tanh u}{2ZL\alpha\vartheta_1} = \frac{b/2}{L}\sqrt{\frac{2\lambda}{\alpha b}}\tanh u = \frac{1}{1.419}\tanh u = 0.626 \qquad (3.57)$$

Note: The quantity $\alpha(b/2)/\lambda$ often shows up in heat conduction problems when heating or cooling by convection occurs. This is a nondimensional quantity and is often called the Biot number and is denoted by Bi.

3.8.5   Fin arrangement

Consider now an optimized rectangular fin. According to the previous section, one has

$$\dot{Q}_1 = Z\vartheta_1 \sqrt{2\alpha\lambda b}\, \tanh\left(\sqrt{\frac{2\alpha}{\lambda b}\frac{A_1}{b}}\right) \qquad (3.52)$$

$$u = \sqrt{\frac{2\alpha}{\lambda b}\frac{A_1}{b}} \qquad (3.58)$$

$$u = 1.419 \qquad (3.59)$$

Equation (3.58) gives

$$b^{3/2} = \frac{A_1}{u}\sqrt{\frac{2\alpha}{\lambda}} \qquad (3.60)$$

Substituting eq. (3.60) in eq. (3.52) gives

$$\dot{Q}_1 = Z\vartheta_1 \sqrt{2\alpha\lambda}\left(\frac{A_1}{u}\sqrt{\frac{2\alpha}{\lambda}}\right)^{1/3}\tanh u$$

or

$$A_1 = \left(\frac{\dot{Q}_1}{Z\vartheta_1}\right)^3 \frac{u}{\tanh^3 u}\frac{1}{4\alpha^2\lambda} \qquad (3.61a)$$

or in eq. (3.59) gives

$$A_1 = \frac{2.017}{4\alpha^2\lambda}\left(\frac{\dot{Q}_1}{Z\vartheta_1}\right)^3 \qquad (3.61b)$$

From eq. (3.61b), a few important conclusions can be drawn. Assume that it is desirable to double the heat transfer rate. Equation (3.61b) then tells that the fin area $A_1 = bL$ must be increased by a factor of 8. In practice this will be more conveniently done by using two fins instead. It is also clear that at a given heat transfer rate $\dot{Q}_1$, the fin area A_1 is proportional to $1/\lambda$. If the fin mass $M = \rho A_1 Z$ is considered, it is evident that this is proportional to ρ/λ. By comparing various materials, it is possible to judge what material being most appropriate.

Consider the data in Table 3.1. It is evident that if aluminum is applied instead of copper as fin material, a 50% reduction in weight can be achieved. Thus despite copper is conducting heat much better than aluminum, the copper material is not favorable.

Table 3.1: Comparison of different fin materials, 20°C.

Material	Thermal conductivity, λ (W/m K)	Density, ρ (kg/m³)	ρ/λ (kg K/W m²)
Copper	386	8890	23.0
Aluminum, pure	228	2700	11.8
Aluminum alloy	122	2770	22.8
Magnesium, pure	172	1750	10.2
Steel (1% C)	42.9	7820	182.3
Stainless steel	16.3	7820	480.0

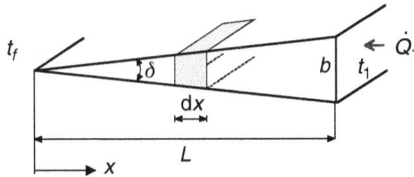

Figure 3.15: Triangular fin.

3.8.6   Straight triangular fin

Consider a triangular fin as depicted in Fig. 3.15.

The heat conduction is assumed to be one-dimensional and the temperature at the base (note $x = L$) is t_1. The fluid temperature is t_f. The convective heat transfer coefficient, α, between the fin surface and the fluid is assumed to be known and uniform across the fin surface. Consider now the heat balance for the element dx in Fig. 3.15.

The amount of heat conducted in through the left surface is (location x)

$$\dot{Q}_x = -\lambda A \frac{dt}{dx}$$

The amount of heat conducted out through the right surface is (location $x + dx$)

$$\dot{Q}_{x+dx} = -\lambda A \frac{dt}{dx} + \frac{d}{dx} \left(-\lambda A \frac{dt}{dx} \right) dx$$

The heat transfer rate due to convection can be written (if the angle δ is so small that the area between the fluid and fin element is $C \, dx$)

$$\dot{Q}_{conv} = \alpha C \, dx(t - t_f)$$

At steady state, the supplied and released heat must be equal and then one has

$$-\frac{d}{dx}\left(-\lambda A\frac{dt}{dx}\right) - \alpha C(t - t_f) = 0$$

Here one has $A = bZx/L$ and $C = 2Z$. If in addition $\vartheta = t - t_f$ is introduced, the equation above becomes

$$\frac{d}{dx}\left(x\frac{d\vartheta}{dx}\right) - \frac{2\alpha L}{\lambda b}\vartheta = 0$$

or

$$\frac{d^2\vartheta}{dx^2} + \frac{1}{x}\frac{d\vartheta}{dx} - \frac{1}{x}\frac{2\alpha L}{\lambda b}\vartheta = 0 \tag{3.62}$$

Commonly the parameter $\beta = 2\alpha L/\lambda b$ is also introduced.

Equation (3.62) is a so-called modified Bessel differential equation and its solution is given by

$$\vartheta = AI_0(2\sqrt{\beta x}) + BK_0(2\sqrt{\beta x}) \tag{3.63}$$

where I_0 and K_0 are the so-called modified Bessel functions of order zero.

In Table 3.2, values of these functions are given for different values of the argument $2\sqrt{\beta x}$. It is found that $K_0 \to \infty$ as $x \to 0$. To make sure that the temperature will be finite at the fin tip, the constant B in eq. (3.63) must be set to zero. The other constant A is found from the condition

$$x = L: \ \vartheta = \vartheta_1 \tag{3.64}$$

The temperature distribution in the triangular fin can now be written as

$$\frac{\vartheta}{\vartheta_1} = \frac{I_0(2\sqrt{\beta x})}{I_0(2\sqrt{\beta L})} \tag{3.65}$$

The heat transfer rate from the fin is found from

$$\dot{Q}_1 = \left(\lambda A\frac{dt}{dx}\right)_{x=L} = \lambda bZ\left(\frac{d\vartheta}{dx}\right)_{x=L}$$

The derivative $dI_0(\xi)d\xi$ can be expressed by the modified Bessel function $I_1(\xi)$, i.e.,

$$\frac{dI_0(\xi)}{d\xi} = I_1(\xi)$$

The heat transfer rate can now be written as

$$\dot{Q}_1 = Z\vartheta_1\sqrt{2\alpha\lambda b}\frac{I_1(2\sqrt{\beta L})}{I_0(2\sqrt{\beta L})} \tag{3.66}$$

Table 3.2: Numerical values for the Bessel functions I_0, I_1, K_0, and K_1.

ζ	$I_0(\zeta)$	$I_1(\zeta)$	$(2/\pi)K_0(\zeta)$	$(2/\pi)K_1(\zeta)$
0.0	1.0000	0.0000	∞	∞
0.2	1.0100	0.1005	1.1158	3.0405
0.4	1.0404	0.2040	0.70953	1.3906
0.6	1.0920	0.3137	0.49498	0.82941
0.8	1.1665	0.4329	0.35991	0.54862
1.0	1.2661	0.5652	0.26803	0.38318
1.2	1.3937	0.7147	0.20276	0.27667
1.4	1.5534	0.8861	0.15512	0.20425
1.6	1.7500	1.0848	0.11966	0.15319
1.8	1.9896	1.3172	0.092903	0.11626
2.0	2.2796	1.5906	0.072507	0.089041
2.2	2.6291	1.9141	0.056830	0.068689
2.4	3.0493	2.2981	0.044702	0.053301
2.6	3.5533	2.7554	0.035268	0.041561
2.8	4.1573	3.3011	0.027896	0.032539
3.0	4.8808	3.9534	0.022116	0.025564
3.2	5.7472	4.7343	0.017568	0.020144
3.4	6.7848	5.6701	0.013979	0.015915
3.6	8.0277	6.7028	0.011141	0.012602
3.8	9.5169	8.1404	0.008891	0.009999
4.0	11.3019	9.7595	0.007105	0.007947
4.2	13.4425	11.7056	0.005684	0.006327
4.4	16.0104	14.0462	0.004551	0.005044
4.6	19.0926	16.8626	0.003648	0.004027
4.8	22.7937	20.2528	0.002927	0.003218
5.0	27.2399	24.3356	0.002350	0.002575
5.2	32.5836	29.2543	0.001888	0.002062
5.4	39.0088	35.1821	0.001518	0.001653
5.6	46.7376	42.3283	0.001221	0.001326
5.8	56.0381	50.9462	0.0009832	0.001064
6.0	67.2344	61.3419	0.0007920	0.0008556
6.2	80.7179	73.8859	0.0006382	0.0006879
6.4	96.9616	89.0261	0.0005146	0.0005534

Table 3.2: Continued.

ζ	$I_0(\zeta)$	$I_1(\zeta)$	$(2/\pi)K_0(\zeta)$	$(2/\pi)K_1(\zeta)$
6.6	116.537	107.305	0.0004141	0.0004455
6.8	140.136	129.378	0.0003350	0.0003588
7.0	168.593	156.039	0.0002704	0.0002891
7.2	202.921	188.250	0.0002184	0.0002331
7.4	244.341	227.175	0.0001764	0.0001880
7.6	294.332	274.222	0.0001426	0.0001517
7.8	354.685	331.099	0.0001153	0.0001225
8.0	427.564	399.873	0.00009325	0.00009891
8.2	515.593	483.048	0.00007543	0.00007991
8.4	621.944	583.657	0.00006104	0.00006458
8.6	750.461	705.377	0.00004941	0.00005220
8.8	905.797	852.663	0.00004000	0.00004221
9.0	1093.59	1030.91	0.00003239	0.00003415
9.2	1320.66	1246.68	0.00002624	0.00002763
9.4	1595.28	1507.88	0.00002126	0.00002236
9.6	1927.48	1824.14	0.00001722	0.00001810
9.8	2329.39	2207.13	0.00001396	0.00001465
10.0	2815.72	2670.99	0.00001131	0.00001187

From L.M.K. Boelter, Heat Transfer Notes, McGraw-Hill, New York (1965).

The function $I_1(\xi)$ is also given in Table 3.2.

In a similar way as for rectangular fins, the heat transfer rate can be optimized for a given fin mass. The fin length L should then be chosen in relation to the fin base b according to

$$\frac{L}{b/2} = 1.309\sqrt{\frac{2\lambda}{\alpha b}} \qquad (3.67)$$

If triangular and rectangular fins are compared for identical heat transfer rate, it can be found that the triangular fin will be 44% lighter than the rectangular one.

3.8.7 Formulas for the fin performance

In the previous section, the fin effectiveness η and the fin efficiency φ were defined. By some simple calculations, the following formulas are found.

Rectangular fins

$$\eta = \sqrt{\frac{2\lambda}{\alpha b}}\tanh mL$$

$$\varphi = \frac{\tanh mL}{mL}$$

Triangular fins

$$\eta = \sqrt{\frac{2\lambda}{\alpha b}}\frac{I_1(2\sqrt{\beta L})}{I_0(2\sqrt{\beta L})}$$

$$\varphi = \frac{I_1(2\sqrt{\beta L})/I_0(2\sqrt{\beta L})}{\sqrt{\beta L}}$$

3.8.8 Circular or annular fins

Fins fixed to tubes are often circular or annular as shown in Fig. 3.16. As the governing differential equation is derived, a similar procedure to that for a triangular fin is applied. The differences are in the expression for the fin cross-sectional area A and the perimeter C. Here the expressions $A = 2\pi r b$ and $C = 4\pi r$ are valid.

The following differential equation is obtained where $\vartheta = t - t_f$

$$\frac{d^2\vartheta}{dr^2} + \frac{1}{r}\frac{d\vartheta}{dr} - \frac{2\alpha}{\lambda b}\vartheta = 0 \qquad (3.68)$$

The boundary conditions to eq. (3.68) are (if the fin is long and thin).

$$r = r_1: \ \vartheta = \vartheta_1; \quad r = r_2: \ \frac{d\vartheta}{dr} = 0$$

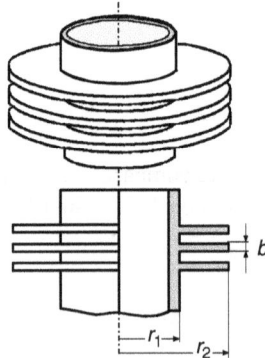

Figure 3.16: Arrangement of circular or annular fins.

Equation (3.68) is also a modified Bessel equation and the solution is (with the boundary conditions above)

$$\frac{\vartheta}{\vartheta_1} = \frac{K_1(r_2\sqrt{\beta})I_0(r\sqrt{\beta}) + I_1(r_2\sqrt{\beta})K_0(r\sqrt{\beta})}{K_1(r_2\sqrt{\beta})I_0(r_1\sqrt{\beta}) + I_1(r_2\sqrt{\beta})K_0(r_1\sqrt{\beta})} \tag{3.69}$$

The function K_1 is available in Table 3.2.

The heat transfer rate from the fin is given by

$$\dot{Q}_1 = 2\pi r_1 b \lambda \sqrt{\beta}\vartheta_1 \frac{I_1(r_2\sqrt{\beta})K_1(r_1\sqrt{\beta}) - K_1(r_2\sqrt{\beta})I_1(r_1\sqrt{\beta})}{I_1(r_2\sqrt{\beta})K_0(r_1\sqrt{\beta}) + K_1(r_2\sqrt{\beta})I_0(r_1\sqrt{\beta})} \tag{3.70}$$

In eqs. (3.69) and (3.70), $\beta = 2\alpha/\lambda b$. (Note β is not the same as for triangular fins.)

In the literature, other geometries of fins are also treated. For further studies of the fin theory, the reader is referred to Gardner [3], Schneider [4], and Kern and Kraus [5].

3.9   Application of the fin efficiency φ in engineering calculations

Consider the finned heat transfer surfaces in Fig. 3.17a and b.

The heat transfer rate can be written as

$$\dot{Q} = \dot{Q}_{\text{unfinned area}} + \dot{Q}_{\text{fins}}$$

or

$$\dot{Q} = \alpha A_b(t_b - t_f) + \phi\dot{Q}_{\text{fins with } \lambda=\infty}$$

where A_b = base area, t_b = temperature of the base area, and $\dot{Q}_{\text{fins with } \lambda=\infty} = \alpha A_{\text{fins}}(t_b - t_f)$.

The heat transfer rate from the finned heat transfer surface can now be written as

$$\dot{Q} = \alpha(t_b - t_f)(A_b + \varphi A_{\text{fins}}) \tag{3.71}$$

In Fig. 3.18, a graph is given and it is possible to determine the fin efficiency φ for circular or annular fins. In the previous section, formulas for rectangular and triangular fins were provided.

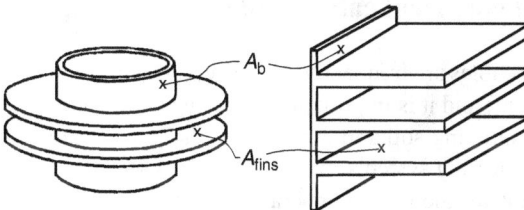

Figure 3.17: Examples of finned heat transfer surfaces.

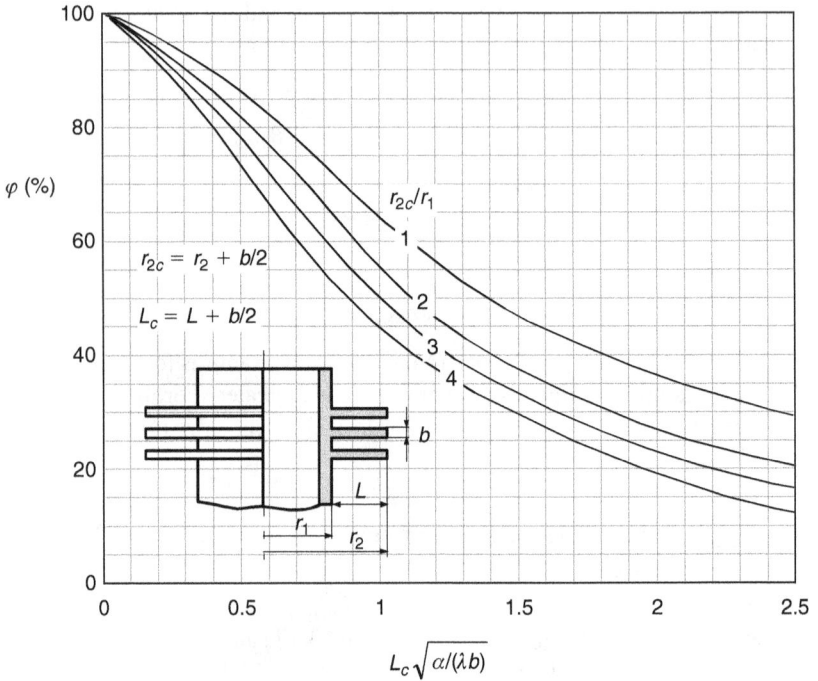

Figure 3.18: The fin efficiency φ for circular or annular fins.

3.10 Limitations in the presented fin theory

In the derivation of the relations for rectangular, triangular, and circular (annular) fins, the heat transfer coefficient was assumed to be constant (uniform) along the fin surface. In reality, the heat transfer coefficient is dependent on the actual flow field around the fins and may vary over the fin surfaces. Investigations, see Refs. [6–11], have shown that the total heat transfer rate and the fin efficiency calculated by using the mean value of the heat transfer coefficient are reasonably accurate. Recent research activities are presented in Ref. [12].

3.11 Buried electrical cables and hot water pipes

Buried electrical cables are frequently appearing. The current in the cable generates heat (Joule heating) and it is important to know what temperature distribution will occur in the surrounding soil and ground. Similarly, hot water carrying pipes are buried in district heating systems as shown in Fig. 3.19.

It is important to know the temperature drop along the water pipe. Another issue is to know how deep the pipe should be buried to keep the heat loss within acceptable limits.

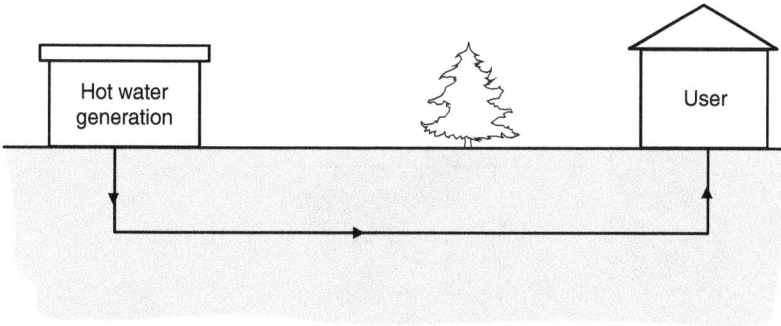

Figure 3.19: Buried water pipe for transporation of hot water.

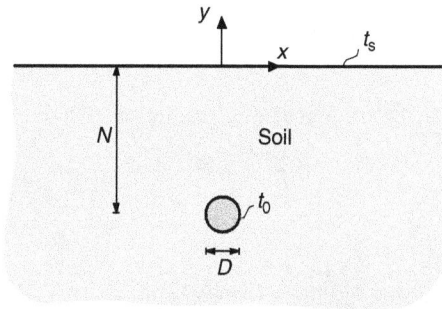

Figure 3.20: Cross section of the buried cable or water pipe.

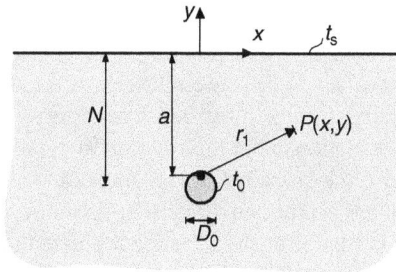

Figure 3.21: Buried cable shown with a heat source for the analysis.

Consider now the cross section of the cable or pipe in Fig. 3.20.

The cable or pipe has a diameter D and its center is placed a distance N below the ground surface. The ground surface is assumed to be horizontal and has the temperature t_s and the cable or pipe surface has the temperature t_0. The temperature differences $\vartheta = t - t_s$ and $\vartheta_0 = t_0 - t_s$ are introduced.

The cable or pipe is now replaced by a heat source with radius r_0 placed at the depth a, see Fig. 3.21.

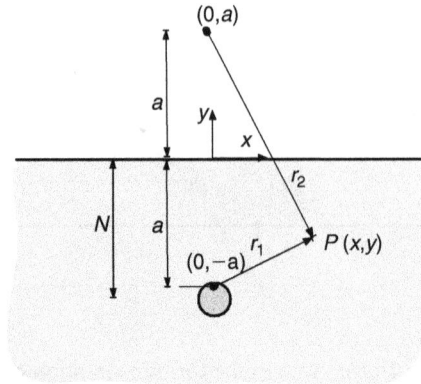

Figure 3.22: Buried cable (pipe) with heat source–heat sink system.

Consider now an arbitrary point $P(x, y)$ in the soil at a distance r_1 from the heat source. According to eq. (3.10), one has

$$\dot{Q} = 2\pi \lambda L \frac{t_0' - t_1}{\ln(r_1/r_0)} = 2\pi \lambda L \frac{\vartheta_0 - \vartheta_1}{\ln(r_1/r_0)} \tag{3.72}$$

where t_0' is the temperature of the heat source.

From eq. (3.72), the temperature at an arbitrary point $P(x, y)$ can be found.

$$\vartheta_1 = \vartheta_0' - \frac{\dot{Q}}{2\pi \lambda L} \ln \frac{r_1}{r_0} \tag{3.73}$$

Equation (3.73) is the temperature at point $P(x, y)$ due to the heat source. The isotherms ($\vartheta =$ constant) of eq. (3.73) are circles. However, according to the boundary conditions, the ground surface must be an isotherm. This is not satisfied by eq. (3.73) and thus this equation is not the solution to the problem considered.

To solve the problem, the so-called mirror method is used (see, for example, Ref. [13]). The heat source is then mirrored in the ground surface and a heat sink with temperature $-\vartheta_0'$ is placed at the distance a above the ground surface. The heat sink receives the same amount of heat as released from the heat source. With the notation in Fig. 3.22, the heat sink introduces the temperature ϑ_2 in the point $P(x, y)$ according to (compare with eq. (3.73))

$$\vartheta_2 = -\vartheta_0' + \frac{\dot{Q}}{2\pi \lambda L} \ln \frac{r_2}{r_0} \tag{3.74}$$

As the heat conduction equation is linear, eqs. (3.72) and (3.74) can be added, and one finds

$$\vartheta = \vartheta_1 + \vartheta_2 = \frac{\dot{Q}}{2\pi \lambda L} \ln \frac{r_2}{r_1} \tag{3.75}$$

r_1 and r_2 now have to be determined. By considering Fig. 3.22, one finds $r_1^2 = x^2 + (y + a)^2$ and $r_2^2 = x^2 + (y - a)^2$. Substituting these into eq. (3.75) gives

$$\vartheta = \frac{\dot{Q}}{4\pi \lambda L} \ln \frac{x^2 + (a - y)^2}{x^2 + (a + y)^2} \tag{3.76}$$

It must now be checked if the solution (76) satisfies the condition that the ground surface is an isotherm. By rewriting eq. (3.76), one has

$$\frac{x^2 + (a - y)^2}{x^2 + (a + y)^2} = \exp\left(\frac{4\pi \lambda L}{\dot{Q}} \vartheta\right) \tag{3.77}$$

The equation for the isotherms, $\vartheta = $ constant, is achieved if the right-hand side of eq. (3.77) is kept constant, say C. One finds

$$x^2 + (a - y)^2 = C(x^2 + (a + y)^2)$$

which can be transformed to

$$x^2 + \left(y - a\frac{1 + C}{1 - C}\right)^2 = \frac{4Ca^2}{(1 - C)^2} \tag{3.78}$$

For $C \neq 1$, eq. (3.78) describes circles with centers given by

$$(x, y) = \left(0, a\frac{1 + C}{1 - C}\right) \quad \text{and} \quad \text{radius } R = \frac{2a\sqrt{C}}{|1 - C|}.$$

Especially for $\vartheta = 0 \Rightarrow t = t_s$ and $C = 1$, one has $R \to \infty$ and $(x, y) \to (0, -\infty)$. Thus $\vartheta = 0$ corresponds to a straight line, which agrees with the condition that the ground surface must be an isotherm.

The solution given by eq. (3.76) is therefore the solution to the considered problem.

An expression for the heat transfer rate $\dot{Q}$ is now requested. To find this expression, the assumption that the pipe or cable surface is an isotherm has to be applied. One then has

$$\left(\frac{D}{2}\right)^2 = \frac{4C_0 a^2}{(1 - C_0)^2} \tag{3.79}$$

$$-N = a\frac{1 + C_0}{1 - C_0} \tag{3.80}$$

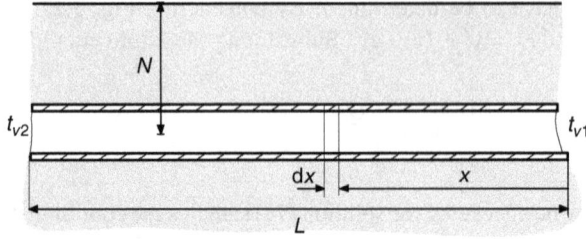

Figure 3.23: Buried hot water pipe.

In eqs. (3.79) and (3.80), two equations and two unknowns (a, C_0) are present. One solves for C_0 and thus for ϑ_0 and $\dot{Q}$. A few calculations result in

$$\dot{Q} = \frac{2\pi \lambda L \vartheta_0}{\ln\left((2N/D) + \sqrt{4((N/D)^2)} - 1\right)} \tag{3.81}$$

In analogy with the previous sections, a thermal resistance is defined as

$$\frac{\ln\left((2N/D) + \sqrt{4((N/D)^2)} - 1\right)}{2\pi \lambda L} \tag{3.82}$$

Especially if $N/D \gg 1$, eq. (3.81) can be written as

$$\dot{Q} = \frac{2\pi \lambda L \vartheta_0}{\ln(4N/D)} \tag{3.83}$$

3.11.1 Calculation of the heat loss from the whole hot water pipe

Consider now a buried hot water pipe according to Fig. 3.23. Attention is now on the increment dx. The outer surface of the pipe is assumed to have the temperature t_0. The heat transfer coefficient between the water and the inner pipe surface is α and the pipe material has a thermal conductivity λ_{pipe} and the pipe inner radius is r_1 while its outer radius is r_2. It is now possible to write

$$t_v - t_s = (t_v - t_0) + (t_0 - t_s) \tag{3.84}$$

With eqs. (3.12) and (3.81), eq. (3.84) can be written as

$$t_v - t_s = d\dot{Q}\frac{R'_{\text{tot}}}{dx} \tag{3.85}$$

where

$$R'_{\text{tot}} = \frac{1}{2\pi}\left(\frac{1}{\alpha r_1} + \frac{\ln(r_2/r_1)}{\lambda_{\text{pipe}}} + \frac{\ln(2N/D) + \sqrt{4(N/D)^2 - 1})}{\lambda_{\text{soil}}}\right) \tag{3.86}$$

The incremental heat transfer rate $\mathrm{d}\dot{Q}$ must be taken from the water and the equation below holds

$$\mathrm{d}\dot{Q} = -\dot{m}c_\mathrm{p}\frac{\mathrm{d}t_\mathrm{v}}{\mathrm{d}x}\mathrm{d}x \qquad (3.87)$$

By introducing $\vartheta = t_\mathrm{v} - t_\mathrm{s}$, and then by using eqs. (3.85) and (3.87), one arrives at

$$\frac{\mathrm{d}\vartheta}{\mathrm{d}x} + \frac{1}{\dot{m}c_\mathrm{p}R'_\mathrm{tot}}\vartheta = 0 \qquad (3.88)$$

The solution to eq. (3.88) with the boundary condition $x = 0$, $\vartheta = \vartheta_1 = t_{\mathrm{v}_1} - t_\mathrm{s}$ gives

$$\vartheta = \vartheta_1 \exp\left(-\frac{x}{\dot{m}c_\mathrm{p}R'_\mathrm{tot}}\right) \qquad (3.89)$$

Especially for $x = L$, one obtains

$$\vartheta_2 = \vartheta_1 \exp\left(-\frac{L}{\dot{m}c_\mathrm{p}R'_\mathrm{tot}}\right) \qquad (3.90)$$

For the heat transfer rate $\dot{Q}$, one has

$$\left|\dot{Q}\right| = \dot{m}c_\mathrm{p}(t_{\mathrm{v}_1} - t_{\mathrm{v}_2}) = \dot{m}c_\mathrm{p}(\vartheta_1 - \vartheta_2)$$

With eq. (3.90), one obtains

$$\dot{Q} = \frac{L}{R'_\mathrm{tot}}\frac{\vartheta_1 - \vartheta_2}{\ln(\vartheta_1/\vartheta_2)} \qquad (3.91)$$

L/R'_tot corresponds to the overall heat transfer coefficient.

3.11.2   The case when the temperature t_s of the ground surface is not given

In practice, the ground surface temperature is not known as was assumed in the derivation above. More commonly the surrounding air temperature is known. If the heat transfer coefficient α_0 between the ground surface and the surrounding air is known or can be estimated, an extra depth ΔN is introduced according to $\Delta N = \lambda/\alpha_0$. Then instead of N in eqs. (3.81)–(3.83) and (3.86), $N + \Delta N$ should be used, see Refs. [14, 15].

3.11.3   Other common cases

In Table 3.3, formulas for calculation of the heat transfer rate from buried or enclosed objects are given for a few cases.

Table 3.3: Formulas for calculation of the heat transfer rate from buried or enclosed objects.

Configuration		$\dot{Q}/(\lambda\,(T_1 - T_2))$
Vertical cylinder in a semi-infinite medium.		$\dfrac{2\pi L}{\ln(4L/D)}$ $L \gg D$
Isothermal sphere in a semi-infinite medium.		$\dfrac{2\pi D}{1 - D/4Z}$ $Z > D/2$
Circular cylinder with length L centered in a quadratic medium.		$\dfrac{2\pi L}{\ln(1.08b/D)}$ $b > D,\ L \gg b$
Heat conduction between two cylinders with length L in an infinite medium		$\dfrac{2\pi L}{\cosh^{-1}\left(\dfrac{4b^2 - D_1^2 - D_2^2}{2D_1 D_2}\right)}$ $L \gg D_1, D_2,\ L \gg b$

References

[1] E.M. Sparrow, Reexamination and correction of the critical radius for radial heat conduction, AIChE J., 16, 149 (1970).
[2] Standard Mathematical Tables, 17th ed., Chemical Rubber Co., Cleveland, OH (1969).
[3] K.A. Gardner, Efficiency of extended surfaces, Trans. ASME, 67, 621–631 (1945).
[4] P.J. Schneider, Conduction Heat Transfer, 6th ed., Addison-Wesley, Reading, MA (1974).
[5] D.Q. Kern and A.D. Kraus, Extended Surface Heat Transfer, McGraw-Hill, New York (1972).
[6] E.M. Sparrow and S. Acharya, A natural convection fin with solution-determined monotonically varying heat transfer coefficient, ASME J. Heat Transfer, 103, 218–225 (1981).
[7] E.M. Sparrow and M.K. Chyu, Conjugate forced convection–conduction analysis of heat transfer in a plate fin, ASME J. Heat Transfer, 104, 204–206 (1982).

[8]   B. Sunden, Conjugate mixed convection heat transfer from a vertical rectangular fin, Int. Comm. Heat Mass Transfer, 10, 267–276 (1983).
[9]   B. Sunden, Conjugate laminar and/or turbulent forced convection heat transfer from rectangular fins, I. Chem. E. Symp. Series No 86, 725–734, Pergamon Press (1984).
[10] B. Sundén, Analysis of conjugated laminar and turbulent forced convection–conduction heat transfer of a plate fin, Int. Comm. Heat Mass Transfer, 16(6), 821–831 (1989).
[11] B. Sunden, The effect of Prandtl number on conjugate heat transfer from rectangular fins, Int. Comm. Heat Mass Transfer, 12, 225–232 (1985).
[12] B. Sundén and P.J. Heggs (Eds.), Recent Advances in Analysis of Heat Transfer for Fin Type Surfaces, WIT Press, Southampton, (1999).
[13] L.M. Milne-Thomson, Theoretical Hydrodynamics, 5th ed., Macmillan, London (1968).
[14] O. Krischer, Das Temperaturfeld in der Umgebung von Rohrleitungen, die in die Erde verlegt sind, Gesundheitsing., 59(37), 537–539 (1936).
[15] K. Elegeti, Der Wärmeverlust einer erdverlegten Rohrleitungen stationären Zustand unter dem Einfluss der Wärmeubergangszahl der Erdoberfläche, Forsch. Ing. Wes., Bd 33, 101–105 (1967).

Further reading

E.R.G. Eckert and R.M. Drake Jr., Analysis of Heat and Mass Transfer, McGraw-Hill, New York (1972).
H.S. Carslaw and J.C. Jaeger, Conduction of Heat in Solids, Oxford University Press, New York (1959).
Y.A. Cengel, Heat transfer—A Practical Approach, 3rd ed., McGraw-Hill (2007).
A.D. Kraus, A. Aziz and J.R. Welty, Extended Surface Heat Transfer, Wiley, New York (2001).

4 Unsteady heat conduction

4.1 Introduction

A process is said to be unsteady or transient if it depends on time. In some cases, the transient time period is very short compared to the overall time for the heat transfer process. Then the time-dependent period is not so important. In other cases, like metal casting, iron hardening, and vulcanization of rubber tires, the complete heat transfer process is transient and the time dependence is very important. In periodically operating equipment and for starting up and shutting down phases the time dependence is also very important. In this chapter, analysis of unsteady heat conduction will be presented.

4.2 Bodies with very high thermal conductivity

In this section, bodies with high thermal conductivity are considered. The body is heated or cooled by convection (Fig. 4.1). Because the body has a very high thermal conductivity, the temperature within it will be almost uniform. A heat balance then gives

$$\rho c V \frac{dt}{d\tau} = -\alpha A(t - t_f) \tag{4.1}$$

where A is the surface area of the body in contact with the fluid and V is its volume.

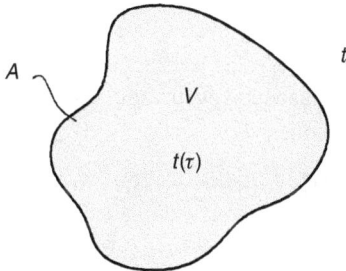

Figure 4.1: Body with very high thermal conductivity.

$\vartheta = t - t_f$ is introduced and then eq. (4.1) can be written as

$$\frac{d\vartheta}{d\tau} + \frac{\alpha A}{\rho c V}\vartheta = 0 \tag{4.2}$$

The solution to eq. (4.2) is

$$\vartheta = C_1 \exp\left(-\frac{\alpha A}{\rho c V}\tau\right) \tag{4.3}$$

As initial condition, the following is set, $t = t_0$, i.e., $\vartheta = \vartheta_0 = t_0 - t_f$, at $\tau = 0$. Equation (4.3) then gives

$$\vartheta = \vartheta_0 \exp\left(-\frac{\alpha A}{\rho c V}\tau\right) \tag{4.4}$$

The quantity $\alpha A/\rho c V$ is commonly rewritten. If $L = V/A$ is introduced as the characteristic length, one may write

$$\frac{\alpha A}{\rho c V}\tau = \frac{\alpha}{\rho c L}\tau = \frac{\alpha L}{a\rho c} \cdot \frac{a\tau}{L^2} = \frac{\alpha L}{\lambda} \cdot \frac{a\tau}{L^2} \tag{4.5}$$

where $a = \lambda/\rho c =$ thermal diffusivity. In eq. (4.5) the dimensionless numbers below are introduced

$$\mathrm{Bi} = \frac{\alpha L}{\lambda}, \quad \text{Biot number}$$

and

$$\mathrm{Fo} = \frac{a\tau}{L^2}, \quad \text{Fourier number}$$

Equation (4.4) can be written as

$$\frac{\vartheta}{\vartheta_0} = \exp(-\mathrm{Bi} \cdot \mathrm{Fo}) \tag{4.6}$$

If the condition $\mathrm{Bi} = \alpha L/\lambda < 0.1$ is satisfied, the assumption that the body temperature is uniform is reasonably well valid, see, for example, Ref. [1].

4.3   Infinite plate with moderate thermal conductivity

Consider an infinite plate with thickness L as shown in Fig. 4.2. Initially, the plate has a temperature distribution $t_0(x)$. The side wall at $x = 0$ is insulated while the side wall at $x = L$ is cooled convectively and the heat transfer coefficient there is α.

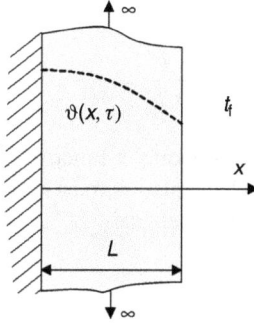

Figure 4.2: Unsteady heat conduction in an infinite plate.

The heat conduction is assumed to be one-dimensional and in the x-direction. Equation (1.18) then gives

$$\rho c \frac{\partial t}{\partial \tau} = \lambda \frac{\partial^2 t}{\partial x^2} \tag{4.7}$$

If $\vartheta = t - t_f$ and $a = \lambda/\rho c$ are introduced, then eq. (4.7) can be written as

$$\frac{\partial \vartheta}{\partial \tau} = a \frac{\partial^2 \vartheta}{\partial x^2} \tag{4.8}$$

The initial condition to eq. (4.8) is

$$\tau = 0: t = t_0(x) \Rightarrow \vartheta(0, x) = \vartheta_0 = t_0 - t_f \tag{4.9}$$

and the boundary conditions are

$$x = 0: \frac{\partial t}{\partial x} = 0 \Rightarrow \frac{\partial \vartheta}{\partial x} = 0 \tag{4.10}$$

$$x = L: -\lambda \frac{\partial t}{\partial x} = \alpha(t - t_f) \Rightarrow -\lambda \frac{\partial \vartheta}{\partial x} = \alpha \vartheta \tag{4.11}$$

To solve eq. (4.8), subject to the conditions (4.9)–(4.11), the so-called method of separating the independent variables is applied. A solution of the following form is then considered

$$\vartheta = F(\tau) \cdot G(x) \tag{4.12}$$

Equation (4.12) inserted into eq. (4.8) gives

$$F'(\tau) \cdot G(x) = aF(\tau) \cdot G''(x)$$

which can be rewritten as

$$\frac{1}{a}\frac{F'}{F} = \frac{G''}{G} \tag{4.13a}$$

The left-hand side in eq. (4.13a) is only a function of time τ, and the right-hand side is only a function of x. As the equation must be valid at all τ and x, eq. (4.13a) must be a constant, say $\pm\beta^2$. One then has

$$\frac{1}{a}\frac{F'}{F} = \frac{G''}{G} = \pm\beta^2 \tag{4.13b}$$

From eq. (4.13b) one can write

$$F' - (\pm)\beta^2 aF = 0 \tag{4.14}$$

$$G'' - (\pm)\beta^2 G = 0 \tag{4.15}$$

The solution to eq. (4.14) is

$$F = C\exp(\pm\beta^2 a\tau) \tag{4.16a}$$

If a cooling process is considered, i.e., $t_0(x) > t_f$, $\vartheta = t - t_f$ will decrease with time. The minus sign in eq. (4.16a) must then be chosen.

$$F = C\exp(-\beta^2 a\tau) \tag{4.16b}$$

As the minus sign in $(\pm)$ is chosen (see, for example, Ref. [2]), the solution of eq. (4.15) becomes

$$G = C_1 \cos\beta x + C_2 \sin\beta x \tag{4.17}$$

Equations (4.16b) and (4.17) are now inserted in eq. (4.12) which gives

$$\vartheta = \exp(-\beta^2 a\tau) \cdot [A\cos\beta x + B\sin\beta x] \tag{4.18}$$

where $A = C \cdot C_1$ and $B = C \cdot C_2$.

In eq. (4.18) there are three unknowns, namely, A, B, and β. These are now to be determined by using the boundary conditions (4.10) and (4.11) as well as the initial condition (4.9).

With eq. (4.10), one obtains

$$\frac{\partial \vartheta}{\partial x} = \exp(-\beta^2 a\tau) \cdot [-A\beta \sin \beta x + B\beta \cos \beta x]$$

$$\left(\frac{\partial \vartheta}{\partial x}\right)_{x=0} = 0 \Rightarrow B = 0$$

Equation (4.18) is now reduced to

$$\vartheta = A \exp(-\beta^2 a\tau) \cdot \cos \beta x \qquad (4.19)$$

Equations (4.19) and (4.11) give

$$-\lambda A \exp(-\beta^2 a\tau) \cdot (-\beta \sin \beta L) = \alpha A \exp(-\beta^2 a\tau) \cdot \cos \beta L$$

or

$$\frac{\beta \lambda}{\alpha} = \cot \beta L \qquad (4.20a)$$

Obviously eq. (4.20a) determines β.
Equation (4.20a) can be written as

$$\beta L \frac{\lambda}{\alpha L} = \cot \beta L \qquad (4.20b)$$

Equation (4.20a) or (4.20b) must be solved graphically or numerically.
Figure 4.3 shows how a graphical solution can be achieved.

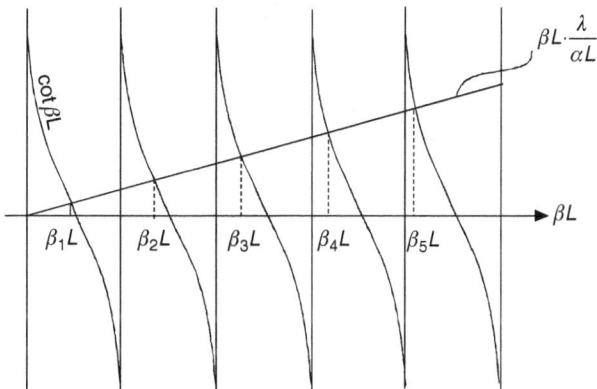

Figure 4.3: Sketch of a graphical solution for finding βL in eq. (4.20).

As is evident from Fig. 4.3, there is an infinite number of β-values satisfying eq. (4.20). The complete solution (4.19) must therefore be written as

$$\vartheta = \sum_{n=1}^{\infty} A_n \exp(-\beta_n^2 a\tau)\cos \beta_n x \tag{4.21}$$

The constants A_n are determined by using the initial condition (4.9). One finds

$$\vartheta_0 = \sum_{n=1}^{\infty} A_n \cos \beta_n x \tag{4.22}$$

Equation (4.22) is multiplied by $\cos \beta_m x$ and then integration from $x = 0$ to $x = L$ is carried out. The following relation is valid

$$\int_0^L \vartheta_0(x)\cos \beta_m x \, dx = A_m \int_0^L \cos^2 \beta_m x \, dx \tag{4.23}$$

because

$$\int_0^L \cos \beta_m x \cos \beta_n x \, dx = 0 \quad \text{if } m \neq n$$

Especially if $\vartheta_0(x) = $ constant (4.23) can be rewritten as

$$A_n = \frac{2\vartheta_0 \sin \beta_n L}{\beta_n L + \sin \beta_n L \cos \beta_n L} \tag{4.24}$$

where n is used to indicate the arbitrary constant.

The solution to the temperature field can now be written as

$$\frac{\vartheta_0}{\vartheta} = \sum_{n=1}^{\infty} \exp(-\beta_n^2 a\tau)\frac{2 \sin \beta_n L \cos \beta_n x}{\beta_n L + \sin \beta_n L \cos \beta_n L} \tag{4.25}$$

The amount of heat (in J) can be calculated from

$$dQ = -\lambda A \left(\frac{\partial \vartheta}{\partial x}\right)_{x=L} d\tau$$

By using eq. (4.25) and integrating from time zero to an arbitrary time τ, the amount of heat as a function of time is found. The result is

$$\frac{Q}{A} = \frac{2\lambda L}{a} \vartheta_0 \sum_{n=1}^{\infty} \frac{\sin^2 \beta_n L (1 - \exp(-\beta_n^2 a\tau))}{(\beta_n L)^2 + \beta_n L \sin \beta_n L \cos \beta_n L} \qquad (4.26)$$

For engineering calculations, eqs. (4.25) and (4.26) are not so handy as they include sums of an infinite number of terms. However, the results may also conveniently be shown as graphs. The main focus will now be on the temperature distribution in eq. (4.25).

β_n from eqs. (4.20a) and (4.20b) can schematically be written as

$$\beta_n L = \text{function}\left(\frac{\lambda}{\alpha L}\right) \qquad (4.27)$$

The exponent $\beta_n^2 a\tau$ in eq. (4.25) can be written as

$$\beta_n^2 a\tau = \frac{a\tau}{L^2} \cdot \beta_n^2 L^2 = \frac{a\tau}{L^2} \cdot \text{function}\left(\frac{\lambda}{\alpha L}\right) \qquad (4.28)$$

$\beta_n x$ in eq. (4.25) can be written as

$$\beta_n x = \frac{x}{L} \cdot \beta_n L = \frac{x}{L} \cdot \text{function}\left(\frac{\lambda}{\alpha L}\right) \qquad (4.29)$$

By using eqs. (4.27)–(4.29), eq. (4.25) can schematically be written as

$$\frac{\vartheta}{\vartheta_0} = \text{function}\left(\frac{\lambda}{aL}, \frac{a\tau}{L^2}, \frac{x}{L}\right) \qquad (4.30)$$

In Fig. 4.4, the temperature distribution (4.25) is shown in the way presented by eq. (4.30). It should be recognized that the modulus $\text{Bi} = \alpha L/\lambda$ and $\text{Fo} = a\tau/L^2$ are appearing again.

4.3.1   Heating or cooling

In the above derivation, the plate was assumed to have an initial temperature $t_0(x) > t_f$, i.e., cooling was considered. For heating, the fluid is warmer and one has $t_0(x) < t_f$. If however ϑ is defined as $\vartheta = t_f - t$, all the results given above are valid.

Plane plate

$$\frac{\vartheta}{\vartheta_0} = \frac{t - t_f}{t_0 - t_f} = \text{Function} \left(\frac{\lambda}{\alpha L}, \frac{\alpha \tau}{L^2}, \frac{x}{L} \right)$$

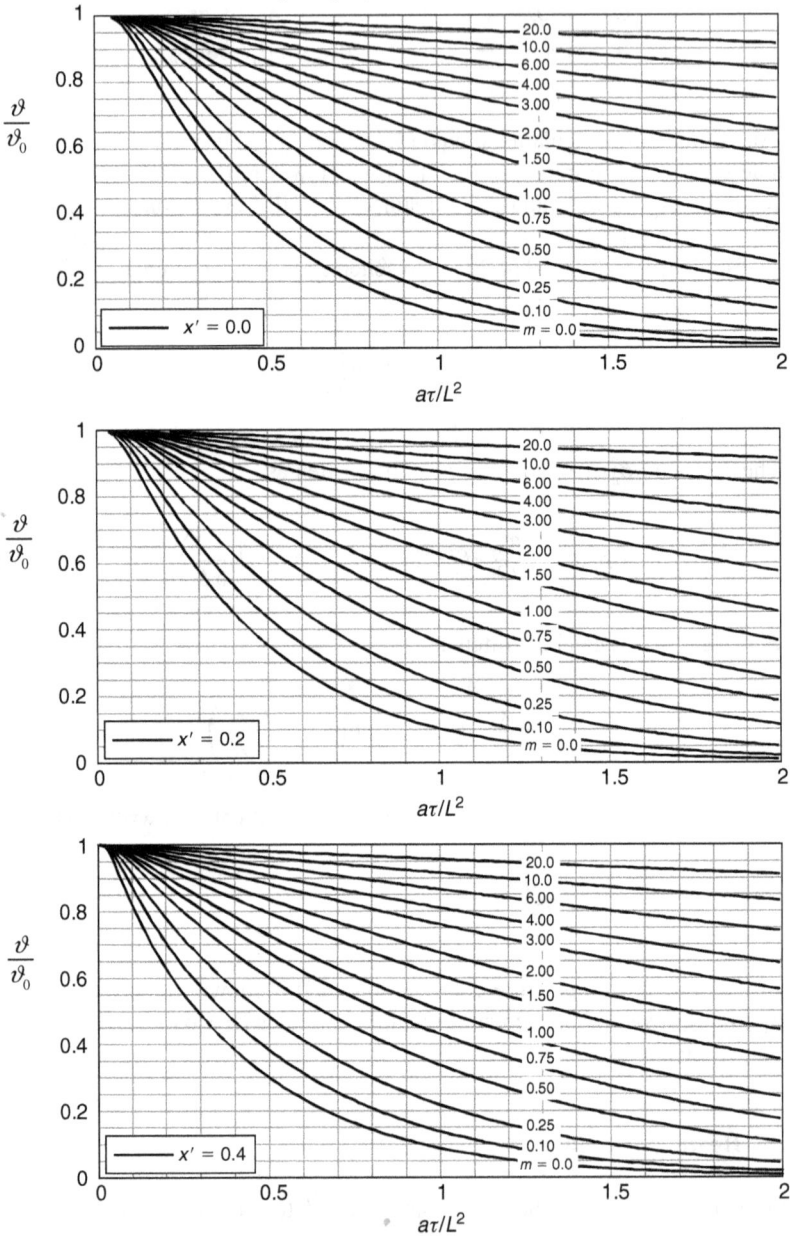

Figure 4.4: Unsteady temperature distribution in a plane plate. $x' = x/L$, $m = \lambda/\alpha L$.

Plane plate

$$\frac{\vartheta}{\vartheta_0} = \frac{t - t_f}{t_0 - t_f} = \text{Function}\left(\frac{\lambda}{\alpha L}, \frac{a\tau}{L^2}, \frac{x}{L}\right)$$

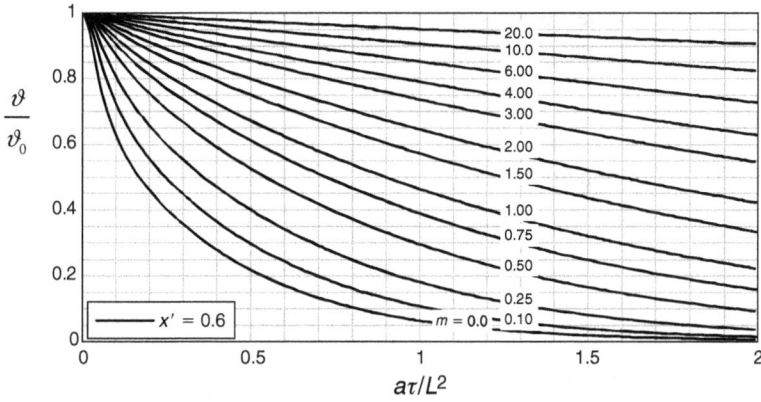

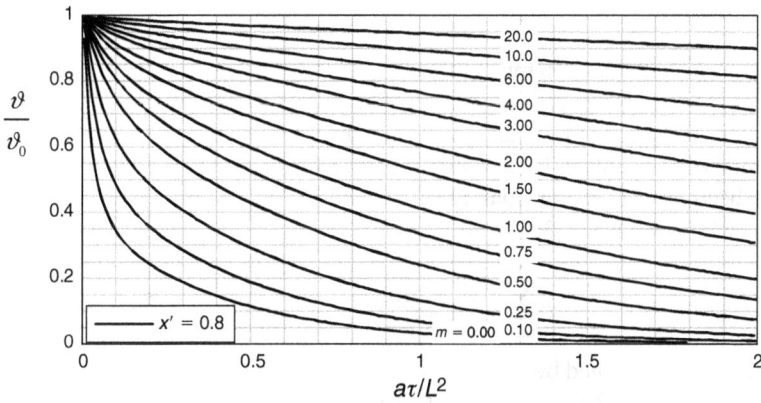

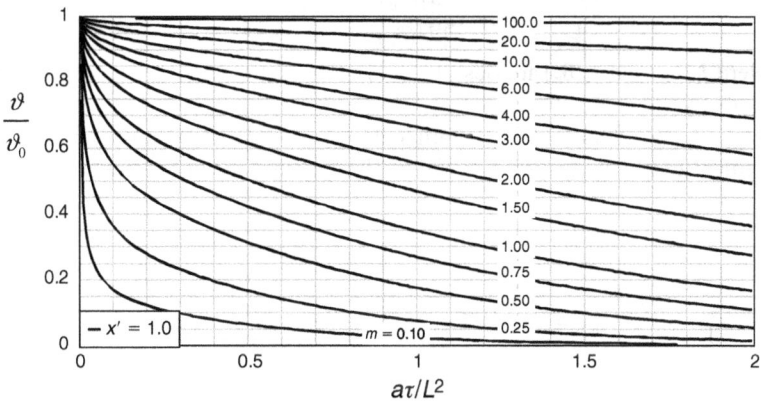

Figure 4.4: Continued.

4.4 Infinite long circular cylinder or cylinder with insulated end surfaces

Consider a long circular cylinder as shown in Fig. 4.5.

Transient radial heat conduction is now investigated. If $\vartheta = t - t_f$ and $a = \lambda/\rho c$ are introduced, then eq. (1.20) can be written as

$$\frac{\partial \vartheta}{\partial \tau} = a \left(\frac{\partial^2 \vartheta}{\partial r^2} + \frac{1}{r} \frac{\partial \vartheta}{\partial r} \right) \tag{4.31}$$

The boundary conditions are

$$r = 0: \frac{\partial \vartheta}{\partial r} = 0 \tag{4.32}$$

$$r = r_0: -\lambda \frac{\partial \vartheta}{\partial r} = \alpha \vartheta \tag{4.33}$$

The initial condition is

$$\tau = 0: \vartheta = \vartheta_0 = t_0 - t_f \tag{4.34}$$

The case with the initial temperature constant, $t(r, \tau = 0) = \text{constant}$, i.e., independent of the radius r, is considered here. The procedure to solve eq. (4.31) with the boundary conditions (4.32) and (4.33) and the initial condition (4.34) is similar to that for the infinite plate. Thus a solution of the form $\vartheta(r, \tau) = F(\tau) \cdot G(r)$ is assumed. $F(\tau)$ becomes an exponential function while $G(r)$ becomes Bessel functions. The result is (see Refs. [3, 4])

$$\frac{\vartheta}{\vartheta_0} = 2 \sum_{n=1}^{\infty} \frac{\exp(-\delta_n^2 a\tau/r_0^2)}{\delta_n} \cdot \frac{J_1(\delta_n) J_0(\delta_n r/r_0)}{J_0^2(\delta_n) + J_1^2(\delta_n)} \tag{4.35}$$

where δ_n is determined by

$$\delta_n \frac{J_1(\delta_n)}{J_0^2(\delta_n)} = \frac{\alpha r_0}{\lambda} \tag{4.36}$$

Here $J_0(\delta_n)$ and $J_1(\delta_n)$ are Bessel functions of the first kind, order zero and unity, respectively.

Equation (4.35) is not so appropriate for engineering calculations and it is more convenient to use the graphs in Fig. 4.6. These give ϑ/ϑ_0 in a schematic form.

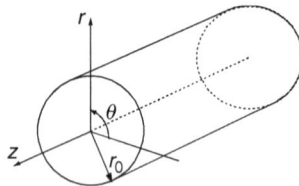

Figure 4.5: Long circular cylinder.

Cylinder

$$\frac{\vartheta}{\vartheta_0} = \frac{t - t_f}{t_0 - t_f} = \text{Function}\left(\frac{\lambda}{\alpha r_0}, \frac{a\tau}{r_0^2}, \frac{r}{r_0}\right)$$

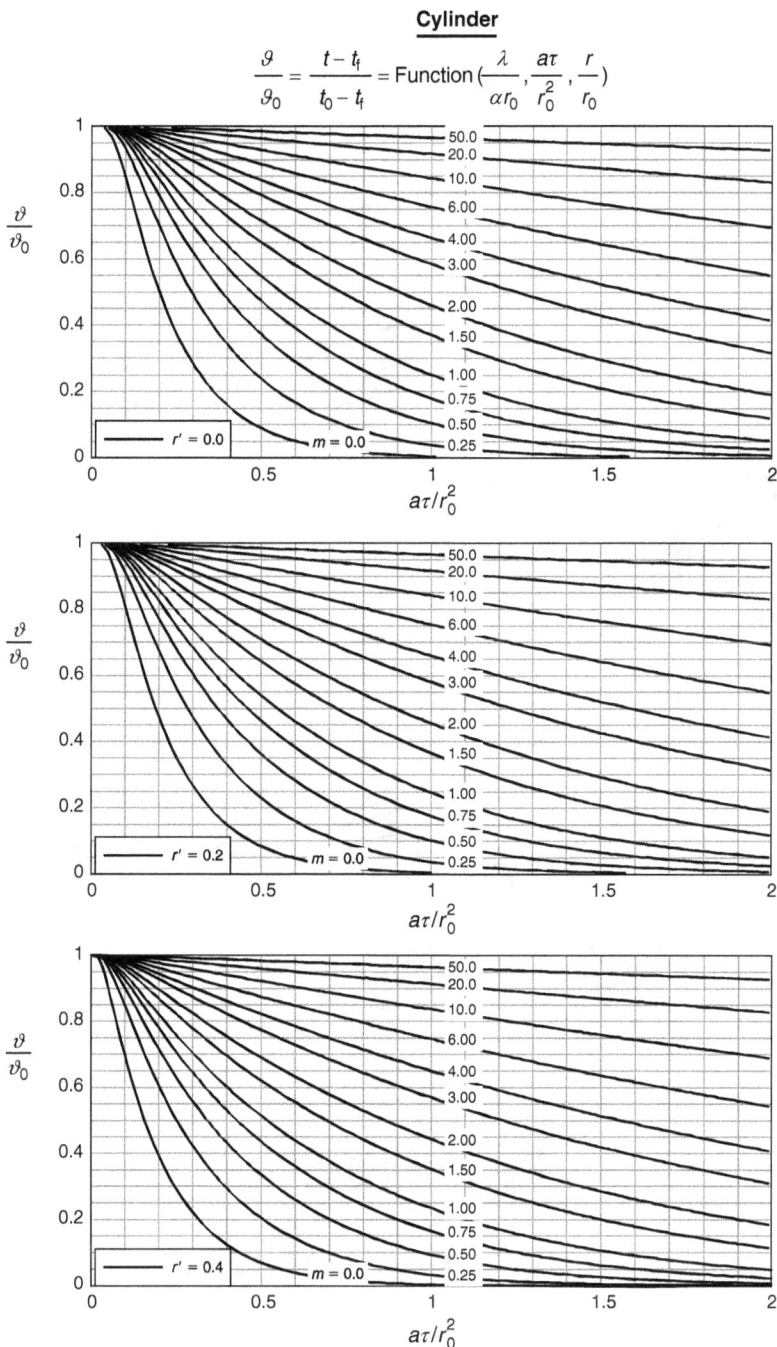

Figure 4.6: Unsteady temperature distribution in a cylinder. $r' = r/r_0$, $m = \lambda/\alpha r_0$.

Cylinder

$$\frac{\vartheta}{\vartheta_0} = \frac{t - t_f}{t_0 - t_f} = \text{Function}\left(\frac{\lambda}{\alpha r_0}, \frac{a\tau}{r_0^2}, \frac{r}{r_0}\right)$$

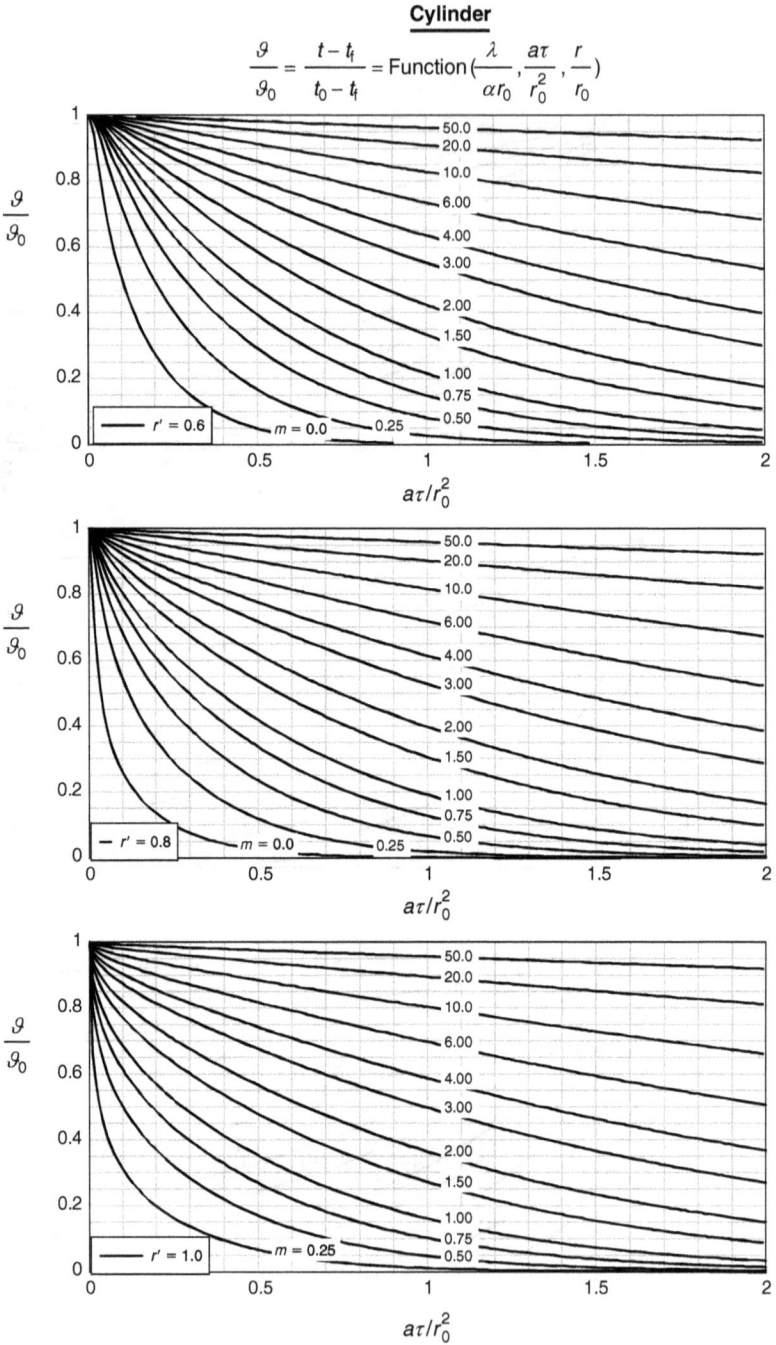

Figure 4.6: Continued.

4.5   Transient radial heat conduction in a sphere

The previous unsteady heat conduction problems for a plate and a circular cylinder can also be solved for a spherical body. The solution is given by

$$\frac{\vartheta}{\vartheta_0} = 4\frac{r_0}{r} \sum_{n=1}^{\infty} \frac{\exp(-\delta_n^2 a\tau/r_0^2)}{\delta_n} \sin\left(\frac{\delta_n r}{r_0}\right) \cdot \frac{\sin\delta_n - \delta_n \cos\delta_n}{2\delta_n - \sin 2\delta_n} \qquad (4.37)$$

where δ_n is determined by the condition

$$1 - \delta_n \cot \delta_n = \frac{a r_0}{\lambda} \qquad (4.38)$$

For engineering calculations, it is convenient to use the graphical solution provided in Fig. 4.7.

4.6   Graphical representation of the amount of heat flow

The dimensionless heat flow Q/Q_i as a function of time τ (i.e., from $\tau = 0$ to τ) for a plate was given in eq. (4.26). In Fig. 4.8, this is presented in a graphical form. Here, Q_i denotes $\lambda L\vartheta_0 A/a$ or $\rho c L\vartheta_0 A$.

The heat flow from a long cylinder is given by [5]

$$\frac{Q}{Q_i} = 4 \sum_{n=1}^{\infty} \left(1 - \exp\left(\frac{-\delta_n^2 a\tau}{r_0^2}\right)\right) \cdot \frac{J_1^2(\delta_n)}{J_0^2(\delta_n) + J_1^2(\delta_n)} \qquad (4.39)$$

where $Q_i = \pi r_0^2 L\rho c\vartheta_0$.

For a sphere, the corresponding heat flow can be written as [5]

$$\frac{Q}{Q_i} = 6\frac{a r_0}{\lambda} \sum_{n=1}^{\infty} \frac{(1 - \exp(-\delta_n^2 a\tau/r_0^2))}{\delta_n^3} \sin(\delta_n) \cdot \frac{\sin \delta_n - \delta_n \cos \delta_n}{\delta_n - \sin \delta_n \cos \delta_n} \qquad (4.40)$$

where $Q_i = 4/3\pi r_0^3 \rho c\vartheta_0$.

Figures 4.9 and 4.10 present the heat flow Q/Q_i for a cylinder and sphere, respectively.

4.7   Two- and three-dimensional solutions

In practice, it is common that transient heat conduction problems occur in rectangular beams or prismatic bodies or in short circular cylinders. In this section, it will be shown how the one-dimensional solutions can be extended to two- and three-dimensional solutions.

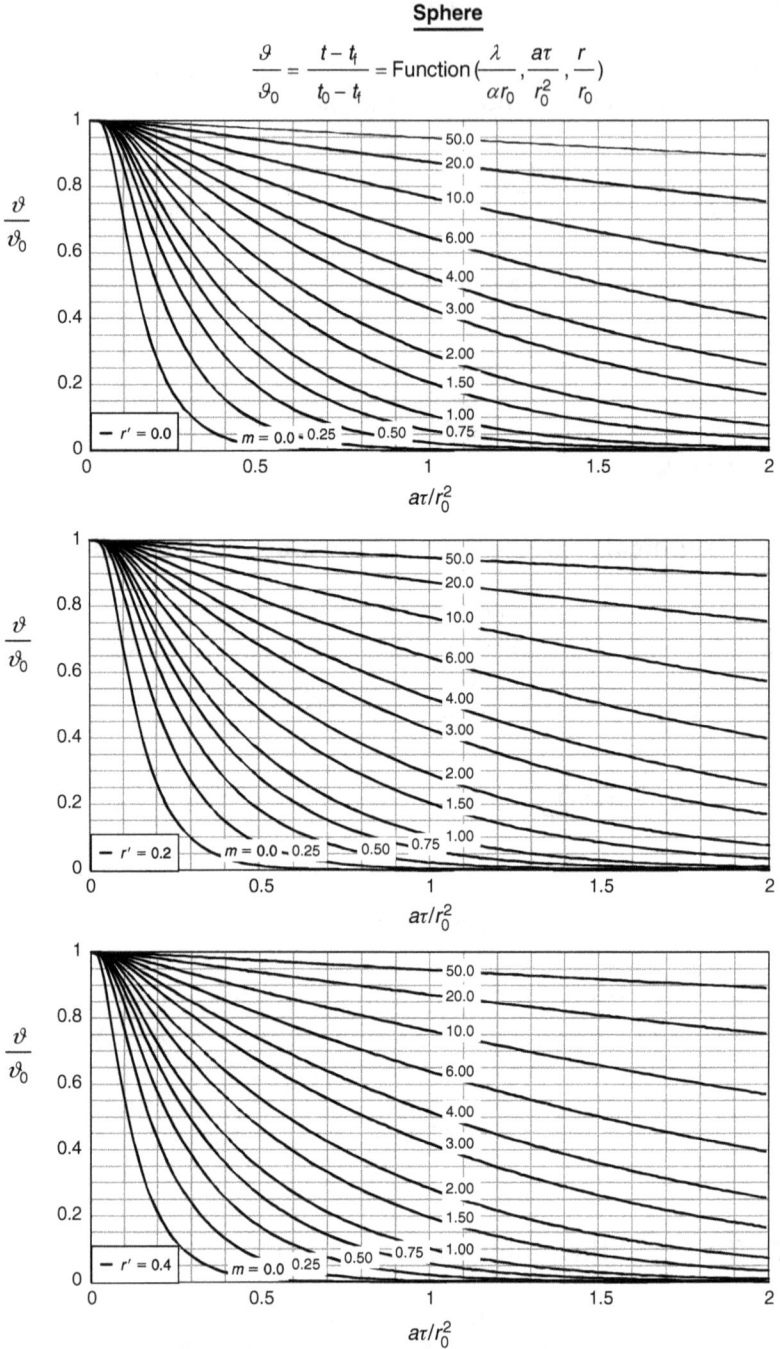

Figure 4.7: Unsteady temperature distribution in a sphere. $r' = r/r_0$, $m = \lambda/\alpha r_0$.

Sphere

$$\frac{\vartheta}{\vartheta_0} = \frac{t - t_f}{t_0 - t_f} = \text{Function}\left(\frac{\lambda}{\alpha r_0}, \frac{a\tau}{r_0^2}, \frac{r}{r_0}\right)$$

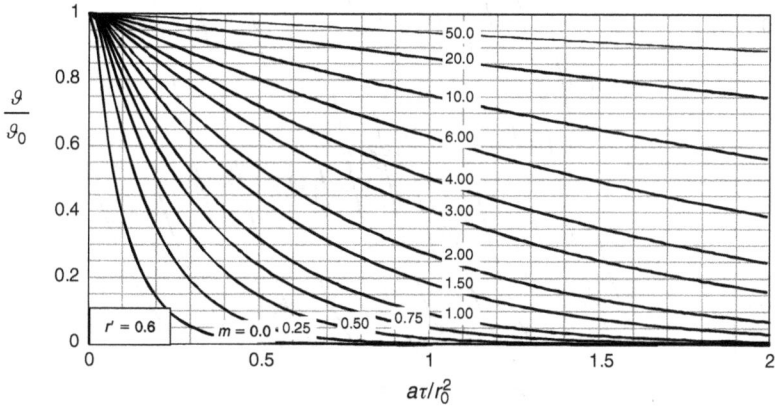

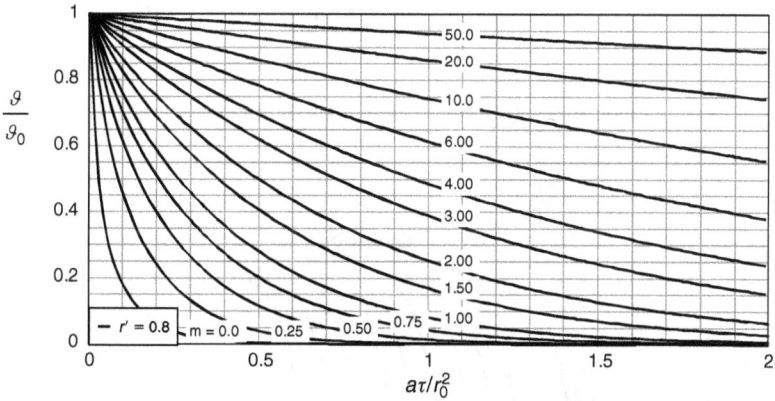

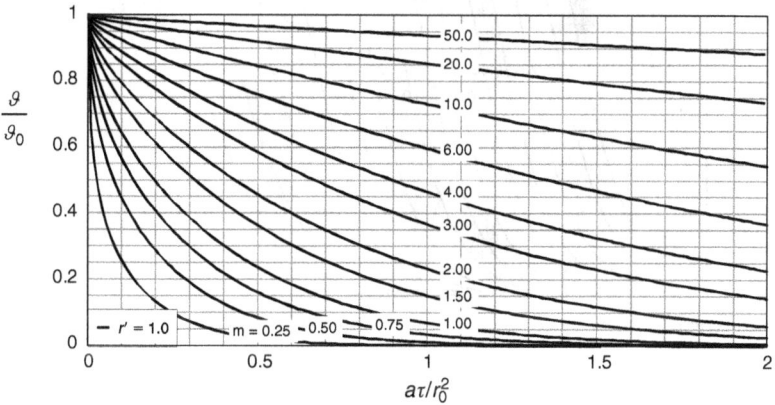

Figure 4.7: Continued.

Plate

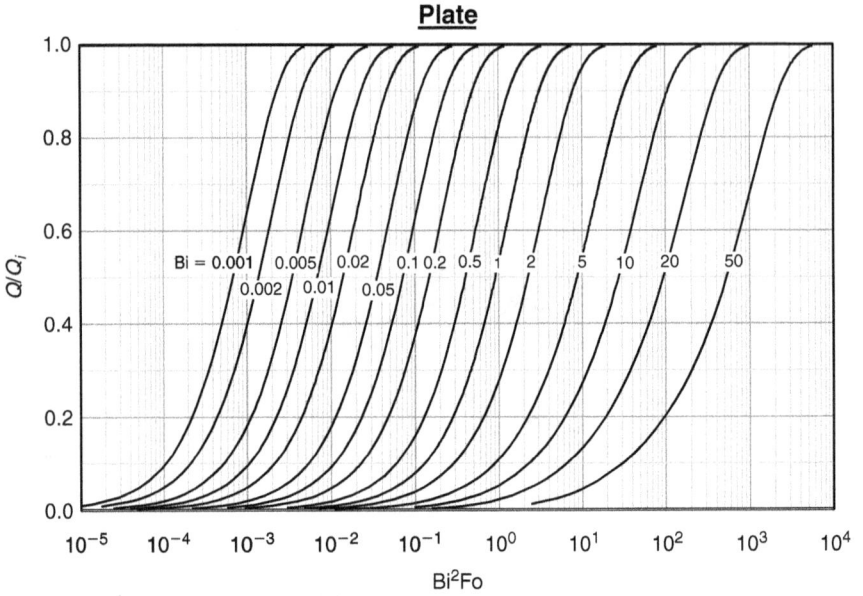

Figure 4.8: The heat flow Q/Q_i for a plane plate as function of $\alpha^2 a\tau/\lambda^2 = \mathrm{Bi}^2\mathrm{Fo}$.

Cylinder

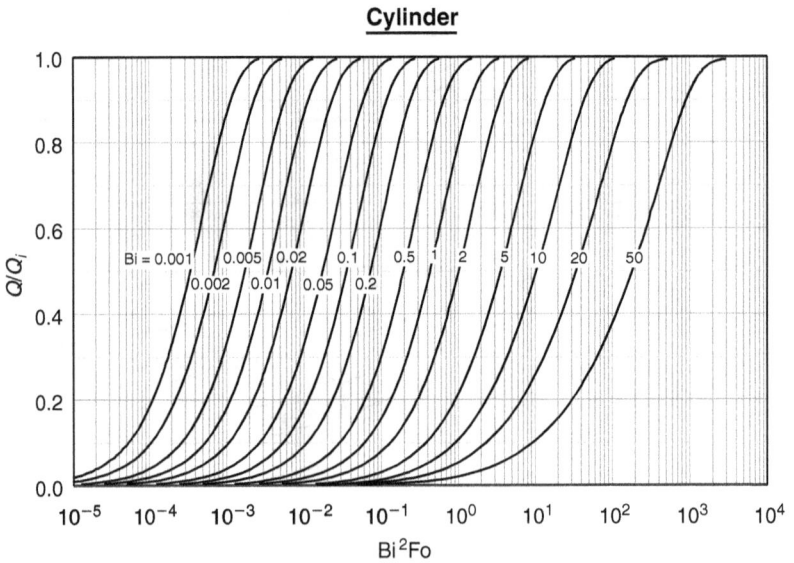

Figure 4.9: The heat flow Q/Q_i for a long cylinder as function of $\alpha^2 a\tau/\lambda^2 = \mathrm{Bi}^2\mathrm{Fo}$.

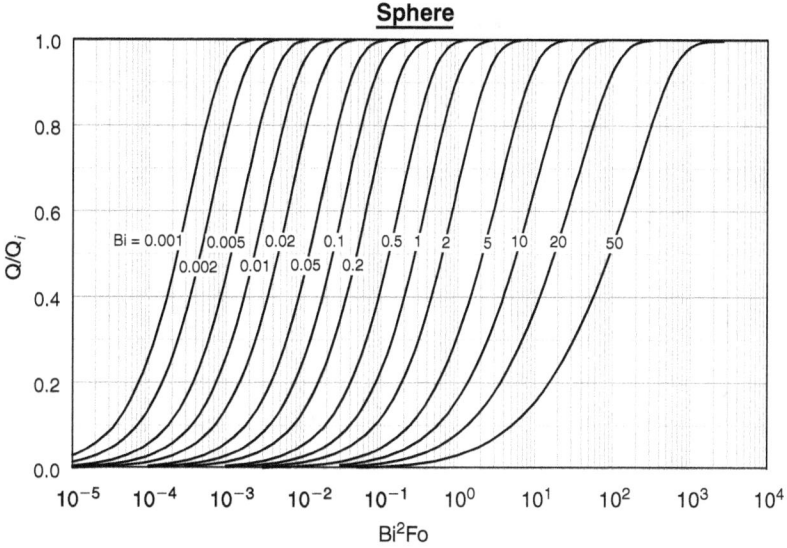

Figure 4.10: The heat flow Q/Q_i for a sphere as function of $\alpha^2 a\tau/\lambda^2 = \text{Bi}^2\text{Fo}$.

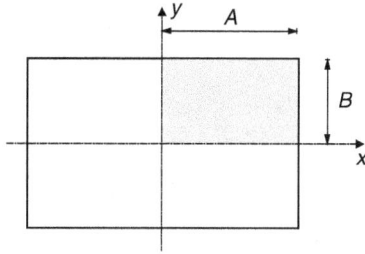

Figure 4.11: Rectangular beam (cross section).

Consider the cross section of a long rectangular beam as shown in Fig. 4.11.

The beam is cooled by convection and the heat transfer coefficient α is assumed to be constant along the boundary in contact with the surrounding fluid. Due to symmetry, it is sufficient to study a quarter of the cross section. As noted previously, when $\vartheta = t - t_f$ and $a = \lambda/\rho c$ are introduced, eq. (1.18) can be written as

$$\frac{\partial \vartheta}{\partial \tau} = a\left(\frac{\partial^2 \vartheta}{\partial x^2} + \frac{\partial^2 \vartheta}{\partial y^2}\right) \tag{4.41}$$

The boundary conditions to eq. (4.41) are

$$x = 0: \frac{\partial \vartheta}{\partial x} = 0$$

$$x = A: -\lambda\frac{\partial \vartheta}{\partial x} = \alpha\vartheta$$

$$y = 0: \frac{\partial \vartheta}{\partial y} = 0$$

$$y = B: -\lambda \frac{\partial \vartheta}{\partial y} = \alpha \vartheta$$

Introduce the nondimensional variables

$$x' = \frac{x}{A}; \quad y' = \frac{y}{B}; \quad \tau'_x = \frac{a\tau}{A^2}; \quad \tau'_y = \frac{a\tau}{B^2}; \quad \vartheta' = \frac{\vartheta}{\vartheta_0}$$

For the derivatives in eq. (4.41) one now has

$$\frac{\partial^2}{\partial x^2} = \frac{1}{A^2}\frac{\partial^2}{\partial x'^2}; \quad \frac{\partial^2}{\partial y^2} = \frac{1}{B^2}\frac{\partial^2}{\partial y'^2}; \quad \frac{\partial}{\partial \tau} = \frac{a}{A^2}\frac{\partial}{\partial \tau'_x} + \frac{a}{B^2}\frac{\partial}{\partial \tau'_y}$$

Equation (4.41) can now be written as

$$\frac{\partial \vartheta'}{\partial \tau'_x} - \frac{\partial^2 \vartheta'}{\partial x'^2} = -\frac{A^2}{B^2}\left(\frac{\partial \vartheta'}{\partial \tau'_y} - \frac{\partial^2 \vartheta'}{\partial y'^2}\right) \qquad (4.42)$$

The boundary conditions are transformed according to

$$x' = 0: \frac{\partial \vartheta'}{\partial x'} = 0$$

$$x' = 1: -\frac{\partial \vartheta'}{\partial x'} = \frac{\alpha A}{\lambda}\vartheta'$$

$$y' = 0: \frac{\partial \vartheta'}{\partial y'} = 0$$

$$y' = 1: -\frac{\partial \vartheta'}{\partial y'} = \frac{\alpha B}{\lambda}\vartheta'$$

Assume that the solution has the form

$$\vartheta' = \vartheta'_x(\tau'_x, x') \cdot \vartheta'_y(\tau'_y, y') \qquad (4.43)$$

Inserting (4.43) in eq. (4.42) gives

$$\frac{1}{\vartheta'_x}\left(\frac{\partial \vartheta'}{\partial \tau'_x} - \frac{\partial^2 \vartheta'}{\partial x'^2}\right) = -\frac{1}{\vartheta'_y}\frac{A^2}{B^2}\left(\frac{\partial \vartheta'}{\partial \tau'_y} - \frac{\partial^2 \vartheta'}{\partial y'^2}\right) \qquad (4.44)$$

Equation (4.44) is of the form

$$\text{function}(x', \tau'_x) = \text{function}(y', \tau'_y) \qquad (4.45)$$

Equation (4.45) must hold for all combinations of x', τ'_x, y', τ'_y, and thus the functions in eq. (4.45) must be a constant. In this case, the constant must however be zero. This is so because if the boundary $y = B$ (or $x = A$) is insulated, one-dimensional heat conduction prevails in the x-direction (y-direction). Equation (4.44) must then give the one-dimensional case of the general heat conduction equation. Thus one has

$$\frac{\partial \vartheta'}{\partial \tau'_x} = \frac{\partial^2 \vartheta'}{\partial x'^2} \tag{4.46}$$

$$\frac{\partial \vartheta'}{\partial \tau'_y} = \frac{\partial^2 \vartheta'}{\partial y'^2} \tag{4.47}$$

By this it has been shown that the solution to the two-dimensional unsteady heat conduction, given by eq. (4.41), can be found as the product of two one-dimensional solutions, eqs. (4.46) and (4.47) through (4.43). In a similar way, it can be shown that the three-dimensional unsteady heat conduction in a finite rectangular or prismatic body can be found as a product of three one-dimensional solutions.

Equations (4.46) and (4.47) are identical to eq. (4.8) but in a dimensionless form. Thus the solutions to eqs. (4.46) and (4.47) are given by the solution of eq. (4.8).

For a short circular cylinder, the solution can be found as a product of the solution for an infinite long cylinder and the solution of the infinite plate.

4.8   Semi-infinite bodies

In many engineering situations, the thickness of the heat conducting material is so big that the transient heat conduction is occurring within a thin surface layer at a boundary of the body. For such cases, the material is assumed to have an infinite thickness and the body is said to be semi-infinite as one surface or boundary is well defined (Fig. 4.12).

Consider now the semi-infinite body as shown in Fig. 4.12. Initially the body has a uniform temperature t_0. The surface temperature (at $x = 0$) is suddenly dropped to t_s and is then kept at this temperature. The task is now to determine the temperature

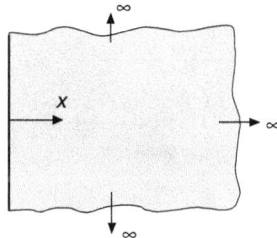

Figure 4.12: Transient heat conduction in a semi-infinite body.

distribution in the body as function of time τ and location x. As the temperature distribution is known, the heat transfer rate can be determined.

If the thermophysical properties are assumed constant, eq. (1.18) gives

$$\frac{\partial t}{\partial \tau} = a \frac{\partial^2 t}{\partial x^2} \tag{4.48}$$

The initial and boundary conditions are given by

$$\tau = 0: t(x, 0) = t_0 \tag{4.49}$$

$$x = 0, (\tau > 0): t(0, \tau) = t_s \tag{4.50}$$

The solution to this problem can be determined by applying Laplacian transformation, see Appendix 1, and is given by

$$\frac{t(x, \tau) - t_s}{t_0 - t_s} = \mathrm{erf}\left(\frac{x}{2\sqrt{a\tau}}\right) \tag{4.51}$$

where erf is the Gaussian error function defined by

$$\mathrm{erf}\left(\frac{x}{2\sqrt{a\tau}}\right) = \frac{2}{\sqrt{\pi}} \int_{0}^{x/2\sqrt{a\tau}} e^{-\eta^2} d\eta \tag{4.52}$$

The Gaussian error function is given in Table 4.1.

Using eqs. (4.51) and (4.52), the heat transfer rate at an arbitrary point can be determined as

$$\frac{\dot{Q}}{A} = \frac{1}{A}\frac{dQ}{d\tau} = -\lambda \frac{\partial t}{\partial x}$$

Finally one finds

$$\frac{\dot{Q}}{A} = -\lambda(t_0 - t_s)\frac{2}{\sqrt{\pi}}\exp\left(\frac{-x^2}{4a\tau}\right)\frac{1}{2\sqrt{a\tau}} = -\frac{\lambda(t_0 - t_s)}{\sqrt{a\pi\tau}}\exp\left(\frac{-x^2}{4a\tau}\right) \tag{4.53}$$

Especially for $x = 0$ one has

$$\left(\frac{\dot{Q}}{A}\right)_{x=0} = \frac{\lambda(t_s - t_0)}{\sqrt{a\pi\tau}} \tag{4.54}$$

4.8.1   Constant heat flux at $x = 0$

The semi-infinite body in Fig. 4.12 with the initial temperature t_0 may also all of a sudden be exposed to a constant heat flux $q_0 = \dot{Q}/A$ at $x = 0$. The solution to this

Table 4.1: $\mathrm{erf}(\phi) = (2/\sqrt{\pi}) \int_0^n e^{-\eta^2} d\eta$.

ϕ	erf ϕ	ϕ	erf ϕ	ϕ	erf ϕ
0.00	0.00000	0.44	0.46623	1.28	0.92973
0.02	0.02256	0.48	0.50275	1.32	0.93807
0.04	0.04511	0.52	0.53790	1.40	0.95229
0.06	0.06762	0.56	0.57162	1.50	0.96611
0.08	0.09008	0.60	0.60386	1.60	0.97635
0.10	0.11246	0.64	0.63459	1.70	0.98379
0.12	0.13476	0.68	0.66378	1.80	0.98909
0.14	0.15695	0.72	0.69143	1.90	0.99279
0.16	0.17901	0.76	0.71754	2.00	0.99532
0.18	0.20094	0.80	0.74210	2.10	0.99702
0.20	0.22270	0.84	0.76514	2.20	0.99814
0.22	0.24430	0.88	0.78669	2.30	0.99886
0.24	0.26570	0.92	0.80677	2.40	0.9993115
0.26	0.28690	0.96	0.82542	2.60	0.9997640
0.28	0.30788	1.00	0.84270	2.80	0.9999250
0.30	0.32863	1.04	0.85865	3.00	0.9999779
0.32	0.34913	1.08	0.87333	3.20	0.9999940
0.34	0.36936	1.12	0.88679	3.40	0.9999985
0.36	0.38933	1.16	0.89910	3.60	0.9999996441
0.38	0.40901	1.20	0.91031	3.80	0.9999999230
0.40	0.42839	1.24	0.92050	4.00	0.9999999846

problem can also be obtained by Laplacian transformation, see Appendix 1, and the solution is

$$t - t_0 = \frac{2q_0}{\lambda}\sqrt{\frac{a\tau}{\pi}}\exp\left(\frac{-x^2}{4a\tau}\right) - \frac{q_0 x}{\lambda}\left(1 - \mathrm{erf}\left(\frac{x}{2\sqrt{a\tau}}\right)\right) \tag{4.55}$$

4.8.2 Convective boundary conditions

In most engineering cases, convective heat transfer occurs along the boundaries of the bodies. For the semi-infinite body in Fig. 4.12, the following conditions apply

$$\tau = 0: t(x,0) = t_0$$

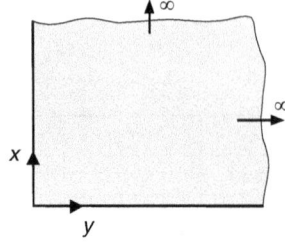

Figure 4.13: Quarter infinite body.

$$x = 0, (\tau > 0): \alpha(t_f - t(0, \tau)) = -\lambda \frac{\partial t}{\partial x}$$

where t_f is the fluid temperature.

The solution for this problem is (see, for example, Schneider [5])

$$\frac{t - t_0}{t_f - t_0} = 1 - \text{erf}\left(\frac{x}{2\sqrt{a\tau}}\right) - \left(1 - \text{erf}\left(\frac{x}{2\sqrt{a\tau}} + \frac{\alpha\sqrt{a\tau}}{\lambda}\right)\right) \cdot \exp\left(\frac{\alpha x}{\lambda} + \frac{\alpha^2 a\tau}{\lambda^2}\right)$$

(4.56)

4.8.3 Two-dimensional quarter infinite bodies

Consider a quarter infinite body as shown in Fig. 4.13.

The body has initially a uniform temperature t_0. The surface temperatures (at $x = 0$ and $y = 0$) are suddenly brought to the temperature t_s. It can be shown, see Ref. [6], that the temperature distribution is given by

$$\vartheta' = \frac{\vartheta}{\vartheta_0} = \frac{t - t_s}{t_0 - t_s} = \text{erf}\left(\frac{x}{2\sqrt{a\tau}}\right) \cdot \text{erf}\left(\frac{y}{2\sqrt{a\tau}}\right)$$

(4.57)

References

[1] J.P. Holman, Heat Transfer, 10th ed., McGraw-Hill, New York (2009).
[2] Standard Mathematical Tables, 17th ed., Chemical Rubber Co., Cleveland, OH (1969).
[3] E.R.G. Eckert and R.M. Drake Jr., Analysis of Heat and Mass Transfer, McGraw-Hill, New York (1972).
[4] J.R. Welty, Engineering Heat Transfer, John Wiley & Sons, New York (1978).
[5] P.J. Schneider, Conduction Heat Transfer, 6th ed., Addison-Wesley, Reading, MA (1974).
[6] Heat Exchanger Design Handbook, Hemisphere Publ. Corp., Washington, DC (1983).

5 Heat conduction with moving boundaries

5.1 Introduction

In many engineering applications, the heat transfer process occurs in a material undergoing a phase transition or in a material in which the composition is changed due to some chemical reaction. In the active zone, thermal energy is released or absorbed. The released or absorbed energy is transported through the material by conduction. Engineering examples are melting and solidification processes, chemical reactions like combustion, penetration of frost in the ground, and so-called ablation cooling in space applications.

An essential and characteristic feature for problems of this type is that an interface separates two regions with different thermal properties and that this interface moves as a function of time. At the interface, heat or energy is released or absorbed. It is very important to be able to predict the movement of the interface.

Only a few analytical solutions exist for this problem type. In this chapter, cases of melting and solidification are considered.

5.2 Solidification process where the solid phase has negligible heat capacity

Consider the formation of ice in a water volume as illustrated in Fig. 5.1.

At time τ the solid phase has the thickness $X(\tau)$. The exposed surface is kept at the temperature t_s while the liquid is at the melting (solidification) temperature t_0, $(t_s < t_0)$. The released energy (latent heat), at the interface between the liquid and solid phases, is conducted across the solid phase.

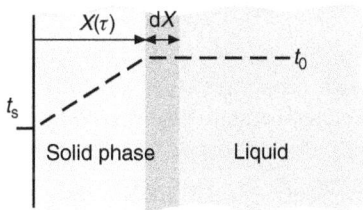

Figure 5.1: Solidification process where the solid phase has negligible heat capacity.

A heat balance now gives

$$\rho A \frac{dX}{d\tau} r_s = \frac{\lambda A}{X}(t_0 - t_s) \tag{5.1}$$

With the initial condition $X = 0$ for $\tau = 0$, the solution to eq. (5.1) is given by

$$X(\tau) = \sqrt{\frac{2\lambda(t_0 - t_s)}{\rho r_s}\tau} \tag{5.2}$$

In eqs. (5.1) and (5.2), r_s (in J/kg) is the latent heat for melting or solidification.

From eq. (5.2) it is obvious that the thickness of the solid phase increases with time proportional to $\sqrt{\tau}$. This is a feature commonly appearing for heat conduction problems including a phase change.

5.3   Melting and solidification taking heat capacity into account

Consider now the solidification process in a liquid. Melting of a solid phase can be considered in a similar way. In Fig. 5.2, the liquid phase occupying the whole region at time $\tau = 0$ is shown. Initially, the liquid has the temperature T_2. The temperature of the liquid surface is suddenly changed to T_1 which is lower than the solidification (melting) temperature T_M. At time τ the interface between the liquid and solid phases is situated at a distance $X(\tau)$ from the original liquid surface.

Heat is conducted from the liquid through the solid phase to the exposed surface. At the interface the latent heat is released.

At time τ the region $x < X(\tau)$ is occupied by the solid phase with the physical properties λ_1, a_1, ρ_1, and c_1. If t_1 is the temperature of the solid phase, one has

$$\frac{\partial t_1}{\partial \tau} = a_1 \frac{\partial^2 t_1}{\partial x_1^2} \tag{5.3}$$

where x_1 is the coordinate in the solid phase. The region $x > X(\tau)$ is occupied by the liquid phase with the physical properties λ_2, a_2, ρ_2, and c_2. Let t_2 be the temperature in the liquid phase and then temperature is governed by

$$\frac{\partial t_2}{\partial \tau} = a_2 \frac{\partial^2 t_2}{\partial x_2^2} \tag{5.4}$$

where x_2 is the coordinate in the liquid phase.

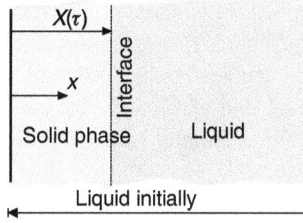

Figure 5.2: Solidification process.

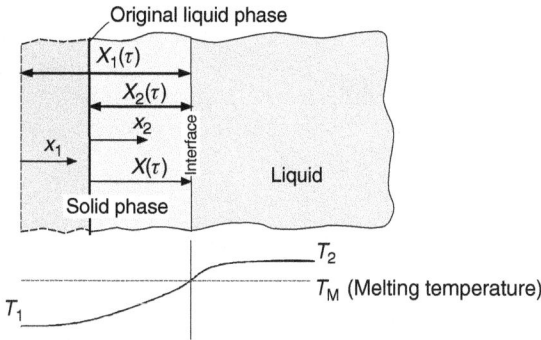

Figure 5.3: Heat conduction process for solidification of water to ice.

When water is solidified, the volume increases (the density decreases). This means that the ice surface does not coincide with the original liquid surface (Fig. 5.3).

A mass balance results in

$$\frac{\rho_1}{\rho_2} = \frac{X_2}{X_1} = \frac{1}{\beta} \tag{5.5}$$

where $X_1(\tau)$ is the thickness of the ice layer and $X_2(\tau)$ is the thickness of the liquid layer which solidifies.

The boundary conditions of eqs. (5.3) and (5.4) are

$$x_1 = 0: t_1 = T_1 \tag{5.6}$$

$$x_2 \rightarrow \infty: t_2 \rightarrow T_2 \tag{5.7}$$

$$\left.\begin{array}{l} x_1 = X_1(\tau) \\ x_2 = X_2(\tau) \end{array}\right\} t_1 = t_2 = T_M \tag{5.8}$$

$$\lambda_1 \left(\frac{\partial t_1}{\partial x_1}\right)_{x_1=X_1} - \lambda_2 \left(\frac{\partial t_2}{\partial x_2}\right)_{x_2=X_2} = \rho_1 r_s \frac{dX_1}{d\tau} = \rho_2 r_s \frac{dX_2}{d\tau} \tag{5.9}$$

Equations (5.8) and (5.9) express the so-called interface conditions; eq. (5.8) means continuity in temperature, and eq. (5.9) tells that the difference in heat flux between the liquid and solid phases is balanced by the released energy.

The solution of eqs. (5.3) and (5.4) with the conditions in eqs. (5.5)–(5.9) will now be presented, see, for example, Carslaw and Jaeger [1] and Eckert and Drake [2].

Introduce

$$\vartheta_1 = t_1 - T_M = (T_1 - T_M) + A\,\mathrm{erf}\frac{x_1}{2\sqrt{a_1\tau}} \tag{5.10}$$

$$\vartheta_2 = t_2 - T_M = (T_2 - T_M) + B\left(1 - \mathrm{erf}\frac{x_2}{2\sqrt{a_2\tau}}\right) \tag{5.11}$$

where A and B are constants. With the results in Chapter 4, one knows that eqs. (5.10) and (5.11) satisfy eqs. (5.3) and (5.4), respectively. By applying condition (5.8) one finds

$$0 = (T_1 - T_M) + A\,\mathrm{erf}\frac{X_1}{2\sqrt{a_1\tau}} \tag{5.12}$$

$$0 = (T_2 - T_M) + B\left(1 - \mathrm{erf}\frac{X_2}{2\sqrt{a_2\tau}}\right) \tag{5.13}$$

Equations (5.12) and (5.13) must be valid for all $X_1(\tau)$ and $X_2(\tau)$. Because T_1, T_2, and T_M are constant, the erf function must have a constant value. This is possible only if X_1 and X_2 are proportional to $\sqrt{\tau}$. With the condition (5.5) one may write

$$X_1 = K\beta\sqrt{\tau} \tag{5.14}$$

$$X_2 = K\sqrt{\tau} \tag{5.15}$$

where K is a constant.

The condition (5.9) is now applied on expressions (5.10) and (5.11). One then finds

$$\lambda_1 A\frac{\exp(-X_1^2/4a_1\tau)}{\sqrt{\pi a_1}} + \lambda_2 A\frac{\exp(-X_2^2/4a_2\tau)}{\sqrt{\pi a_2}} = \rho_1 r_s\frac{\mathrm{d}X_1}{\mathrm{d}\tau}$$

Applying eqs. (5.12), (5.13), and (5.14), eq. (5.15) results in

$$\frac{(T_M - T_1)\lambda_1\exp(-K^2\beta^2/4a_1)}{\sqrt{\pi a_1}\,\mathrm{erf}(K\beta/2\sqrt{a_1})} - \frac{(T_2 - T_M)\lambda_2\exp(-K^2/4a_2)}{\sqrt{\pi a_2}(1 - \mathrm{erf}(K/2\sqrt{a_2}))} = \rho_1 r_s\frac{K\beta}{2} \tag{5.16}$$

From eq. (5.16), K can be solved numerically as a function of T_1, T_2, and T_M as well as of the physical properties. As K is known, A and B can be determined from eqs. (5.12) and (5.13), respectively, with eqs. (5.14) and (5.15), respectively, inserted.

The temperature distribution in the solid phase can be written as

$$\frac{t_1 - T_M}{T_1 - T_M} = 1 - \frac{\mathrm{erf}(x/2\sqrt{a_1\tau})}{\mathrm{erf}(K\beta/2\sqrt{a_1})} \tag{5.17}$$

and the temperature distribution in the liquid phase can be written as

$$\frac{t_2 - T_M}{T_2 - T_M} = 1 - \frac{1 - \mathrm{erf}(x/2\sqrt{a_2\tau})}{1 - \mathrm{erf}(K/2\sqrt{a_2})} \tag{5.18}$$

One observes that eq. (5.17) satisfies the condition $t_1 = T_1$ for $x = 0$ and that eq. (5.18) satisfies the condition $t_2 \to T_2$ as $x \to \infty$.

From eq. (5.16), K has been determined for values of T_1 and T_2 close to the solidification temperature $0°C$. The value of β was set to $\beta = 1$ and $\beta = 1.09$. The latter value corresponds to the density ratio between water and ice. The result for K is

$$K_{\beta=1} = \sqrt{\frac{2(T_M - T_1)\lambda_1}{\rho_1 r_s}} \tag{5.19}$$

$$K_{\beta=1.09} = \sqrt{\frac{2(T_M - T_1)\lambda_1}{\rho_1 r_s \beta^2}} \tag{5.20}$$

From eqs. (5.20) and (5.14), it is evident that for $\beta > 1$, the time to reach a certain thickness of the solid phase is longer than for $\beta = 1$.

References

[1] H.S. Carslaw and J.C. Jaeger, Conduction of Heat in Solids, 2nd ed., Oxford University Press, New York (1959).
[2] E.R.G. Eckert and R.M. Drake Jr., Analysis of Heat and Mass Transfer, McGraw-Hill, Tokyo (1972).

6 Convection—general theory

6.1 Introduction

In the preceding chapters, heat conduction in solid bodies was treated extensively. Often convective cooling or heating with an adjacently moving fluid occurred along the boundary of the body. In the fluid the heat transport is due to a combination of molecular heat conduction and transport of internal energy due to the macroscopic motion of the fluid. If the fluid movement is created by external devices such as pumps, fans, and compressors one speaks about *forced convection*. If the fluid motion is created as a result of inequalities in the density (due to temperature differences) one speaks about *natural* or *free convection*.

In Fig. 6.1, the solid body as well as in the fluid a temperature field is created. Especially at the surface of the body, the fluid velocity must be zero which means that the heat transfer process there only can occur due to conduction. The heat flux can then be written as

$$\frac{\dot{Q}}{A} = q = -\lambda_f \left(\frac{\partial t}{\partial y} \right)_{y=0^+} = \left\{ -\lambda \left(\frac{\partial t}{\partial y} \right)_{y=0^-} \right\}_{body} \qquad (6.1)$$

Commonly a heat transfer coefficient α is introduced according to

$$q = \alpha(t_w - t_\infty) \qquad (6.2)$$

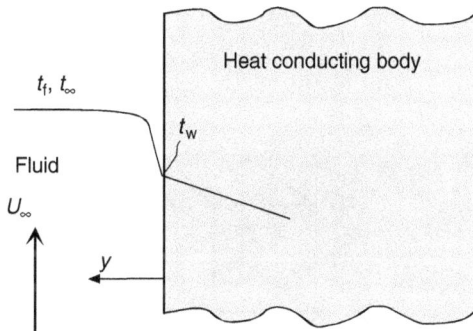

Figure 6.1: Illustration of convective heat transfer.

Equations (6.2) and (6.1) give

$$\alpha = \frac{q}{t_w - t_\infty} = \frac{-\lambda_f \left(\dfrac{\partial t}{\partial y} \right)_{y=0^+}}{t_w - t_\infty} \tag{6.3}$$

The heat transfer coefficient α in eq. (6.3) is a complex and non-constant variable which depends on the geometry of the solid body, the flow field, and the physical properties of the fluid. To determine α is the main issue in the theory of convective heat transfer. (In what follows the notation t_∞ will often be used instead of t_f.)

The overall objective of all chapters on convective heat transfer in this book is to determine α for either a prescribed surface temperature distribution t_w or a prescribed surface heat flux q_w. To enable this, one needs to analyze the flow and temperature fields in the fluid. The tools for this are:

a) mass conservation equation (continuity equation)
b) Navier–Stokes' equations
c) energy equation (from first law of thermodynamics).

6.2   Continuity equation (mass conservation equation)

The continuity equation (K.E.) expresses that mass is constant or not destroyable. If one considers a mass balance for the volume element in Fig. 6.2, the following equation appears

$$\frac{\partial \rho}{\partial \tau} + \frac{\partial(\rho u)}{\partial x} + \frac{\partial(\rho v)}{\partial y} + \frac{\partial(\rho w)}{\partial z} = 0 \tag{6.4}$$

where ρ is the density of the fluid and u, v, and w are the velocity components in the x-, y-, and z-directions.

Especially for steady state ($\partial/\partial \tau \equiv 0$), incompressible flow ($\rho = \text{constant}$) and two-dimensional ($w \equiv 0$, $\partial/\partial z \equiv 0$) flow, one has

$$\frac{\partial u}{\partial x} + \frac{\partial v}{\partial y} = 0 \tag{6.5}$$

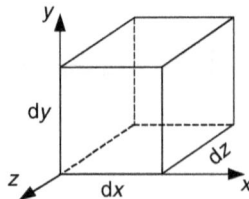

Figure 6.2: Volume element dx dy dz.

The derivation of eqs. (6.4) and (6.5) can be found in White [1].

Equation (6.4) tells that the sum of the mass within the volume element $dx\,dy\,dz$ in Fig. 6.2 and the net in- or outflowing mass is constant. This mass can therefore be considered as a system from a thermodynamics point of view.

6.3   Navier–Stokes' equations

The equations of motion are derived from Newton's second law which says that mass times acceleration in a certain direction is equal to the net external forces acting in the same direction. The external forces acting on a fluid element are splitted up into surface forces and volume or body forces. The volume or body forces act uniformly over the fluid element and are commonly generated by the gravity acceleration, electrical or magnetic fields. The surface forces act on the boundary surfaces of the fluid element and are acting as normal forces or shear forces. Newton's second law can now be written

$$\text{mass} \times \text{acceleration}_i = \text{volume forces}_i + \text{surface forces}_i \qquad (6.6)$$

Index i means direction i, i.e., x, y, or z.

The volume or body forces are calculated per unit mass and denoted by F_i.

$$F_i = (F_x, F_y, F_z)$$

The surface forces are calculated per unit area and are rather called stresses. As a stress is acting perpendicular to a surface (in the direction of the surface normal) it is called a normal stress and if it is acting along the surface it is called a shear stress. In Fig. 6.3 the stresses on a two-dimensional element $dx\,dy$ are shown. σ_{xx} and σ_{yy} are normal stresses in the x- and y-directions, respectively. The shear stresses are denoted as σ_{xy} and σ_{yx}, where the first index indicates the axis to which the considered surface is directed perpendicularly and the second index gives the direction of the stress.

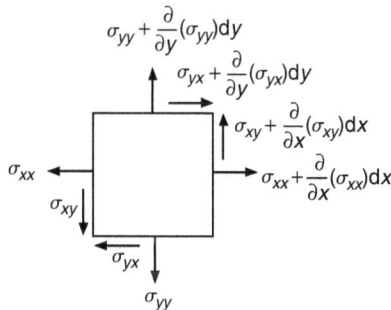

Figure 6.3: Stresses on the element $dx\,dy$.

The schematic picture in Fig. 6.3 can easily be developed to the three-dimensional case but then the normal stress σ_{zz} and the shear stresses σ_{xz}, σ_{zx} and σ_{yz}, σ_{zy} have to be added.

If the various terms in eq. (6.6) are set up for the x-, y-, and z-directions one finds, see Refs. [1–3],

$$\hat{x}: \rho \left(\frac{\partial u}{\partial \tau} + u\frac{\partial u}{\partial x} + v\frac{\partial u}{\partial y} + w\frac{\partial u}{\partial z} \right) = \rho F_x + \frac{\partial \sigma_{xx}}{\partial x} + \frac{\partial \sigma_{yx}}{\partial y} + \frac{\partial \sigma_{zx}}{\partial z} \tag{6.7}$$

$$\hat{y}: \rho \left(\frac{\partial v}{\partial \tau} + u\frac{\partial v}{\partial x} + v\frac{\partial v}{\partial y} + w\frac{\partial v}{\partial z} \right) = \rho F_y + \frac{\partial \sigma_{xy}}{\partial x} + \frac{\partial \sigma_{yy}}{\partial y} + \frac{\partial \sigma_{zy}}{\partial z} \tag{6.8}$$

$$\hat{z}: \rho \left(\frac{\partial w}{\partial \tau} + u\frac{\partial w}{\partial x} + v\frac{\partial w}{\partial y} + w\frac{\partial w}{\partial z} \right) = \rho F_z + \frac{\partial \sigma_{xz}}{\partial x} + \frac{\partial \sigma_{yz}}{\partial y} + \frac{\partial \sigma_{zz}}{\partial z} \tag{6.9}$$

6.3.1   The stress tensor σ_{ij}

The nine stresses σ_{xx}, σ_{xy},..., σ_{zz} can be brought together in the tensor σ_{ij} (matrix with three rows and three columns) according to

$$\sigma_{ij} = \begin{bmatrix} \sigma_{xx} & \sigma_{xy} & \sigma_{xz} \\ \sigma_{yx} & \sigma_{yy} & \sigma_{yz} \\ \sigma_{zx} & \sigma_{zy} & \sigma_{zz} \end{bmatrix} \tag{6.10}$$

The mean value of the normal stresses is defined as the fluid pressure as

$$-p = \frac{\sigma_{xx} + \sigma_{yy} + \sigma_{zz}}{3} = \frac{\sigma_{ii}}{3} \tag{6.11}$$

where the negative sign is necessary due to sign convention. This definition of the pressure is the same as the one being used for fluids at rest. For the case with a moving fluid, the definition is arbitrary but convenient. The quantity p according to eq. (6.11) is however not depending on the chosen coordinate system but has the same value in a Cartesian x-, y-, z-coordinate system as in a cylindrical coordinate system r, θ, z. A quantity having this property is called an *invariant*.

The indices i and j in σ_{ij} can be x, y, or z. If an index is repeated as in σ_{ii} (eq. (6.11)), this means that summation over the repeated index should be carried out, i.e., $\sigma_{ii} = \sigma_{xx} + \sigma_{yy} + \sigma_{zz}$.

The stress tensor σ_{ij} is usually written as

$$\sigma_{ij} = -p\delta_{ij} + d_{ij} \tag{6.12}$$

where

d_{ij} is called the deviatoric-stress tensor

and

$$\delta_{ij} \text{ is the Kronecker delta } \begin{cases} 0 & i \neq j \\ 1 & i = j \end{cases}$$

From the definitions in eqs. (6.11) and (6.12), it follows (note $\delta_{ii} = 3$)

$$d_{ii} = d_{xx} + d_{yy} + d_{zz} = 0 \qquad (6.13)$$

Especially for a Newtonian fluid one writes

$$d_{ij} = 2\mu \left(\frac{e_{ij} - \Delta \delta_{ij}}{3} \right) \qquad (6.14)$$

where $\mu =$ dynamic viscosity (kg/m s) and

$$e_{ij} = \text{strain} - \text{rate} - \text{tensor} = \frac{1}{2} \left(\frac{\partial u_i}{\partial x_j} + \frac{\partial u_j}{\partial x_i} \right) \qquad (6.15)$$

$$\Delta = e_{ii} = \frac{\partial u_i}{\partial x_i} = \frac{\partial u}{\partial x} + \frac{\partial v}{\partial y} + \frac{\partial w}{\partial z} \qquad (6.16)$$

The derivation of the equations above can be found in Batchelor [2] and Young et al. [3]. Originally the expression (6.14) comes from Navier (1822), Poisson (1829), Saint-Venant (1843), and Stokes (1845).

For water and most gases the assumption of a Newtonian fluid corresponding to eq. (6.14) is a valid approach.

The pressure p is defined as $-p = \sigma_{ii}/3$ and that it is not self-evident that this pressure is equal to that one in thermodynamic relations. However, if the flow is incompressible, the defined pressure is equal to the thermodynamic pressure, see Batchelor [2] and Kestin [4].

If eqs. (6.12), (6.14), and (6.15) are inserted in eqs. (6.7)–(6.9) one finds the Navier–Stokes' equations (N.S.).

6.3.2   Navier–Stokes' equations for two-dimensional and incompressible flow

If the flow is incompressible one has $\rho =$ constant and $\Delta = 0$.

For the stresses, one finds by using eqs. (6.12), (6.14), and (6.15):

$$\sigma_{xx} = -p + 2\mu e_{xx} = -p + 2\mu \frac{\partial u}{\partial x} \qquad (6.17)$$

$$\sigma_{xy} = \sigma_{yx} = 2\mu e_{xy} = \mu \left(\frac{\partial u}{\partial y} + \frac{\partial v}{\partial x} \right) \qquad (6.18)$$

$$\sigma_{yy} = -p + 2\mu e_{yy} = -p + 2\mu \frac{\partial v}{\partial y} \qquad (6.19)$$

By inserting eqs. (6.5), (6.17) and (6.18) in eq. (6.7), and eqs. (6.5), (6.18) and (6.19) in eq. (6.8), one finds

$$\hat{x}: \rho\left(\frac{\partial u}{\partial \tau} + u\frac{\partial u}{\partial x} + v\frac{\partial u}{\partial y}\right) = \rho F_x - \frac{\partial p}{\partial x} + \mu\left(\frac{\partial^2 u}{\partial x^2} + \frac{\partial^2 u}{\partial y^2}\right) \qquad (6.20)$$

$$\hat{y}: \rho\left(\frac{\partial v}{\partial \tau} + u\frac{\partial v}{\partial x} + v\frac{\partial v}{\partial y}\right) = \rho F_y - \frac{\partial p}{\partial y} + \mu\left(\frac{\partial^2 v}{\partial x^2} + \frac{\partial^2 v}{\partial y^2}\right) \qquad (6.21)$$

In eqs. (6.20) and (6.21) the dynamic viscosity μ has been assumed constant.

6.4 Derivation of the temperature field equation

It is here assumed that steady state prevails and that the thermophysical properties are constant. Consider a volume element according to Fig. 6.2.

The energy equation (first law of thermodynamics for an open system) for steady flow conditions through the volume element reads, see Refs. [5] and [6],

$$d\dot{Q} - d\dot{W}_t = d\dot{H} + d\left(\frac{\dot{m}w^2}{2}\right) + d(\dot{m}gr_h) \qquad (6.22)$$

where $d\dot{Q}$ is the heat, $d\dot{W}_t$ the engineering work, $d\dot{H}$ the change in enthalpy, $d(\dot{m}w^2/2)$ the change in kinetic energy $d(\dot{m}gr_h)$ the change in potential energy and r_h is a position in relation to a reference position.

If the work by the friction forces can be neglected, which is true at low flow velocities, one has $d\dot{W}_t = 0$.

The changes in kinetic and potential energy are generally negligible and then eq. (6.22) can be written

$$d\dot{Q} = d\dot{H} \qquad (6.23)$$

The heat flow is calculated as in Chapter 1 and one has for the x-direction

$$\Delta\dot{Q}_x = -\lambda A_x\frac{\partial t}{\partial x} - \frac{\partial}{\partial x}\left(\lambda A_x\frac{\partial t}{\partial x}\right)dx + \lambda A_x\frac{\partial t}{\partial x} = -\frac{\partial}{\partial x}\left(\lambda\frac{\partial t}{\partial x}\right)dx\,dy\,dz \qquad (6.24)$$

In a similar way one obtains in the y- and z-directions

$$\Delta\dot{Q}_y = -\frac{\partial}{\partial y}\left(\lambda\frac{\partial t}{\partial y}\right)dx\,dy\,dz \qquad (6.25)$$

$$\Delta\dot{Q}_z = -\frac{\partial}{\partial z}\left(\lambda\frac{\partial t}{\partial z}\right)dx\,dy\,dz \qquad (6.26)$$

In eq. (6.23) the heat is positive, if it is supplied to the volume element and negative if it is rejected. One then has

$$d\dot{Q} = -(\Delta\dot{Q}_x + \Delta\dot{Q}_y + \Delta\dot{Q}_z) = \left(\frac{\partial}{\partial x}\left(\lambda\frac{\partial t}{\partial x}\right) + \frac{\partial}{\partial y}\left(\lambda\frac{\partial t}{\partial y}\right) + \frac{\partial}{\partial z}\left(\lambda\frac{\partial t}{\partial z}\right) \right) dx\,dy\,dz$$

(6.27)

The procedure to determine the change in enthalpy is as follows. In the x-direction, the enthalpy flow is

$$\dot{H}_x = \dot{m}_x h = \rho u\,dy\,dz\,h$$

$$\Rightarrow d\dot{H}_x = \rho h\frac{\partial u}{\partial x}dx\,dy\,dz + \rho u\frac{\partial h}{\partial x}dx\,dy\,dz$$

(6.28)

Similarly one obtains in the y- and z-directions

$$d\dot{H}_y = \rho h\frac{\partial v}{\partial y}dx\,dy\,dz + \rho v\frac{\partial h}{\partial y}dx\,dy\,dz$$

(6.29)

$$d\dot{H}_z = \rho h\frac{\partial w}{\partial z}dx\,dy\,dz + \rho w\frac{\partial h}{\partial z}dx\,dy\,dz$$

(6.30)

By using eqs. (6.28)–(6.30), $d\dot{H}$ can be expressed as

$$d\dot{H} = d\dot{H}_x + d\dot{H}_y + d\dot{H}_z$$

$$= \rho h\left(\frac{\partial u}{\partial x} + \frac{\partial v}{\partial y} + \frac{\partial w}{\partial z} \right) dx\,dy\,dz + \rho\left(u\frac{\partial h}{\partial x} + v\frac{\partial h}{\partial y} + w\frac{\partial h}{\partial z} \right) dx\,dy\,dz$$

(6.31)

In eq. (6.31) the first term is zero according to the mass conservation equation because steady flow conditions and constant physical properties are assumed. Equations (6.28) and (6.31) in eq. (6.23) give

$$\frac{\partial}{\partial x}\left(\lambda\frac{\partial t}{\partial x}\right) + \frac{\partial}{\partial y}\left(\lambda\frac{\partial t}{\partial y}\right) + \frac{\partial}{\partial z}\left(\lambda\frac{\partial t}{\partial z}\right) = \rho\left(u\frac{\partial h}{\partial x} + v\frac{\partial h}{\partial y} + w\frac{\partial h}{\partial z} \right)$$

(6.32)

For the enthalpy h the following relations are generally valid

$$h = h(p, t)$$

$$\Rightarrow dh = \left(\frac{\partial h}{\partial p}\right)_t dp + \left(\frac{\partial h}{\partial t}\right)_p dt$$

(6.33)

By definition one has

$$c_p = \left(\frac{\partial h}{\partial t}\right)_p$$

(6.34)

For ideal gases the enthalpy is independent of pressure, i.e., $(\partial h/\partial p)_t \equiv 0$. For liquids, one commonly assumes that the derivative $(\partial h/\partial p)_t$ is small and/or that the pressure variation dp is small compared to the change in temperature. Then generally one states

$$dh = c_p dt \qquad (6.35)$$

Equation (6.32) can now be written if λ also is assumed constant:

$$u\frac{\partial t}{\partial x} + v\frac{\partial t}{\partial y} + w\frac{\partial t}{\partial z} = \frac{\lambda}{\rho c_p}\left(\frac{\partial^2 t}{\partial x^2} + \frac{\partial^2 t}{\partial y^2} + \frac{\partial^2 t}{\partial z^2}\right) \qquad (6.36)$$

For the unsteady case, a term $\partial t/\partial \tau$ has to be included in the left-hand side of eq. (6.36).

In Appendix 2, a more complete derivation of the temperature field equation, eq. (6.36), including the work by friction forces is given.

6.5   Basic equations in cylindrical coordinates

For incompressible and steady flow conditions the following equations are valid in cylindrical coordinates (see Fig. 6.4) if the physical properties are assumed constant.

K.E.

$$\frac{1}{r}\frac{\partial(ru_r)}{\partial r} + \frac{1}{r}\frac{\partial(u_\theta)}{\partial \theta} + \frac{\partial(u_z)}{\partial z} = 0 \qquad (6.37)$$

N.S.

r-direction:

$$u_r\frac{\partial u_r}{\partial r} + u_\theta\frac{1}{r}\frac{\partial u_r}{\partial \theta} + u_z\frac{\partial u_r}{\partial z} - \frac{u_\theta^2}{r}$$

$$= -\frac{1}{\rho}\frac{\partial p}{\partial r} + \frac{\mu}{\rho}\left(\frac{1}{r}\frac{\partial}{\partial r}\left(r\frac{\partial u_r}{\partial r}\right) + \frac{1}{r^2}\frac{\partial^2 u_r}{\partial \theta^2} + \frac{\partial^2 u_r}{\partial z^2} - \frac{u_r}{r^2} - \frac{2}{r^2}\frac{\partial u_\theta}{\partial \theta}\right)$$

$$(6.38)$$

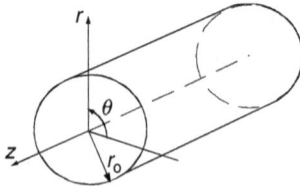

Figure 6.4: Cylindrical coordinaets r, θ, z.

θ-direction:

$$u_r \frac{\partial u_\theta}{\partial r} + u_\theta \frac{1}{r} \frac{\partial u_\theta}{\partial \theta} + u_z \frac{\partial u_\theta}{\partial z} + \frac{u_r u_\theta}{r}$$

$$= -\frac{1}{\rho} \frac{\partial p}{\partial \theta} + \frac{\mu}{\rho} \left(\frac{1}{r} \frac{\partial}{\partial r} \left(r \frac{\partial u_\theta}{\partial r} \right) + \frac{1}{r^2} \frac{\partial^2 u_\theta}{\partial \theta^2} + \frac{\partial^2 u_\theta}{\partial z^2} + \frac{2}{r^2} \frac{\partial u_r}{\partial \theta} - \frac{u_\theta}{r^2} \right)$$

$$(6.39)$$

z-direction:

$$u_r \frac{\partial u_z}{\partial r} + u_\theta \frac{1}{r} \frac{\partial u_z}{\partial \theta} + u_z \frac{\partial u_z}{\partial z} = -\frac{1}{\rho} \frac{\partial p}{\partial z} + \frac{\mu}{\rho} \left(\frac{1}{r} \frac{\partial}{\partial r} \left(r \frac{\partial u_z}{\partial r} \right) + \frac{1}{r^2} \frac{\partial^2 u_z}{\partial \theta^2} + \frac{\partial^2 u_z}{\partial z^2} \right)$$

$$(6.40)$$

The energy equation (temperature field equation)

$$\rho c_p \left(u_r \frac{\partial t}{\partial r} + u_\theta \frac{1}{r} \frac{\partial t}{\partial \theta} + u_z \frac{\partial t}{\partial z} \right) = \lambda \left(\frac{1}{r} \frac{\partial}{\partial r} \left(r \frac{\partial t}{\partial r} \right) + \frac{1}{r^2} \frac{\partial^2 t}{\partial \theta^2} + \frac{\partial^2 t}{\partial z^2} \right) \quad (6.41)$$

The basic equations expressed in other coordinate systems can be found in Ref. [7].

6.6   Boundary layer equations for the laminar case

The mathematical difficulties in solving the flow field equations (6.4), (6.7)–(6.9), and the temperature field equation (6.36), have forced researchers and engineers to develop ideas which are simplifying the equations. The so-called *boundary layer theory* according to Prandtl (1904) [8], was found to be successful and has then been applied for many engineering problems. In the boundary layer theory, the flow field is splitted up into two regions: (a) a thin layer adjacent to the surface of the body where the friction forces are strong and high gradients in the flow and temperature fields exist; (b) the region outside the boundary layer where the friction forces are negligible. This region is called the potential flow field or the outer region. The gradients in the flow and temperature fields are relatively small in that region.

A basic assumption in the boundary layer theory is that the fluid immediately at the surface is at rest in relation to the body. This assumption is valid except at very low pressures as then the mean free path of the molecular motion is of the same order of magnitude as the overall dimensions of the body. Thus the boundary layer can be defined as that region where the flow velocity is changed from the potential flow value to zero at the body surface (see Fig. 6.5).

In a similar way the thermal boundary layer or the temperature boundary layer can be defined as that region where the fluid temperature is changed from the value in the potential flow to that of the surface (see Fig. 6.6).

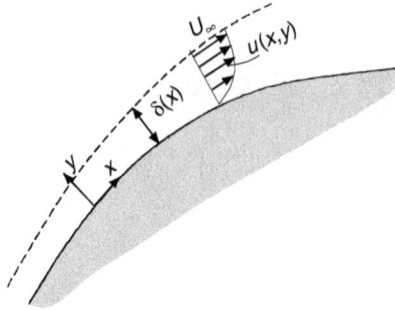

Figure 6.5: Flow or velocity boundary layer along a body surface.

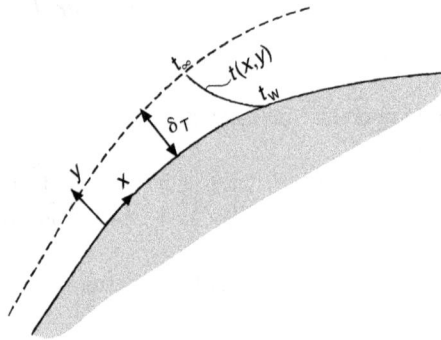

Figure 6.6: The thermal or temperature boundary layers along a body surface.

If the boundary layer thicknesses δ and δ_T, respectively, are very small compared to other dimensions, one has for a two-dimensional boundary layer, see Schlichting [9].

$$u \gg v \tag{6.42}$$

$$\frac{\partial u}{\partial y} \gg \frac{\partial u}{\partial x}, \frac{\partial v}{\partial x}, \frac{\partial v}{\partial y} \tag{6.43}$$

$$\frac{\partial t}{\partial y} \gg \frac{\partial t}{\partial x} \tag{6.44}$$

From Navier–Stokes' equation in the y-direction, eq. (6.21), it is found that the pressure p is independent of y, i.e.,

$$p = p(x) \tag{6.45}$$

Navier–Stokes' equation in the x-direction is simplified to

$$\rho \left(u \frac{\partial u}{\partial x} + v \frac{\partial u}{\partial y} \right) = \rho F_x - \frac{dp}{dx} + \mu \frac{\partial^2 u}{\partial y^2} \tag{6.46}$$

The temperature field equation (6.36) is simplified to

$$u\frac{\partial t}{\partial x} + v\frac{\partial t}{\partial y} = \frac{\lambda}{\rho c_p}\frac{\partial^2 t}{\partial y^2} \qquad (6.47)$$

For the inviscid flow, the potential flow, outside the boundary layer the Bernoulli equation is valid, i.e.,

$$p + \frac{1}{2}\rho U^2 = \text{constant} \qquad (6.48)$$

where U is the potential flow velocity. With eq. (6.48) the pressure term in eq. (6.46) can be written as

$$\frac{dp}{dx} = -\rho U\frac{dU}{dx} \qquad (6.49)$$

Commonly the Prandtl number Pr is introduced as

$$\text{Pr} = \frac{\nu\rho c_p}{\lambda} = \frac{\mu c_p}{\lambda} \qquad (6.50)$$

In eq. (6.50) ν is the kinematic viscosity (m^2/s) defined as $\nu = \mu/\rho$.

In summary, the following equations are valid for a two-dimensional laminar boundary layer at steady flow conditions and by neglecting the mass or body forces:

K.E.

$$\frac{\partial u}{\partial x} + \frac{\partial v}{\partial y} = 0 \qquad (6.51)$$

N.S.

$$u\frac{\partial u}{\partial x} + v\frac{\partial u}{\partial y} = U\frac{dU}{dx} + \frac{\mu}{\rho}\frac{\partial^2 u}{\partial y^2} \qquad (6.52)$$

T.E.

$$u\frac{\partial t}{\partial x} + v\frac{\partial t}{\partial y} = \frac{\mu}{\rho\,\text{Pr}}\frac{\partial^2 t}{\partial y^2} \qquad (6.53)$$

Equations (6.51)–(6.53) are only valid for *laminar cases* because the flow and temperature fields are assumed to be independent of time. If the curvature of the body surface is modest, the equations will be valid if x is along the surface and y is normal to the surface, see Ref. [9].

The boundary layer can be either laminar or turbulent depending on the flow conditions. Figure 6.7 illustrates these two types of the velocity boundary layer along a plane plate. At the leading edge a laminar boundary layer is developing. At a certain distance, say x_c, disturbances appear and these are growing within the boundary layer and then transition from laminar to turbulent boundary layer flow takes place. The distance beyond which the flow field no longer is laminar is determined by the so-called critical Reynolds number $\text{Re}_c = U_\infty x_c/\nu$. A typical

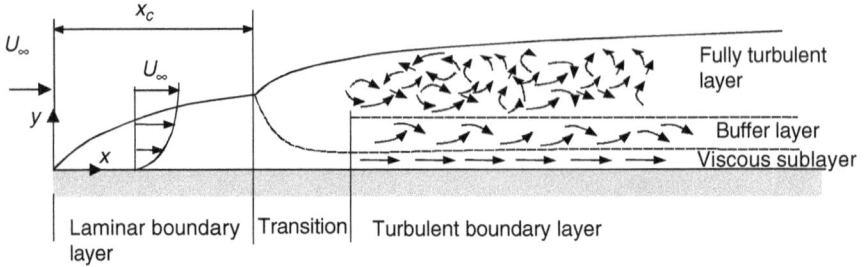

Figure 6.7: Laminar and turbulent boundary layer over a plane plate.

value is Re_c equal to $5 \cdot 10^5$. This value is assumed in this book if not otherwise stated.

If the plate surface is rough the transition may occur at a much lower Reynolds number, e.g., 10^5. In other cases, the plate surface might be extremely smooth and for ideal situations the transition might be delayed and the laminar boundary layer can survive up to $\mathrm{Re}_c = 5 \cdot 10^6$.

6.7 Dimensionless groups and rules of similarity

As stated previously it is extremely difficult to solve the governing equations for convective flow and heat transfer. Experimental investigations have dominated in the past and results have been presented in terms of empirical correlations which include dimensionless numbers. The advantage of using dimensionless groups is that many variables can be combined to a limited number of dimensionless numbers. Thus it is important to find the dimensionless groups being relevant for a certain heat transfer problem. Two different methods are commonly used. In one of the methods all variables which affect the heat transfer process are set up and then the number of independent dimensionless groups is determined by the so-called Buckingham Π-theorem, see Ref. [10]. The procedure is simple but the analysis can result in errors if any of the variables are missing. In the other method, the dimensionless groups are determined from the dimensionless form of the governing equations. Here a principle description based on the second method is provided.

If the following non-dimensional variables are introduced

$$\bar{x} = \frac{x}{L}, \qquad \bar{y} = \frac{y}{L}, \qquad \bar{z} = \frac{z}{L},$$

$$\bar{u} = \frac{u}{U_0}, \qquad \bar{v} = \frac{v}{U_0}, \qquad \bar{w} = \frac{w}{U_0}, \qquad (6.54)$$

$$\bar{p} = \frac{p}{\rho U_0^2}, \qquad \bar{\tau} = \frac{\tau U_0}{L}, \qquad \bar{t} = \frac{t - t_w}{t_\infty - t_w}$$

where U_0 is a characteristic velocity and L is a characteristic dimension (length), it is possible to show that the solutions to the flow and temperature fields have the forms

$$\bar{u} = f_1(\bar{x}, \bar{y}, \bar{z}, \bar{\tau}, \text{Re}) \tag{6.55}$$

$$\bar{v} = f_2(\bar{x}, \bar{y}, \bar{z}, \bar{\tau}, \text{Re}) \tag{6.56}$$

$$\bar{w} = f_3(\bar{x}, \bar{y}, \bar{z}, \bar{\tau}, \text{Re}) \tag{6.57}$$

$$\bar{p} = f_4(\bar{x}, \bar{y}, \bar{z}, \bar{\tau}, \text{Re}) \tag{6.58}$$

$$\bar{t} = f_5(\bar{x}, \bar{y}, \bar{z}, \bar{\tau}, \text{Re}, \text{Pr}) \tag{6.59}$$

where

$$\text{Re} = \frac{U_0 L}{\nu} \tag{6.60}$$

Equations (6.55)–(6.59) are valid if the mass or body force is excluded which means that only forced convection is considered.

From eqs. (6.55)–(6.58) it is possible to find that the drag coefficient c_D and the lift coefficient c_L only depend on Reynolds number, Re, see, for example, Ref. [1].

For the heat transfer coefficient α according to eq. (6.3), one obtains by using eqs. (6.54) and (6.59)

$$\alpha = \frac{1}{L} \lambda_f \left(\frac{\partial \bar{t}}{\partial \bar{y}} \right)_{y=0^+}$$

or:

$$\frac{\alpha L}{\lambda_f} = \left(\frac{\partial \bar{t}}{\partial \bar{y}} \right)_{y=0^+} \tag{6.61}$$

The group $\alpha L / \lambda_f$ is called the Nusselt number Nu. By applying eqs. (6.55)–(6.59) one can write

$$\text{Nu} = \frac{\alpha L}{\lambda_f} = f_6(\bar{x}, \bar{z}, \bar{\tau}, \text{Re}, \text{Pr}) \tag{6.62}$$

If steady flow conditions are considered, a mean value of the heat transfer coefficient can be written as:

$$\text{Nu} = f_7(\text{Re}, \text{Pr}) \tag{6.63}$$

The so-called rules of similarity can be formulated as:

- The flow field around or inside geometrically similar bodies are identical if the Reynolds number is equal and if dimensionless variables are applied. (The boundary conditions need to be the same.)

- The temperature field around or inside geometrically similar bodies are identical if the Reynolds and Prandtl numbers are equal and if dimensionless variables are applied. (The boundary conditions need to be the same.)

References

[1] F.M. White, Fluid Mechanics, 6th ed., McGraw-Hill, New York (2008).

[2] G.K. Batchelor, An Introduction to Fluid Dynamics, 2nd ed., Cambridge University Press, Cambridge (1970).

[3] D.F. Young, B.R. Munson and T.H. Okiishi, A Brief Introduction to Fluid Mechanics, 4th ed., John Wiley Sons (2007).

[4] J. Kestin, A Course in Thermodynamics, vol. II, Blaisdell Publ. Co., Waltham, Massachusetts (1966).

[5] Y.A. Cengel and M.A. Boles, Thermodynamics: An Engineering Approach, 7th ed., McGraw-Hill, New York (2011).

[6] J. Kestin, A Course in Thermodynamics, vol. 1, Blaisdell Publ. Co., Waitham, Massachusetts (1966).

[7] W.F. Hughes and E.W. Gaylord, Basic Equations of Engineering Science, Schaum's outline series, McCraw-Hill, New York (1964).

[8] L. Prandtl, Über Flüssigkeitsbewegung bei sehr kleiner Reibung, Proc. Third Int. Math. Congr., 484–491, Heidelberg (1904).

[9] H. Schlichting, Boundary Layer Theory, 7th ed., McGraw-Hill, New York (1979).

[10] G.W. Bluman and J.D. Gole, Similarity Methods for Differential Equations, Applied Mathematical Sciences, vol. 13, Springer-Verlag, New York (1974).

7 Similarity solutions for laminar boundary layer flow

7.1 Introduction

In this chapter, the so-called similarity solutions for the boundary layer equations are considered. In such solutions the independent variables u, v, and T can be transformed to a function of only one independent variable and the partial differential equations can be transformed to ordinary differential equations.

The presented solutions are idealized; however, they show how different parameters and phenomena influence the heat transfer.

It is assumed that the flow is two-dimensional, steady, incompressible and that the physical properties are constant. It is also assumed that the body forces are negligible.

7.2 Derivation of flow and temperature distribution equations

The boundary layer equations simplified for the situation mentioned above read as follows:

$$\frac{\partial u}{\partial x} + \frac{\partial v}{\partial y} = 0 \tag{7.1}$$

$$u\frac{\partial u}{\partial x} + v\frac{\partial u}{\partial y} = U\frac{dU}{dx} + v\frac{\partial^2 u}{\partial y^2} \tag{7.2}$$

$$u\frac{\partial T}{\partial x} + v\frac{\partial T}{\partial y} = \frac{v}{Pr}\frac{\partial^2 T}{\partial y^2} \tag{7.3}$$

where $v = \mu/\rho$.

The following boundary conditions apply

$$y = 0; \quad u = v = 0; \quad t = t_w(x) \tag{7.4}$$

$$y \to \infty; \quad u \to U; \quad t \to t_\infty \tag{7.5}$$

where $y \to \infty$ indicates the edge of the boundary layer.

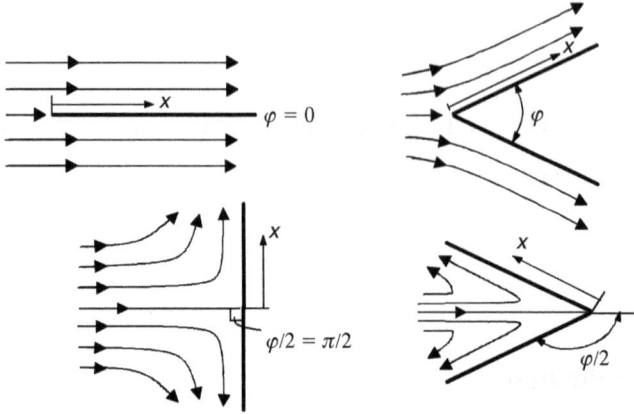

Figure 7.1: Bodies for which the potential flow at the surface is given by $U = cx^m$.

Now, consider the case when the potential flow (just outside the boundary layer) is given by

$$U = cx^m \tag{7.6}$$

This potential flow is referred as wedge-type flow (see Fig. 7.1).

The exponent m in eq. (7.6) is related to the opening angle φ as (see Refs. [1] and [2])

$$\varphi = \frac{2m}{m+1}\pi \tag{7.7}$$

For $m = 0 \Rightarrow \varphi = 0$, one has tangential flow along a flat plate, and with $m = 1 \Rightarrow \varphi = \pi$, one has flow towards a flat plate (impinging flow).

The continuity of eq. (7.1) can be eliminated by introducing the stream function ψ as

$$u = \frac{\partial \psi}{\partial y} \tag{7.8}$$

$$v = -\frac{\partial \psi}{\partial x} \tag{7.9}$$

Introduce the independent variable η as in eq. (7.10), and a new stream function $f(\eta)$ as in eq. (7.11). The dimensionless temperature $\theta(\eta)$ is introduced by eq. (7.12), and the temperature difference between the plate and the free stream is described by eq. (7.13)

$$\eta = \sqrt{\frac{m+1}{2}} \cdot \sqrt{\frac{U}{vx}} y \tag{7.10}$$

$$\psi = \sqrt{\frac{2}{m+1}} \cdot \sqrt{vxU} f(\eta) \tag{7.11}$$

$$T = T_\infty + (T_w - T_\infty)\theta(\eta) \tag{7.12}$$

$$T_w - T_\infty = c_t x^\gamma \tag{7.13}$$

By introducing eqs. (7.10)–(7.13), the boundary layer equations (7.2) and (7.3) can be transformed to ordinary differential equations for the functions $f(\eta)$ and $\theta(\eta)$, respectively. In order to find these differential equations from eqs. (7.2) and (7.3), one needs to derive expressions for the velocities, the velocity derivatives, and the temperature derivatives.

$$u = \frac{\partial \psi}{\partial y} = \frac{\partial \psi}{\partial \eta} \cdot \frac{\partial \eta}{\partial y} = Uf'(\eta) = cx^m f' \tag{7.14}$$

$$\frac{\partial u}{\partial y} = \frac{\partial u}{\partial \eta} \cdot \frac{\partial \eta}{\partial y} = cx^m f'' \sqrt{\frac{m+1}{2}} \cdot \sqrt{\frac{cx^m}{vx}} \tag{7.15}$$

$$\frac{\partial^2 u}{\partial y^2} = cx^m f''' \frac{m+1}{2} \frac{cx^m}{vx} = \frac{m+1}{2} \frac{c^2}{v} x^{2m-1} f''' \tag{7.16}$$

$$\frac{\partial u}{\partial x} = mcx^{m-1} f' + cx^m f'' \frac{\partial \eta}{\partial x}$$

$$\frac{\partial \eta}{\partial x} = \frac{\partial}{\partial x}\left(\sqrt{\frac{m+1}{2}\frac{c}{v}x^{m-1}} y\right) = \frac{m-1}{2}\frac{\eta}{x}$$

$$\Rightarrow \frac{\partial u}{\partial x} = cx^{m-1}\left\{ mf' + \frac{m-1}{2}\eta f''\right\} \tag{7.17}$$

$$v = -\frac{\partial \psi}{\partial x} = -\sqrt{\frac{2}{m+1}}\sqrt{cv}\left\{\frac{m+1}{2}x^{(m-1)/2}f + x^{(m+1)/2}f' \frac{m-1}{2}\frac{\eta}{x}\right\}$$

$$= -\sqrt{\frac{2}{m+1}}\sqrt{cv}x^{(m-1)/2}\left\{\frac{m+1}{2}f + \frac{m-1}{2}\eta f'\right\} \tag{7.18}$$

$$\frac{\partial t}{\partial x} = \frac{\partial}{\partial x}(t_\infty + c_t x^\gamma \theta) = c_t \gamma x^{\gamma-1}\theta + c_t x^\gamma \theta' \frac{m-1}{2}\frac{\eta}{x}$$

$$= c_t x^{\gamma-1}\left\{\gamma\theta + \frac{m-1}{2}\eta\theta'\right\} \tag{7.19}$$

$$\frac{\partial t}{\partial y} = c_t x^\gamma \theta' \sqrt{\frac{m+1}{2}} \sqrt{\frac{c}{\nu} x^{m-1}}$$ (7.20)

$$\frac{\partial^2 t}{\partial y^2} = c_t x^\gamma \theta'' \frac{m+1}{2} \frac{c}{\nu} x^{m-1}$$ (7.21)

Substitution of eqs. (7.14)–(7.18) in eq. (7.2) yields

$$cx^m f' \cdot cx^{m-1} \left\{ mf' + \frac{m-1}{2} \eta f'' \right\} - \sqrt{\frac{2}{m+1}} \sqrt{c\nu} x^{(m-1)/2} \left\{ \frac{m+1}{2} f + \frac{m-1}{2} \eta f' \right\}$$

$$\cdot cx^m f'' \sqrt{\frac{m+1}{2}} \sqrt{\frac{cx^m}{\nu x}} = cx^m \cdot cmx^{m-1} + \nu \frac{m+1}{2} \frac{c^2}{\nu} x^{2m-1} f'''$$

$$\Rightarrow mf'^2 - \frac{m+1}{2} ff'' = m + \frac{m+1}{2} f'''$$

$$f''' + ff'' - \frac{2m}{m+1}(f'^2 - 1) = 0$$ (7.22)

Often the parameter β is introduced as

$$\beta = \frac{2m}{m+1}$$ (7.23)

Equation (7.22) is an ordinary differential equation for the function $f(\eta)$. It is valid for the boundary layer for all coordinates along the plate. The boundary conditions for eq. (7.22) read as

$$\eta = 0: \ f' = f = 0$$ (7.24)

$$\eta \to \infty: \ f' \to 1$$ (7.25)

Equation (7.22) was first derived by Falkner and Skan in 1931, and is called the Falkner–Skan equation (see Ref. [3]). For the case $m = 0$ (flow along a flat plate), Blasius presented a solution as early as 1908 (see Ref. [4]). However, in Blasius' work, the coordinate η was introduced as

$$\eta = \sqrt{\frac{U}{\nu x}} y$$

and the stream function as

$$\psi = \sqrt{\nu x U} f$$

The equation for the temperature function $\theta(\eta)$ is derived by substitution of eqs. (7.14) and (7.18)–(7.21) in eq. (7.3).

$$cx^m f' \cdot c_t x^{\gamma-1} \left(\gamma\theta + \frac{m-1}{2}\eta\theta' \right)$$

$$- \sqrt{\frac{2}{m+1}} \sqrt{c\nu} x^{(m-1)/2} \left(\frac{m+1}{2}f + \frac{m-1}{2}\eta f' \right) c_t x^\gamma \theta' \sqrt{\frac{m+1}{2}} \sqrt{\frac{c}{\nu}} x^{m-1}$$

$$= \frac{\nu}{\Pr} c_t x^\gamma \theta'' \frac{m+1}{2} \frac{c}{\nu} x^{m-1}$$

$$\Rightarrow \gamma f'\theta - \frac{m+1}{2} f\theta' = \frac{1}{\Pr} \frac{m+1}{2}\theta''$$

$$\Rightarrow \theta'' + \Pr\left(f\theta' - \frac{2\gamma}{m+1}f'\theta \right) = 0$$

If $\beta = 2m/(m+1)$ is introduced

$$\frac{2}{m+1} = 2 - \beta$$

$$\Rightarrow \theta'' + \Pr f\theta' - (2 - \beta)\gamma \Pr f'\theta = 0 \qquad (7.26)$$

Equation (7.26) is an ordinary differential equation for the function $\theta(\eta)$, which describes the temperature distribution.

The boundary conditions read as

$$\eta = 0: \ \theta = 1 \qquad (7.27)$$

$$\eta \rightarrow \infty: \ \theta = 0 \qquad (7.28)$$

7.3   Results for the flow and temperature fields

Equations (7.22) and (7.26) with the boundary conditions (7.24)–(7.25) and (7.27)–(7.28), respectively, can be solved numerically. Runge–Kuttas method (see Ref. [5]) is suitable.

Figure 7.2 shows the velocity distribution across the boundary layer. The solution is adopted from Hartree [6]. The shape of the velocity profile depends solely on the parameter β and is independent of x. All velocity distributions are therefore similar in shape independent of x. Such solutions are called *similarity solutions*.

$\beta = 0$ (or $m = 0$) corresponds to flow along a flat plate (tangential flow) with a constant free-stream velocity (potential velocity) U (see Fig. 7.1). For this kind of flow, the velocity at first increases almost linearly with the wall distance y and later turns asymptotically into the potential U. Positive values of β correspond to flows where the potential velocity is zero for $x = 0$ and increases with increasing x. Figure 7.2 shows that the velocity gradient at the wall ($\eta = 0$) increases with

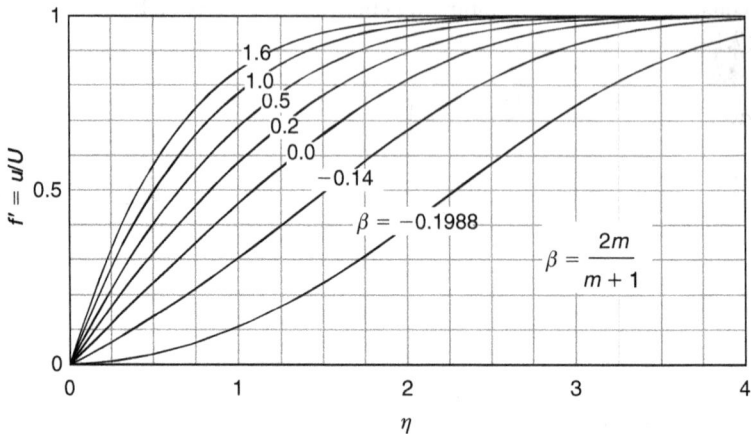

Figure 7.2: Velocity profiles for laminar wedge-type flow, where $U = cx^m$ applies. (Based on Ref. [6].)

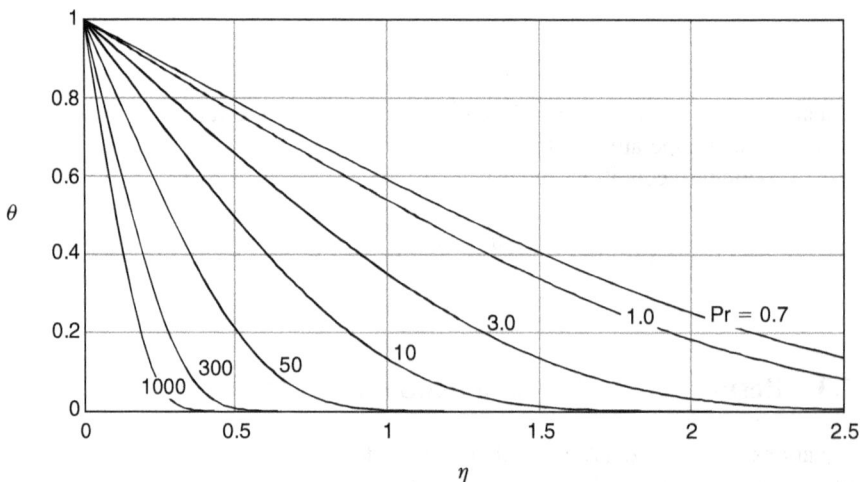

Figure 7.3: Influence of the Prandtl number on the temperature distribution, $\beta = 0$ and $\gamma = 0$. (Based on Ref. [7].)

increasing β. The boundary layer thickness decreases with increased β. For negative values of β, the velocity profile takes on an S-shape. Separation (the wall shear stress becomes zero) occurs for $\beta = -0.1988$ ($m = -0.091$). For $\beta < -0.1988$, no solutions are found.

Figures 7.3–7.5 show solutions for the temperature distribution, eq. (7.26), with the boundary conditions (7.27) and (7.28). Figure 7.3 shows the influence of the Prandtl number for $\beta = 0$ and with constant wall temperature, $\gamma = 0$.

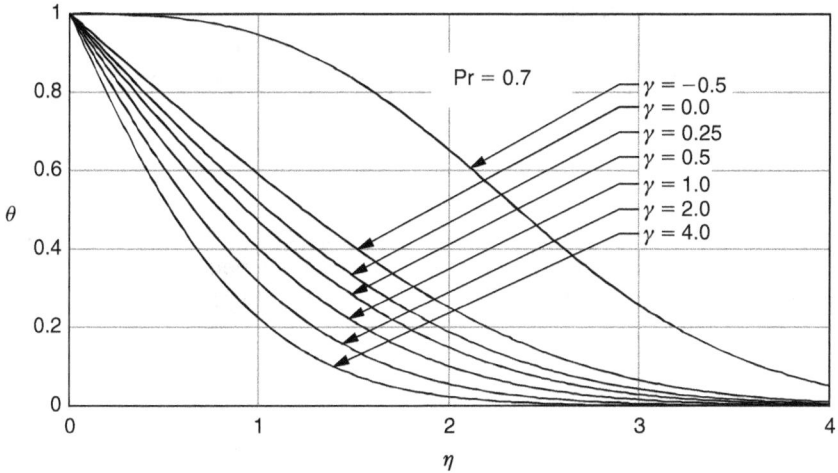

Figure 7.4: Temperature distribution in the boundary layer for non-constant wall temperature, $\gamma \neq 0$ and $\beta = 0$. (Based on Ref. [8].)

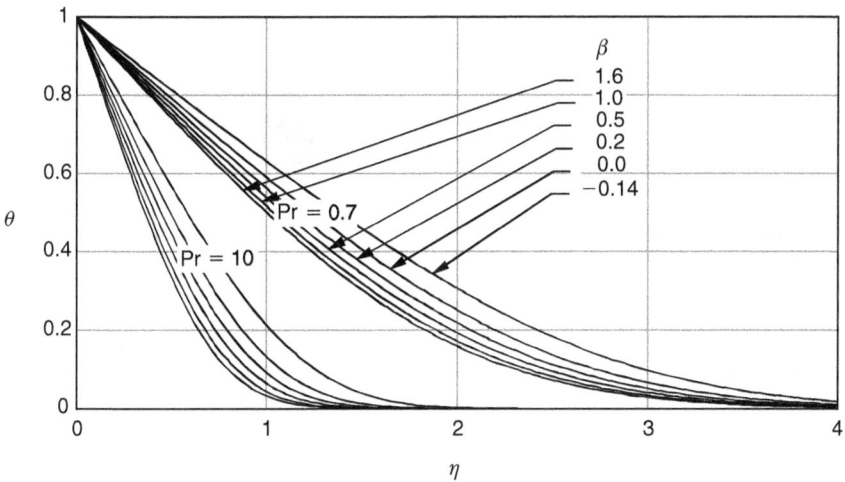

Figure 7.5: Influence of β on the thermal boundary layer $\gamma = 0$. (Based on Ref. [9].)

High Pr number (oils) gives a high temperature gradient at the wall and therefore also a high wall heat transfer. The thermal boundary layer also becomes thinner with increasing Pr number. For $Pr = 1$, the velocity profile and temperature profile are identical (see eqs. (7.22) and (7.26)). Very small Pr number (liquid metals) gives a very thick thermal boundary layer.

Figure 7.4 shows the influence of the parameter γ on the thermal boundary layer for $\beta = 0$. Positive values of γ (increasing wall temperature) give a higher temperature gradient at the wall and a thinner thermal boundary layer compared to the case $\gamma = 0$ (constant wall temperature).

For $\gamma = -0.5$, the temperature gradient is zero at the wall, hence the heat transfer becomes zero although $T_w - T_\infty \neq 0$. Physically, this can be explained in the following way. Assume that $T_w - T_\infty > 0$. For $\gamma = -0.5$ or $T_w - T_\infty = c_t x^{-0.5}$ the wall temperature decreases in the x-direction. This means that fluid particles close to the wall are arriving from a place where the wall temperature was higher and will therefore have convected their enthalpy along and will protect the wall surface from the cooler fluid in the outer region of the boundary layer.

When $\gamma < -0.5$, the temperature gradient at the wall becomes positive, and heat flows into the wall even when the wall temperature is higher than the stream temperature.

Figure 7.5 shows the influence on the temperature distribution by the parameter β, for $\gamma = 0$, for two different Prandtl numbers. It is obvious that the influence of β is much weaker than the influence of the Prandtl number.

7.4   The wall shear stress and the heat transfer coefficient

It is of great engineering interest to determine the wall shear stress σ_w, and the heat transfer coefficient α or the Nusselts number, Nu.

The definition of the wall shear stress yields

$$\sigma_w = \mu \left(\frac{\partial u}{\partial y} \right)_w = \mu c x^m f''(0) \sqrt{\frac{m+1}{2}} \sqrt{\frac{cx^m}{\nu x}} \tag{7.29}$$

The friction factor (or shear stress coefficient) is defined as

$$C_F = \frac{\sigma_w}{\left(\frac{\rho U^2}{2} \right)} \tag{7.30}$$

Equations (7.6) and (7.29) in eq. (7.30) give

$$C_F = \frac{\mu c x^m f''(0) \sqrt{\frac{m+1}{2}} \sqrt{\frac{cx^m}{\nu x}}}{\frac{\rho c^2 x^{2m}}{2}} = \nu \sqrt{2(m+1)} f''(0) \sqrt{\frac{cx^m}{\nu x c^2 x^{2m}}}$$

$$= \sqrt{2(m+1)} f''(0) \sqrt{\frac{\nu}{cx^m x}} \tag{7.31}$$

Introduce

$$Re_x = \frac{Ux}{\nu} = \frac{cx^m x}{\nu} \tag{7.32}$$

$$\Rightarrow C_F = \frac{\sqrt{2(m+1)} f''(0)}{\sqrt{Re_x}} \tag{7.33}$$

Table 7.1: Numerical values of $f''(0)$ and $\theta'(0)$.

$\beta = 2m/(m+1)$	$f''(0)$	β	$\gamma = 0$ $\theta'(0)$	Pr
1.6	1.5210	0	−0.4696	1
1.0	1.2326		−0.4208	0.72
0.5	0.9276	0.5	−0.4756	0.72
0.2	0.6866		−0.5390	1
0	0.4696	1	−0.4959	0.7
−0.14	0.2397			
−0.1988	0.0			

The heat transfer coefficient is defined as

$$\alpha = \frac{q}{T_{\text{w}} - t_\infty} = \frac{-\lambda_f \left(\frac{\partial T}{\partial y}\right)_{0+}}{T_{\text{w}} - T_\infty}$$

and the Nusselt number as

$$\text{Nu}_x = \frac{\alpha x}{\lambda_f}$$

$$\Rightarrow \text{Nu}_x = -\frac{\left(\frac{\partial T}{\partial y}\right)_{\text{w}} x}{(T_{\text{w}} - T_\infty)}$$

Equations (7.13) and (7.20) give

$$\text{Nu}_x = -\frac{c_t x^\gamma \theta'(0) \sqrt{\frac{m+1}{2}} \sqrt{\frac{c}{v} x^{m-1}} x}{c_t x^\gamma}$$

$$\Rightarrow \text{Nu}_x = -\sqrt{\frac{m+1}{2}} \theta'(0) \sqrt{\text{Re}_x} \qquad (7.34)$$

From eqs. (7.33) and (7.34) it can be seen that the wall shear stress is determined by $f''(0)$ and the heat transfer coefficient by $\theta'(0)$. Some values of these parameters are given in Table 7.1.

For $\gamma \neq 0$, Fig. 7.6 can be used.

Figure 7.7 shows the ratio $\text{Nu}_\beta/\text{Nu}_{\beta=0}$ versus β for different Pr ($\gamma = 0$).

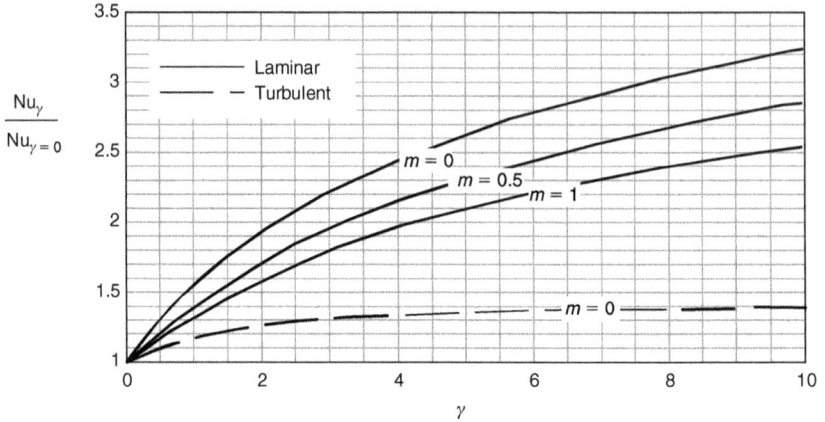

Figure 7.6: The ratio $\mathrm{Nu}_\gamma/\mathrm{Nu}_{\gamma=0}$ $(\theta'_\gamma/\theta'_{\gamma=0})$ versus γ for different m. Valid for all Pr, $T_\mathrm{w} - T_\infty = c_t x^\gamma$, $U = cx^m$. (Based on Ref. [10].)

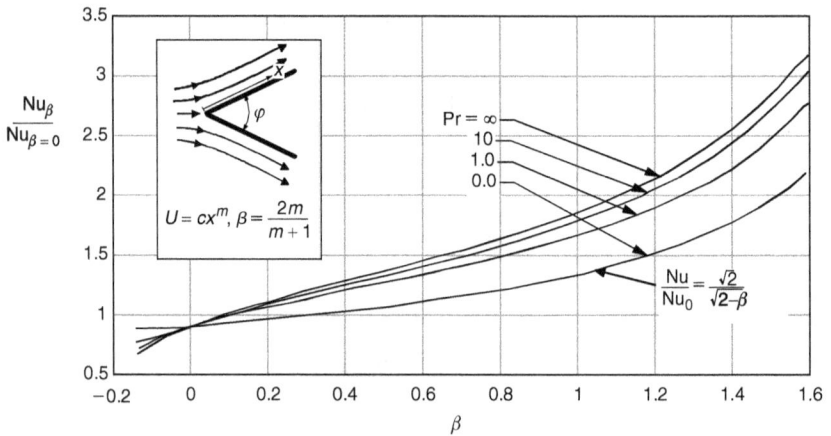

Figure 7.7: The ratio $\mathrm{Nu}_\beta/\mathrm{Nu}_{\beta=0}$ versus β for different Pr. Isothermal surface ($\gamma = 0$). (From Ref. [10].)

7.5 Analytical expressions for the influence of the Prandtl number

Consider eqs. (7.26) and (7.34) for the case $\gamma = 0$ and $m = 0$, i.e., flow along a flat plate with constant wall temperature.

$$\theta'' + \mathrm{Pr}f\theta' = 0 \tag{7.35}$$

$$\mathrm{Nu}_x = -\theta'(0)\sqrt{\frac{\mathrm{Re}_x}{2}} \tag{7.36}$$

Equation (7.35) can be integrated in a straightforward way. The solution which satisfies the boundary conditions (7.27) and (7.28) is

$$1 - \theta = \frac{\int\limits_0^\eta \exp\left(-\int\limits_0^\eta \Pr f \, d\eta\right) d\eta}{\int\limits_0^\infty \exp\left(-\int\limits_0^\eta \Pr f \, d\eta\right) d\eta} \tag{7.37}$$

The derivative $\theta'(0)$ in eq. (7.36) is

$$\theta'(0) = \frac{-1}{\int\limits_0^\infty \exp\left(-\int\limits_0^\eta \Pr f \, d\eta\right) d\eta} \tag{7.38}$$

For decreasing Prandtl number, the thermal boundary layer becomes larger and larger compared to the velocity boundary layer (see Figs. 7.2 and 7.3). The velocity is U in almost the entire thermal boundary layer, i.e., $f'(\eta) = u/U = 1 \Rightarrow f(\eta) = \eta$. The solution to eq. (7.38) then gives

$$\theta'(0) = -\sqrt{\frac{2\Pr}{\pi}} \tag{7.39}$$

For large Prandtl numbers on the other hand, the thermal boundary layer is thin compared to the velocity boundary layer (see Figs. 7.2 and 7.3). The velocity in the thermal boundary layer can then be approximated as $f'(\eta) = u/U = C\eta \Rightarrow f(\eta) = C\eta^2/2$, and the solution to (7.38) becomes

$$\theta'(0) = K\Pr^{1/3} \tag{7.40}$$

Within these limiting results, eqs. (7.39) and (7.40), the Nusselt number for flow over a flat plate ($\beta = 0$) with constant wall temperature ($\gamma = 0$) can be well approximated by the following correlations:

$$Nu_x = 0.332\sqrt{Re_x}\Pr^{1/3} \quad \text{for } \Pr > 0.6 \tag{7.41a}$$

$$Nu_x = 0.564\sqrt{Re_x}\Pr \quad \text{for } 0.005 < \Pr < 0.05 \tag{7.41b}$$

The mean value of the heat transfer coefficient over a plate with the length L, is obtained from $\bar{\alpha} = (1/L)\int_0^L \alpha \, dx$.

Applying this on eq. (7.41a) gives

$$\overline{Nu} = \frac{\bar{\alpha}L}{\lambda_f} = 0.664\sqrt{Re_L}\,\Pr^{1/3} \tag{7.41c}$$

where $Re_L = U_\infty L/\nu$.

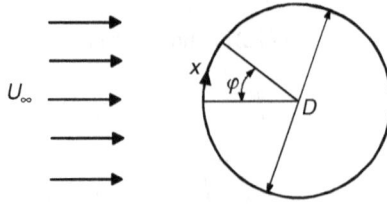

Figure 7.8: Stagnation point flow across a circular cylinder.

7.6   The Stanton number

Consider the expression for the friction factor C_F according to eq. (7.33) with $m = 0$ and the expression for the Nusselt number Nu_x according to eq. (7.41a)

$$C_F = \frac{\sqrt{2} \times 0.4696}{\sqrt{Re_x}}$$

$$Nu_x = 0.332\sqrt{Re_x}\, Pr^{1/3}$$

The Stanton number is defined as

$$St = \frac{\alpha}{\rho c_p U} = \frac{Nu_x}{Re_x Pr} \qquad (7.42)$$

Substitution gives

$$St = \frac{C_F}{2}Pr^{-2/3} \qquad (7.43)$$

The relation (7.43) is sometimes referred to as Reynolds analogy because it relates the wall shear stress to the heat transfer.

7.7   The constant c in $U = cx^m$

For wedge-type flows as in Fig. 7.1, the potential velocity U can be written as

$$U = cx^m \qquad (7.6)$$

The exponent m decides the opening angle of the wedge according to eq. (7.7). The question is now how to determine the constant c? For flow along a flat plate, one has $m = 0 \Rightarrow U = c$, i.e., c is equal to the free-stream velocity, U_∞.

For a circular cylinder in cross flow, see Fig. 7.8, one has $m = 1$ ($\beta = 1$) at the stagnation point ($\varphi = 0$) and $U = cx$.

For potential flow at the surface around a circular cylinder, the velocity U is given by

$$U = 2U_\infty \sin \varphi \qquad (7.44)$$

Close to the stagnation point, Fig. 7.8, the angle can be approximated by $\varphi = 2x/D$, and the velocity can be written as

$$U = 2U_\infty \sin \frac{2x}{D} = 2U_\infty \left(\frac{2x}{D} - \frac{1}{3!} \left(\frac{2x}{D} \right)^3 + \cdots \right)$$

For small x, one has

$$U \approx 4 \frac{U_\infty x}{D} \qquad (7.45)$$

The constant c in $U = cx$ becomes

$$c = 4 \frac{U_\infty}{D} \qquad (7.46)$$

In most other cases, the constant c has to be determined from experimental studies. The case $m = 1$ is sometimes referred to as a Hiemenz flow.

7.8   Linear superposition

The differential equation for the temperature field is linear, and the principle of superposition can be used to some extent. The temperature variation along the surface can in general be expressed by the series

$$T_w - T_\infty = a_0 + a_1 x + a_2 x^2 + a_3 x^3 + \cdots = \sum_{i=0}^{\infty} a_i x^i \qquad (7.47)$$

The heat flux at the wall is then described by

$$q_w = -\lambda_f \left(\frac{\partial T}{\partial y} \right)_w = -\lambda_f \sum_{i=0}^{\infty} \left(\frac{\partial T_i}{\partial y} \right)_w \qquad (7.48)$$

The heat transfer coefficient α is defined as

$$\alpha_i (T_w - t_\infty) = -\lambda_f \left(\frac{\partial T_i}{\partial y} \right)_w \qquad (7.49)$$

The heat flux can now be written as

$$q_w = \sum_{i=0}^{\infty} \alpha_i (T_w - T_\infty)_i$$

or

$$q_w = \alpha_0 a_0 + \alpha_1 a_1 x + \alpha_2 a_2 x^2 + \alpha_3 a_3 x^3 + \cdots$$

or

$$q_w = \alpha_0 \left(a_0 + \frac{\alpha_1}{\alpha_0} a_1 x + \frac{\alpha_2}{\alpha_0} a_2 x^2 + \frac{\alpha_3}{\alpha_0} a_3 x^3 + \cdots \right) \qquad (7.50)$$

By using the definition of the Nusselt number, $Nu = \alpha x / \lambda_f$, eq. (7.50) can be written as

$$q_w = \frac{Nu_0 \lambda_f}{x} \left(a_0 + \frac{Nu_1}{Nu_0} a_1 x + \frac{Nu_2}{Nu_0} a_2 x^2 + \frac{Nu_3}{Nu_0} a_3 x^3 + \cdots \right)$$

$$= \frac{Nu_0 \lambda_f}{x} \sum_{i=0}^{\infty} a_i x^i \frac{Nu_i}{Nu_0} \qquad (7.51)$$

Nu_0 is the Nusselt number for the constant wall temperature case, i.e., $\gamma = 0$, and Nu_i / Nu_0 can be taken from Fig. 7.6 with $\gamma = i$ for a given m.

For the laminar case, Nu_0 is given by

$$Nu_0 = -\sqrt{\frac{m+1}{2}} \theta'(0) \sqrt{Re_x} \qquad (7.52)$$

The above also applies for turbulent cases, only the expression for Nu_0 differs. The practical use of the method is limited to such cases where the wall temperature can be described as a series with relatively few terms.

7.9   Blowing and suction at the surface

Similarity solutions can also be obtained when blowing or suction of fluid at the surface occur. So-called film cooling, where a coolant is blown through a series of openings (holes or pores) in a direction nearly tangential to the surface, is an application. Consider the expression for the velocity v, eq. (7.18):

$$v = -\frac{\partial \psi}{\partial x} = -\sqrt{\frac{2cv}{m+1}} x^{(m-1)/2} \left\{ \frac{m+1}{2} f + \frac{m-1}{2} \eta f' \right\}$$

Especially for $\eta = 0$ one has:

$$v_w = -\sqrt{\frac{2cv}{m+1}} x^{(m-1)/2} \frac{m+1}{2} f_w$$

or

$$v_w = -\sqrt{\frac{cv(m+1)}{2}} x^{(m-1)/2} f_w \qquad (7.53)$$

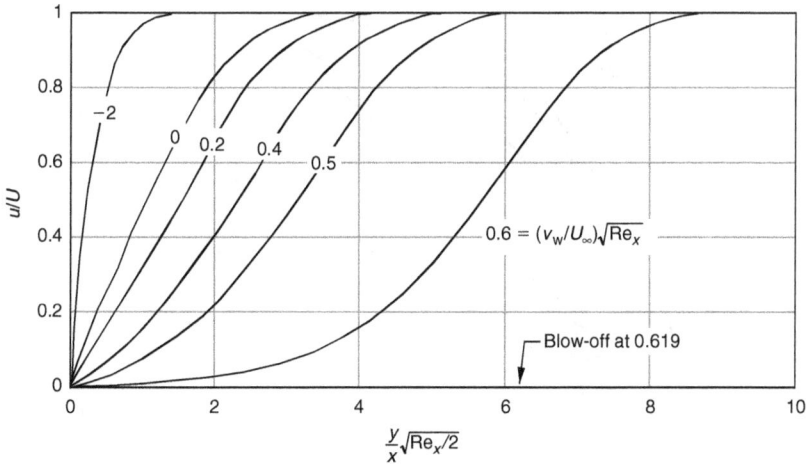

Figure 7.9: Velocity boundary layer when blowing or suction occur, $m = 0$. (Based on Ref. [11].)

For $m = 0$, similarity solutions can be obtained if $v_w \sim x^{-1/2}$, while with $m = 1$, the velocity v_w must be constant. Figure 7.9 depicts results for the velocity distribution in the boundary layer. Suction will decrease the boundary layer thickness while blowing increases the thickness.

Blowing gives the velocity profile an S-shape. At $(v_w/U_\infty)\sqrt{Re_x} = 0.619$ the velocity gradient becomes zero at $y = 0$ and u becomes zero for finite values of y. This means that the wall shear stress is zero and the fluid will no longer be attached to the wall. This phenomenon is called blow-off.

7.10   Temperature distribution and heat transfer coefficient when blowing or suction occur at the surface

Temperature profiles are presented in Fig. 7.10 for similarity solutions for the laminar boundary layer on a flat plate, $m = 0$, with constant wall temperature, $\gamma = 0$, when blowing or suction occur at the surface. The influence of the blowing parameter $(v_w/U_\infty)\sqrt{Re_x}$, is obvious ($v_w > 0$ blowing, $v_w < 0$ suction).

With blowing, $v_w > 0$, the thickness of the thermal boundary layer increases and the gradient $\partial T/\partial \eta$ at the wall decreases, which means that the heat transfer coefficient also decreases. For suction $v_w < 0$ the opposite applies.

Figure 7.11 shows $Nu_x/\sqrt{Re_x}$ versus the blowing parameter $(v_w/U_\infty)\sqrt{Re_x}$.

Blowing decreases the heat transfer coefficient. This is often used for cooling (protection of the surface) of gas turbine blades and nozzles for rockets. On the other hand, suction ($v_w < 0$) will increase the heat transfer coefficient.

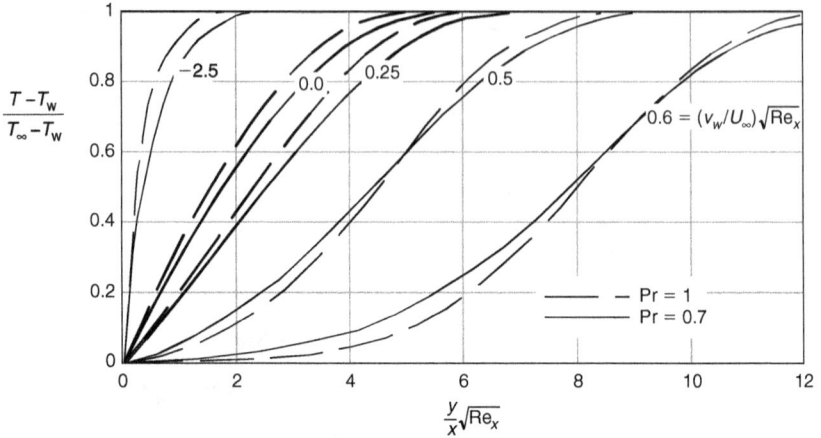

Figure 7.10: Temperature distribution in the boundary layer, for blowing and suction, $m = 0$, $\gamma = 0$. (Based on Ref. [12].)

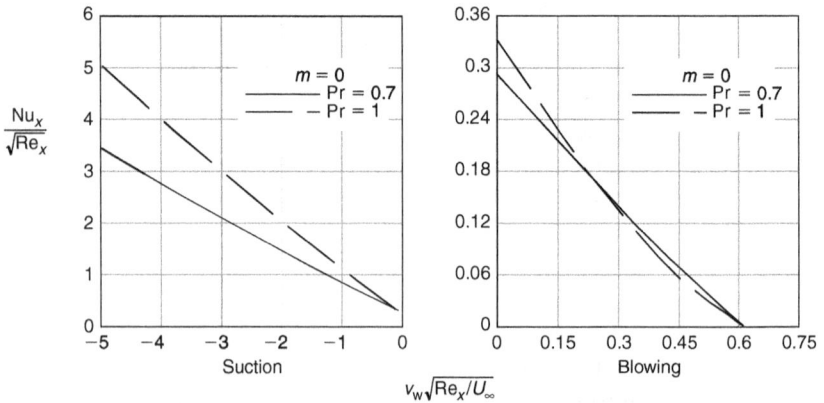

Figure 7.11: Dimensionless heat transfer coefficient for $m = 0$, $\gamma = 0$ with blowing and suction. (Based on Ref. [12].)

Figure 7.12 gives $Nu_x/\sqrt{Re_x}$ versus $(v_w/U_\infty)\sqrt{Re_x}$ for $m = 1$, i.e., flow impinging on a plate or the stagnation point flow of a circular cylinder in cross flow.

Figure 7.13 shows the influence of blowing for flow along a flat plate, for laminar and turbulent flow, respectively. α_0 is the heat transfer coefficient without blowing, i.e., $v_w = 0$. It is observed that blowing can considerably reduce the heat transfer, especially for a laminar boundary layer.

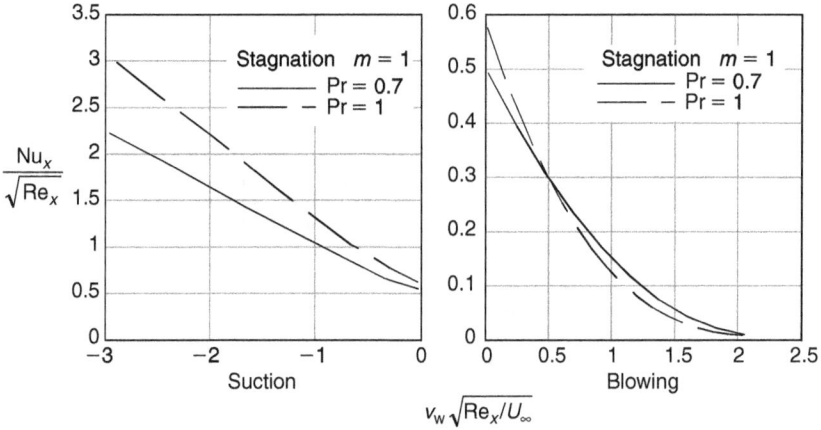

Figure 7.12: Dimensionless heat transfer coefficient when $m = 1$ and $\gamma = 0$ with blowing or suction. (Based on Ref. [12].)

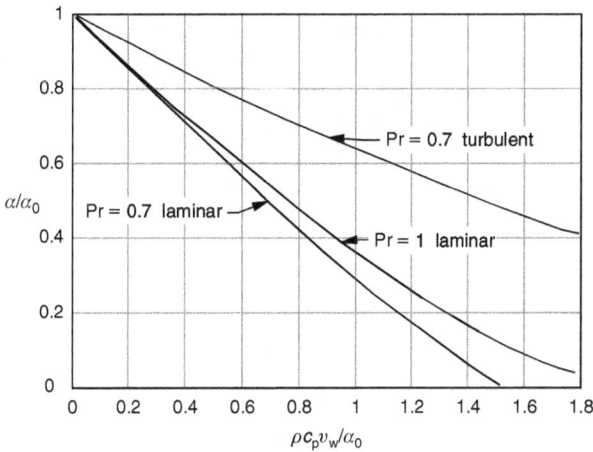

Figure 7.13: The ratio $\alpha(v_w > 0)/\alpha(v_w = 0)$ versus $\rho c_p v_w / \alpha_0$, $m = 0$. (Based on Ref. [12].)

7.11   Physical properties

In the derivation of the equations it was assumed that the physical properties were constant. However, the physical properties are usually dependent on temperature. When there is an appreciable variation between the wall and free-stream conditions, it is recommended that the properties are evaluated at the so-called film temperature. This is defined as the arithmetic mean between the wall and free-stream temperature, i.e.,

$$T_{ref} = \frac{T_w + T_\infty}{2} \tag{7.54}$$

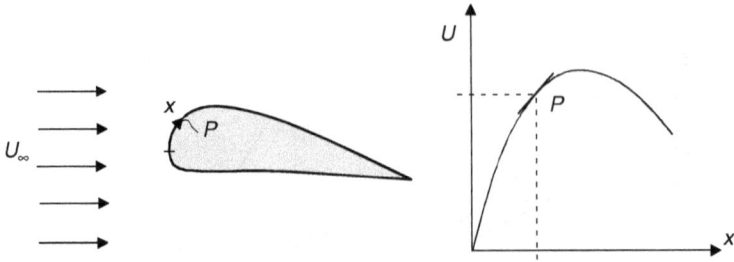

Figure 7.14: Local similarity.

7.12 Local similarity

The presented solutions are based on the fact that the velocity and temperature distributions at different x-positions were similar, so-called similarity solutions. For cases where the potential velocity U outside the boundary layer cannot be described by $U = cx^m$, the solutions are not directly applicable. To determine the velocity and temperature distributions for such cases, the principle of local similarity can be used. It is then assumed that the boundary layer development is controlled by local flow parameters as the pressure gradient and boundary layer thickness. Now, consider the case depicted in Fig. 7.14.

The potential velocity $U(x)$ is assumed to be given (calculated or measured). At an arbitrary point the velocity can be described by

$$U = cx^m \tag{7.55}$$

Differentiation yields

$$\frac{dU}{dx} = cmx^{m-1} = m\frac{U}{x} \tag{7.56}$$

or

$$m = \frac{x}{U}\frac{dU}{dx} \tag{7.57}$$

At the point P, m is determined by eq. (7.57), and thereafter c can be determined from eq. (7.55). For every arbitrary point one obtains values of m and c. In that way one can locally apply the solutions that have been presented.

The development of numerical methods during the last decades has made it possible to solve the boundary layer equations by direct numerical solutions, and corresponding differential equations for other cases (see Refs. [11, 13–16]).

References

[1] H. Schlichting, Boundary Layer Theory, 7th ed., McGraw-Hill, New York (1979).
[2] G.K. Batchelor, An Introduction to Fluid Dynamics, 2nd ed., Cambridge University Press, Cambridge (1970).

[3] V.M. Falkner and S.W. Skan, Some approximative solutions boundary layer equations, Phil. Mag., 12, 865–869 (1931).

[4] H. Blasius, Grenzschichten in Flussigkeiten mit kleiner Reibung, Z. Math. Phys., 56, 1–37 (1908).

[5] J.A. Adams and D.F. Rogers, Computer-Aided Heat Transfer Analysis, McGraw-Hill, New York (1973).

[6] D.R. Hartree, On an equation occurring in Falkner and Skan's approximate treatment of the equations of the boundary layer, Proc. Camb. Phil. Soc., 33, 223–239 (1937).

[7] E. Pohlhausen, Der Wärmeaustausch zwischen festen Körpern und Flussigkeiten mit kleiner Reibung und kleiner Wärmeleitung, Z. Angew. Math. Mech., 1, 115–121 (1921).

[8] S. Levy, Heat transfer to constant property laminar boundary layer flows with power function free-stream velocity and wall temperature variation, J. Aeron. Sci., 19, 341–348 (1952).

[9] E.R.G. Eckert, Die Berechnung des Wärmeüberganges in der laminaren Grenzschicht umströmter Körper, VDI-Forschungsheft 416 (1942).

[10] E.R.G. Eckert and R.M. Drake Jr., Analysis of Heat and Mass Transfer, McGraw-Hill, New York (1972).

[11] F.M. White, Viscous Fluid Flow, 2nd ed., McGraw-Hill, New York (1991).

[12] J.P. Hartnett and E.R.G. Eckert, Mass transfer cooling in a laminar boundary layer with constant fluid properties, Trans. ASME, 79, 247–254 (1957).

[13] S.V. Patankar, Numerical Heat Transfer and Fluid Flow, McGraw-Hill, New York (1980).

[14] H.K. Versteeg and W. Malalasekera, An Introduction to Computational Fluid Dynamics: The Finite Volume Method, 2nd ed., Pearson–Prentice Hall, Glasgow (2007).

[15] D.A. Anderson, J.C. Tannehill and R.H. Pletcher, Computational Fluid Mechanics and Heat Transfer, 2nd ed., Taylor & Francis, New York (1997).

[16] J.H. Ferziger and M. Peric, Computational Methods for Fluid Dynamics, Springer-Verlag, Berlin (1996).

8 Forced convection in channels—laminar case

8.1 Introduction

In this chapter, analysis of forced convective heat transfer and fluid flow in channels and pipes for laminar conditions will be presented. Despite that turbulent flow conditions prevail in many engineering applications, there are still a number of engineering applications where laminar flow prevails. This is true in heat exchangers for highly viscous fluids like oils. It may also be an advantage to reduce the pumping power by reducing the flow velocity although the heat transfer in laminar flow is usually lower. In the so-called compact heat exchangers with small channel dimensions (small hydraulic diameter), the flow is often laminar.

8.2 Flow conditions at the entrance region of channels

As a fluid enters a channel, boundary layers are built up along the channel walls. Figure 8.1 shows the situation in a channel consisting of two semi-infinite parallel plates, and a typical flow velocity profile is shown in Fig. 8.2. The fluid is assumed to have a uniform velocity as it enters the channel. At the walls the velocity is zero, and at the boundary layer edge the core velocity is reached.

The deceleration of the fluid in the boundary layers is of great significance. This implies that a velocity in the y-direction must be developed. This is so because the velocity in the core region must be increased to maintain the mass flow balance in the main flow direction. The boundary layers are growing in thickness and gradually they are merging and the velocity field is then said to be fully developed. For the

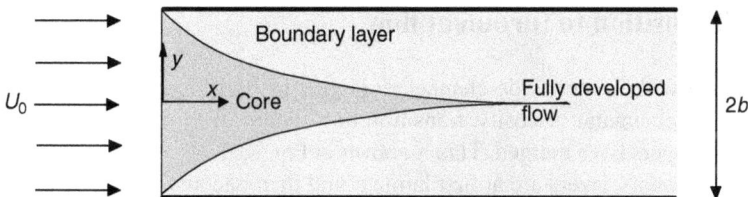

Figure 8.1: Entrance region in a channel.

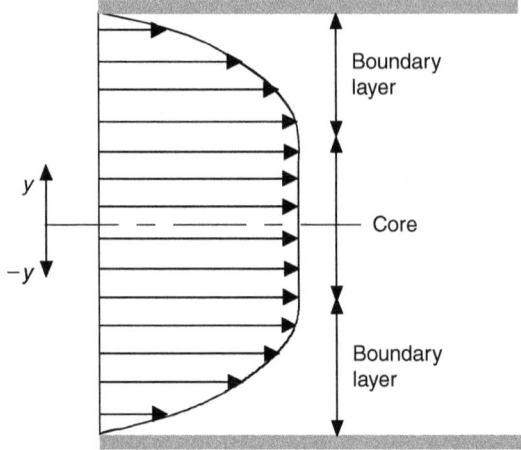

Figure 8.2: Typical flow velocity profile in the entrance region.

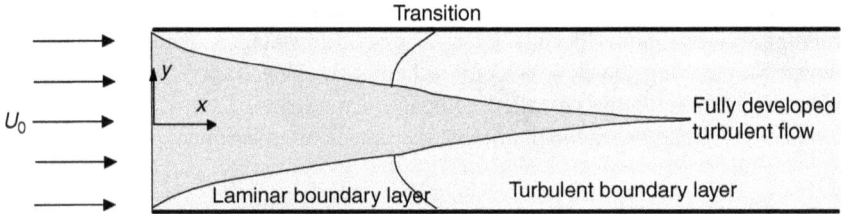

Figure 8.3: Development of a turbulent velocity profile.

case in Fig. 8.1, the velocity profile in the fully developed region is given by (see Ref. [1])

$$u = u_{\max} \left(1 - \left(\frac{y}{b} \right)^2 \right) \tag{8.1}$$

where $u_{\max}$ is the maximum velocity occurring at the center of the channel. $u_{\max}$ is given by

$$u_{\max} = -\frac{b^2}{2\mu} \frac{dp}{dx} \tag{8.2}$$

8.3   Transition to turbulent flow

For a case with a very wide channel (b large), or high inlet velocity, or a fluid with small kinematic viscosity, transition to turbulent flow may occur before the boundary layers have merged. This is shown in Fig. 8.3.

The boundary layers are at first laminar and then become turbulent, and in the fully developed regime, a turbulent velocity profile is reached. Turbulent flow and heat transfer will be discussed in Chapter 9.

8.4   Circular pipes or tubes

The previous discussion for a channel consisting of two parallel plates also holds, in principle, for circular pipes or tubes. The boundary layers are however occupying the whole perimeter and merge at the centerline of the pipe.

If r denotes the radial coordinate from the pipe center and R is the pipe radius, the velocity profile in the fully developed region is given by (see Refs. [1, 2])

$$u = 2u_m \left(1 - \left(\frac{r}{R} \right)^2 \right) \tag{8.3}$$

where u_m is average flow velocity.

8.5   Entrance length in circular pipes (tubes)

The entrance length, i.e., the distance from the inlet to the position where the velocity profile is fully developed, may be estimated by a correlation suggested by Langhaar [3]. Because the boundary layer thickness asymptotically reaches the pipe radius, the entrance length will not be exactly determined. The entrance length L_i can be defined as the distance from the pipe inlet to the position where the center velocity is 99% of the velocity for fully developed flow, and then the Langhaar equation reads

$$\frac{L_i}{D} = 0.0575 \mathrm{Re}_D \tag{8.4}$$

where Re_D is the Reynolds number based on the mean velocity and pipe diameter.

$$\mathrm{Re}_D = \frac{u_m D}{\nu} \tag{8.5}$$

Equation (8.4) is only valid for laminar cases, i.e., $\mathrm{Re}_D < 2300$.

8.6   The pressure drop over the entrance length

The pressure drop over the entrance length is bigger than the corresponding one for fully developed flow. The pressure drop Δp over the entrance region can be calculated with results given by Langhaar [3] (Fig. 8.4).

Another way to present the pressure drop over the entrance region is given in Ref. [4].

8.7   The pressure drop for fully developed flow

For fully developed flow in channels, as a balance (equilibrium) between pressure forces and friction forces prevails, the pressure drop over a certain pipe length L can be determined analytically if the flow is laminar.

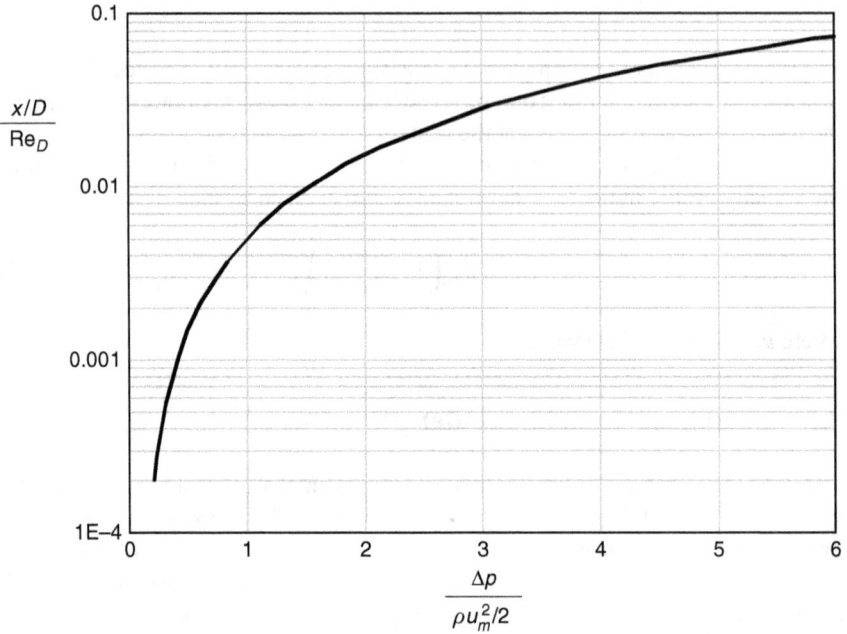

Figure 8.4: The pressure drop over the entrance length for a circular pipe. (Based on Ref. [3].)

The pressure drop is written as

$$\Delta p = f \frac{L}{D} \frac{\rho u_m^2}{2} \qquad (8.6)$$

For a circular pipe, the following expression for the friction factor f is valid (see Refs. [1, 2])

$$f = \frac{64}{\mathrm{Re}_D} \qquad (8.7)$$

For noncircular pipes, a hydraulic diameter D_h is introduced according to

$$D_h = \frac{4 \times \text{cross} - \text{sectional area}}{\text{wetted perimeter}} \qquad (8.8)$$

The friction factor f is given by

$$f = \frac{C}{\mathrm{Re}_D} \qquad (8.9)$$

where $\mathrm{Re}_D = u_m D_h / \nu$. The constant C depends on the channel geometry, and in Table 8.1, C is given for some cases.

Table 8.1: The constant C in eq. (8.9).

Configuration		Hydraulic diameter, D_h	C
	$a/b = 0.7$	$1.17a$	65
	0.5	$1.3a$	68
	0.3	$1.44a$	73
	0.2	$1.5a$	78
	0.1	$1.55a$	79
	$b/a = 1.0$	a	57
	1.25	$1.11a$	57.6
	2.0	$1.33a$	62
	3.0	$1.5a$	69
	5.0	$1.67a$	78
	10.0	$1.82a$	85
	∞	$2a$	96
		$0.58a$	53
	$d_i/d_o = 0.1$	$d_o - d_i$	89.2
	$d_i/d_o = 0.25$		94
	$0.5 < d_i/d_o < 1$		96

The pressure drop Δp is a key parameter for calculation of the necessary pumping power to transport the fluid through the channel.

8.8 Heat transfer for a circular pipe (tube)

In this section, the convective heat transfer between the fluid and the channel walls will be considered. The velocity field is assumed to be fully developed. The cases of uniform wall temperature and uniform wall heat flux will be considered.

8.8.1 Uniform wall temperaure

Consider the circular pipe in Fig. 8.5. The pipe wall has the temperature t_w for $x > 0$ and t_0 for $x < 0$. The fluid has a uniform temperature t_0 for $x < 0$. The task

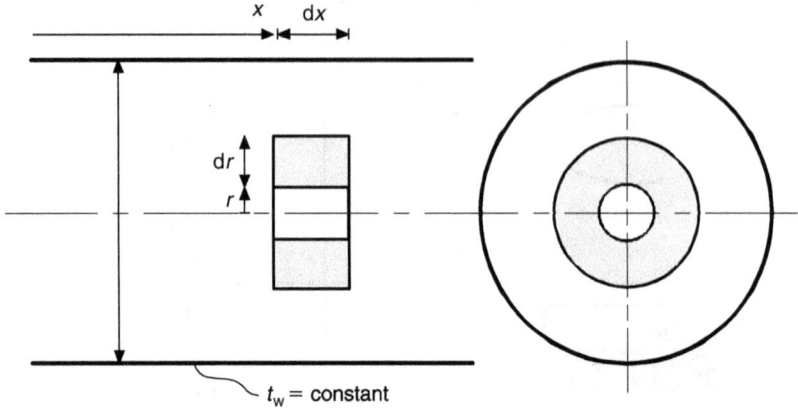

Figure 8.5: Circular pipe with uniform wall temperature.

is now to determine the temperature distribution in the pipe and the heat transfer coefficient.

An energy balance for the annular element in Fig. 8.5 is established. In the radial direction, heat is transported by conduction and one has

$$\dot{Q}_r = -\lambda 2\pi r \, dx \frac{\partial t}{\partial r}$$

$$\dot{Q}_{r+dr} = -\lambda 2\pi r \, dx \frac{\partial t}{\partial r} + \frac{\partial}{\partial r}\left(-\lambda 2\pi r \, dx \frac{\partial t}{\partial r}\right) dr$$

The net heat transfer rate in the radial direction is $\Delta \dot{Q}_r = \dot{Q}_{r+dr} - \dot{Q}_r$

$$\Delta \dot{Q}_r = -2\pi \lambda \, dx \, dr \frac{\partial}{\partial r}\left(r \frac{\partial t}{\partial r}\right) \tag{8.10}$$

In the axial direction, a change in the enthalpy flow rate occurs. This change is given by

$$d\dot{H} = \dot{m}c_p \frac{\partial t}{\partial x} \, dx = \rho 2\pi r \, dr \, uc_p \frac{\partial t}{\partial x} dx \tag{8.11}$$

For steady state, the energy balance equation (6.41) gives (taking the sign convention into account)

$$ur\rho c_p \frac{\partial t}{\partial x} = \lambda \frac{\partial}{\partial r}\left(r \frac{\partial t}{\partial r}\right)$$

or

$$u \frac{\partial t}{\partial x} = \frac{\lambda}{\rho c_p} \frac{1}{r} \frac{\partial}{\partial r}\left(r \frac{\partial t}{\partial r}\right) \tag{8.12}$$

Because the flow field is assumed to be fully developed, the velocity distribution (8.3) is valid. If this is inserted in eq. (8.12), one obtains

$$2u_m \frac{\partial t}{\partial x} = \frac{\lambda}{\rho c_p} \frac{1}{(1 - (r/R)^2)} \frac{1}{r} \frac{\partial}{\partial r} \left(r \frac{\partial t}{\partial r} \right) \tag{8.13}$$

Equation (8.13) describes the temperature distribution. The boundary conditions for the case in Fig. 8.5 are

$$x = 0: t = t_0 \tag{8.14}$$

$$r = R: t = t_w \tag{8.15}$$

$$r = 0: \frac{\partial t}{\partial r} = 0 \tag{8.16}$$

The problem to solve eq. (8.13) with the boundary conditions (8.14)–(8.16) is commonly called the Graetz problem as Graetz already, in the year 1883 [6], studied this problem before anybody else. In 1910 Nusselt [7] presented, without the knowledge of the work by Graetz, a solution to this problem. In what follows, the solution by Nusselt is presented.

It is now convenient to introduce

$$\vartheta = t - t_w \tag{8.17}$$

$$x' = \frac{x}{R} = \frac{2x}{D} \tag{8.18}$$

$$r' = \frac{r}{R} \tag{8.19}$$

With eqs. (8.17)–(8.19), eq. (8.13) can be written as

$$2u_m \frac{1}{R} \frac{\partial \vartheta}{\partial x'} = \frac{\lambda}{\rho c_p} \frac{1}{(1 - r'^2)} \frac{1}{R^2} \frac{1}{r'} \frac{\partial}{\partial r'} \left(r' \frac{\partial \vartheta}{\partial r'} \right)$$

If $Re_D = u_m D/\nu$ and $Pr = \mu c_p/\lambda = \rho \nu c_p/\lambda$ are introduced, one has

$$Re_D Pr \frac{\partial \vartheta}{\partial x'} = \frac{1}{r'(1 - r'^2)} \frac{\partial}{\partial r'} \left(r' \frac{\partial \vartheta}{\partial r'} \right) \tag{8.20}$$

The boundary conditions (8.14)–(8.16) are transformed as

$$x' = 0: \vartheta = \vartheta_0 = t_0 - t_w \tag{8.21}$$

$$r' = 1: \vartheta = 0 \tag{8.22}$$

$$r' = 0: \frac{\partial \vartheta}{\partial r'} = 0 \tag{8.23}$$

The solution to eq. (8.20) with the boundary conditions (8.21)–(8.23) is derived in a similar manner as was adopted for unsteady heat conduction in Chapter 4, i.e., the method of separating the independent variables is used. Here one assumes

$$\vartheta = F(x')G(r') \tag{8.24}$$

Introducing eq. (8.24) into eq. (8.20) gives

$$Re_D Pr G F' = \frac{F}{r'(1-r'^2)} \frac{\partial}{\partial r'}(r'G')$$

or

$$Re_D Pr \frac{F'}{F} = \frac{1}{r'(1-r'^2)} \frac{1}{G} \frac{\partial}{\partial r'}(r'G') \tag{8.25}$$

In eq. (8.25) the left-hand side is a function of x' and the right-hand side is a function of r'. As eq. (8.25) must be valid for all x' and r', eq. (8.25) must be a constant. If cooling of the fluid is considered the constant is called $-\beta^2$, and the following relations are valid

$$Re_D Pr \frac{F'}{F} = -\beta^2 \tag{8.26}$$

$$\frac{1}{r'(1-r'^2)} \frac{1}{G} \frac{\partial}{\partial r'}(r'G') = -\beta^2 \tag{8.27}$$

The solution to eq. (8.26) is

$$F = C \exp\left(-\beta^2 \frac{x'}{Re_D Pr}\right) \tag{8.28}$$

To solve eq. (8.27) is however more cumbersome. Nusselt solved eq. (8.27) numerically. It was found that the condition $G=0$ at $r'=1$ (condition (8.22)) can be satisfied by an infinite number of β-values. Thus also infinite numbers of the functions G and F are true. The solution to the temperature distribution must therefore be written as

$$\vartheta = \sum_i C_i G_i(r') \exp\left(-\beta_i^2 \frac{x'}{Re_D Pr}\right) \tag{8.29}$$

The constants C_i are determined by the condition (8.21), i.e., $\vartheta = \vartheta_0$ for $x' = 0$.

The details of determining G_i, β_i, and C_i will not be given here, but instead the solution (8.29) is presented as depicted in Fig. 8.6.

For large β_i-values, the exponential function in eq. (8.29) is decaying rapidly. At a certain distance downstream the entrance, the term including the smallest β-value will dominate the solution and the temperature distribution is then given by

$$\vartheta = C_0 G_0 \left(\frac{r}{R}\right) \exp\left(-\beta_0^2 \frac{2x/D}{Re_D Pr}\right) \tag{8.30}$$

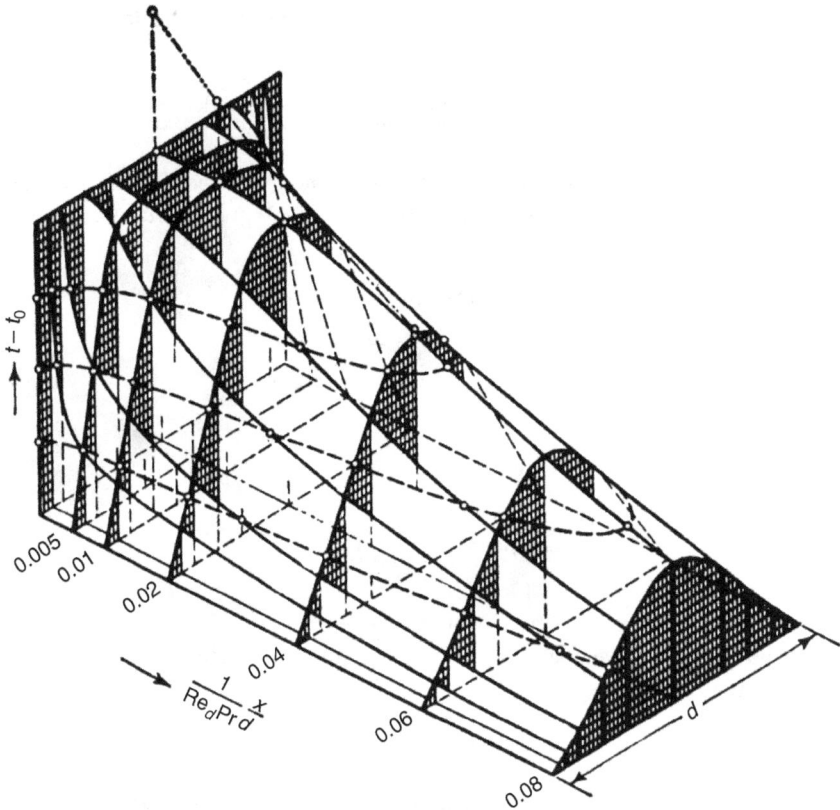

Figure 8.6: Temperature profile in the thermal entrance region for a circular pipe with uniform wall temperature. Fully developed laminar flow. (From Ref. [4].)

It is obvious that the radial dependence will be the same for all x positions. At the position this occurs, the temperature field is said to be thermally fully developed. Figure 8.6 indicates that this occurs at

$$\frac{1}{\text{Re}_D \text{Pr}} \frac{x}{D} \approx 0.05 \qquad (8.31)$$

From eq. (8.31) it is evident that the thermal entrance length is proportional to the product of the Reynolds and Prandtl numbers.

8.8.2   The heat transfer coefficient ($t_w = $ constant)

For channel flow, the heat transfer coefficient is defined as

$$q_w = \alpha(t_B - t_w) \qquad (8.32)$$

where t_B is the so-called bulk temperature. The bulk temperature is defined by

$$t_B = \frac{\int \rho c_p u t \, dA}{\int \rho c_p u \, dA} \tag{8.33}$$

and is a temperature representative of the total energy of the flow at a certain location. Especially for a circular pipe or tube, eq. (8.33) gives if ρc_p is constant

$$t_B = \frac{\int_0^R u t 2\pi r \, dr}{\int_0^R u 2\pi r \, dr} = \frac{\int_0^R u t r \, dr}{\int_0^R u r \, dr} \tag{8.34}$$

The heat flux q_w in eq. (8.32) can be written as

$$q_w = -\lambda \left(\frac{\partial t}{\partial r} \right)_{r=R} \tag{8.35}$$

where the temperature derivative is found from eq. (8.29).

In the thermally developed region, the heat transfer coefficient α and the Nusselt number are found from

$$\mathrm{Nu}_D = \frac{\alpha D}{\lambda} = 3.656 \tag{8.36}$$

In Fig. 8.7, the Nu_D is shown as a function of $(x/D)/\mathrm{Re}_D\mathrm{Pr}$. As x/D increases, the value $\mathrm{Nu}_D = 3.656$ reaches asymptotically.

For engineering calculations, a mean value is often requested, and this should also include effects of the entrance region. Hausen [8] developed a formula which fairly well agrees with the lowest curve in Fig. 8.7:

$$\overline{\mathrm{Nu}_D} = 3.656 + \frac{0.0668 \mathrm{Re}_D \mathrm{Pr} D/x}{1 + 0.04 (\mathrm{Re}_D \mathrm{Pr} D/x)^{2/3}} \tag{8.37}$$

For oils and cases with large temperature differences, it has been found that the influence of the dynamic viscosity (strongly temperature dependent) can be accommodated by multiplying the right-hand side in eq. (8.37) by the factor $(\mu_B/\mu_w)^{0.14}$, where μ_B and μ_w are the dynamic viscosities at bulk temperature t_B and wall temperature t_w, respectively. The thermophysical properties in Re, Pr, and Nu should be taken at the temperature t_B.

For gases it is recommended that eq. (8.37) is used without the viscosity correction. The thermophysical properties are evaluated at $t_{ref} = (t_w + \bar{t}_B)/2$ where $\bar{t}_B$ is the mean value of the inlet and outlet bulk temperatures [4].

If the velocity field is not fully developed at the inlet of the heated (or cooled) pipe section, the average or mean heat transfer coefficient will be higher than that

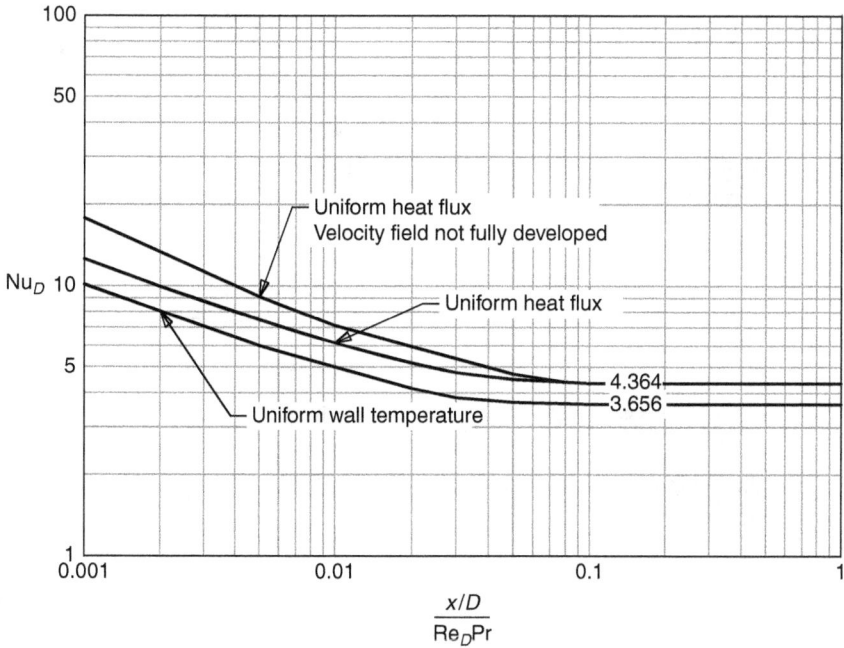

Figure 8.7: Local heat transfer coefficient for a circular pipe.

according to eq. (8.37) because the boundary layers are still developing. Equation (8.38) (developed by Sieder and Tate [9]) can be applied for such cases:

$$\overline{\mathrm{Nu}}_D = 1.86 \left(\frac{\mathrm{Re}_D \mathrm{Pr}}{L/D} \right)^{1/3} \left(\frac{\mu_{\mathrm{B}}}{\mu_{\mathrm{w}}} \right)^{0.14} \tag{8.38}$$

Equation (8.38) is valid for $(L/D)/\mathrm{Re}_D\mathrm{Pr} < 0.1$.

The heat transfer rate is determined by

$$\dot{Q} = \alpha A \Delta t_m$$

where

$$\Delta t_m = \frac{t_{\mathrm{B1}} + t_{\mathrm{B2}}}{2} - t_{\mathrm{w}}$$

8.8.3   Uniform heat flux at the pipe wall

Consider now the circular pipe in Fig. 8.8. At the pipe wall, a uniform heat flux q_{w} is supplied. For $x < 0$ the fluid has a temperature t_0.

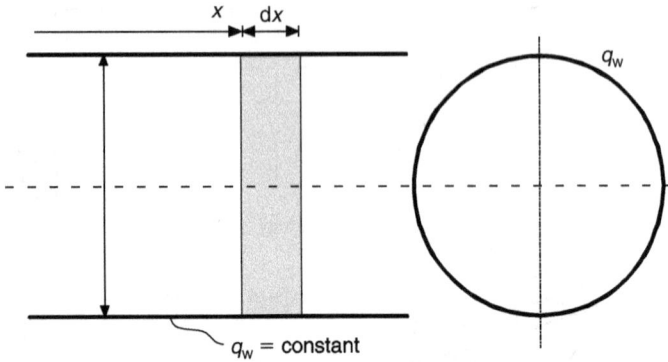

Figure 8.8: Circular pipe with uniform heat flux at the walls.

For the temperature field in the pipe, eq. (8.13) is valid. The boundary conditions are

$$x = 0: t = t_0 \tag{8.39}$$

$$r = 0: \frac{\partial t}{\partial r} = 0 \tag{8.40}$$

$$r = R: -\lambda \frac{\partial t}{\partial r} = -q_w \tag{8.41}$$

An energy balance for the element dx in Fig. 8.8 gives:

$$d\dot{Q} = \dot{m}c_p \frac{dt_B}{dx} dx = q_w \pi D \, dx$$

or

$$\frac{dt_B}{dx} = q_w \frac{\pi D}{\dot{m}c_p} \tag{8.42}$$

As the right-hand side in eq. (8.42) is constant, the bulk temperature will increase linearly with x, i.e., in the main flow direction.

For a thermally developed temperature field, this will be valid at every position r as well [10]. One then has

$$\frac{\partial t}{\partial x} = \text{constant} = C = q_w \frac{\pi D}{\dot{m}c_p} \tag{8.43}$$

The temperature field equation (8.13) can thus be written as

$$\frac{\partial}{\partial r}\left(r\frac{\partial t}{\partial r}\right) = \frac{2u_m C \rho c_p}{\lambda} r \left(1 - \left(\frac{r}{R}\right)^2\right) \tag{8.44}$$

Equation (8.44) can be integrated directly. One then finds

$$\frac{\partial t}{\partial r} = \frac{2u_m C \rho c_p}{\lambda}\left(\frac{r}{2} - \frac{1}{4}\frac{r^3}{R^2}\right) + \frac{C_1}{r}$$

Another integration gives

$$t = \frac{2u_m C \rho c_p}{\lambda}\left(\frac{r^2}{4} - \frac{1}{16}\frac{r^4}{R^2}\right) + C_1 \ln r + C_2 \tag{8.45}$$

C_1 is found from the condition (8.40) and the result is

$$C_1 = 0 \tag{8.46}$$

C is found from eq. (8.43) and can be written as

$$C = q_w \frac{\pi D}{\dot{m}c_p} = q_w \frac{4}{\rho c_p u_m D} \tag{8.47}$$

C_2 is found from the condition at $r = R$ where $t = t_w$. One has

$$t_w = \frac{2u_m C \rho c_p}{\lambda}\frac{3}{16}R^2 + C_2 \tag{8.48}$$

If eqs. (8.46)–(8.48) are inserted in eq. (8.45), the final form of the temperature distribution becomes

$$t - t_w = \frac{4q_w}{\lambda}R\left(\frac{1}{4}\left(\frac{r}{R}\right)^2 - \frac{1}{16}\left(\frac{r}{R}\right)^4 - \frac{3}{16}\right) \tag{8.49}$$

The heat transfer coefficient was defined in eq. (8.32). The bulk temperature t_B is determined by considering eqs. (8.34), (8.49), and (8.3).

After a few calculations, one finds

$$Nu_D = \frac{\alpha D}{\lambda} = \frac{48}{11} = 4.364 \tag{8.50}$$

Figure 8.7 shows Nu_D as a function of $(1/Re_D Pr)(x/D)$ for the case $q_w = $ constant. For increasing values of x, $Nu_D = 4.364$ reaches asymptotically.

Similar to the case with $t_w = $ constant, a mean value which also includes the entrance effect is needed in engineering calculations. Shah [11] has developed the following equations which are reasonably accurate

$$\overline{Nu_D} = \begin{cases} 1.953\left(\frac{1}{Re_D Pr}\frac{x}{D}\right)^{-1/3} & \text{if } \frac{x/D}{Re_D Pr} < 0.03 \\ 4.364 + \frac{0.0722}{x/D}Re_D Pr & \text{if } \frac{x/D}{Re_D Pr} > 0.03 \end{cases} \tag{8.51}$$

Table 8.2: Nusselt number, $\mathrm{Nu} = \alpha D_\mathrm{h}/\lambda$, for channels of different cross section.

Configuration		Nu	
		$t_\mathrm{w} = \text{constant}$	$q_\mathrm{w} = \text{constant}$
		3.66	4.36
	$b/a = 1.0$	2.98	3.63
	2.0	3.39	4.12
	4.0	4.44	5.33
	8.0	5.60	6.49
	∞	7.54	8.23
		2.47	3.11

Fully developed flow and temperature fields prevail. From Refs. [5] and [12].

8.9   Heat transfer for noncircular channels

In the literature, one can find solutions for the convective heat transfer in noncircular channels, e.g., channels with rectangular, triangular, or prismatic cross sections. In Table 8.2 Nusselt numbers, Nu, are presented for some cases. Values for uniform wall temperature and uniform heat flux are provided. Fully developed flow and temperature fields prevail.

8.10   Heat transfer when the velocity field is not fully developed

The so far presented results (except eq. (8.38)) were valid only if the velocity field was fully developed. If this is not the case, the analysis will be more complicated. Figure 8.7 shows the heat transfer coefficient as the flow field and the temperature field are not fully developed. Asymptotically the value $\mathrm{Nu}_D = 4.364$ is however reached. For the mathematical analysis, the book by Shah and London [12] should be studied.

Sometimes the conditions $t_w = $ constant or $q_w = $ constant are not valid. This is the case if heat conduction in the channel walls is important. In Ref. [12], heat transfer in channels for a variety of wall boundary conditions is analyzed.

8.11   Final remarks

In this chapter, analysis and results for laminar forced convection in channels were presented. More details and complementary information can be found in Ref. [13].

References

[1] H. Schlichting, Boundary Layer Theory, 7th ed., McGraw-Hill, New York (1979).
[2] F.M. White, Fluid Mechanics, 6th ed., McGraw-Hill, New York (2008).
[3] H.L. Langhaar, Steady flow in the transition length in a straight tube, J. Appl. Mech., 9, A55–A58 (1942).
[4] E.R.G. Eckert and R.M. Drake Jr., Analysis of Heat and Mass Transfer, McGraw-Hill, Tokyo (1972).
[5] W.M. Kays, Convective Heat and Mass Transfer, McGraw-Hill, New York (1966).
[6] L. Graetz, Über die Wärmeleitungsfähigkeit von Flüssigkeiten, Part 1, Ann. Phys. Chem., 18, 79–94 (1883).
[7] W. Nusselt, Die Abhängigkeit der Wärmeübergangszahl von der Rohrlänge, Z. Ver. Deutsch. Ing., 514, 1154–1158 (1910).
[8] H. Hausen, Darstellung des Wärmeüberganges in Rohren durch verallgemeinerte Potenzbeziehungen, Z. Ver. Deutsch. Ing., 14, 91–98 (1943).
[9] E.N. Sieder and C.E. Tate, Heat transfer and pressure liquids in tubes, Ind. Eng. Chem., 28, 1429–1435 (1936).
[10] F.P. Incropera, D.P. DeWitt, T. Bergman and A. Lavine, Introduction to Heat Transfer, 5th ed., John Wiley & Sons, New York (2007).
[11] R.K. Shah, Thermal entry length solutions for the circular tube and parallel plates, vol. 1, Proceedings of 3rd National Heat Mass Transfer Conference, Indian Institute of Technology, Bombay (1975).
[12] R.K. Shah and A.L. London, Laminar flow forced convection in ducts, in Advances in Heat Transfer, Supplement 1, Academic Press, New York (1978).
[13] R.K. Shah and M.S. Bhatti, Laminar convective heat transfer in ducts, in Handbook of Single-Phase Convective Heat Transfer (Eds. S. Kakac, R.K. Shah and W. Aung), Wiley-Interscience Publication, New York (1987).

9 Forced convection—turbulent flow

9.1 Introduction

Most flow fields in nature and in engineering applications are turbulent. So, for instance, the boundary layer in the earth atmosphere and the boundary layer on airplane wings are turbulent.

Water flow in rivers, pipes, and conduits are often turbulent. The wake flow behind ships, road vehicles, submarines, and airplanes is commonly turbulent.

In this chapter, a qualitative discussion of turbulence is presented, parts of the existing theory will be outlined, and then application of the theory is shown to determine friction factors (pressure drop) and heat transfer coefficients in ducts and over plane surfaces.

9.2 Properties of turbulence

It is very hard to present a precise definition of turbulence. An idea about it can be acquired by considering the smoke from a chimney or from a cigarette. A summary of the most important properties of a turbulent flow field is presented below.

1. The flow field is unsteady.
2. The fluid motion is irregular and shows random variations in time and space concerning the detailed flow structure. This means that a detailed description of the motion as a function of time and the space coordinates x, y, z is impossible. However, the turbulent motion is random in such a manner that statistical methods of analysis can be used. It has been found that it is possible to define mean values (time-averaged values) of properties like velocity, pressure, and temperature.
3. The flow field is characterized by a high eddy activity, and unsteady eddies of various size are present at the same time in the flow field.
4. Turbulent flow exists at high Reynolds number. Often the turbulence is created from an instability of a laminar flow if the Reynolds number is sufficiently high.
5. The turbulence motion is three-dimensional and has a certain amount of vorticity. High fluctuations in the vorticity are present.
6. The turbulence is a continuous phenomenon, i.e., the smallest turbulent motions are bigger than the mean free path for the molecular motion.

7. The turbulence is diffusive, i.e., the velocity fluctuations are spread through the surrounding fluid. This implies that the heat transfer is enhanced, the momentum transfer is increased, and the resistance against separation is increased as well.
8. The turbulent motion is always dissipative. Due to the action of viscous forces, turbulent energy is converted into internal energy in the fluid. The turbulent motion therefore needs supply of energy to compensate for the viscous losses. If no energy is supplied, the turbulent motion will disappear.
9. Turbulence is not a property of fluids but a way fluids may flow.

9.3 Methods of analysis

Turbulent flow fields have been investigated for many years, but no general solution to the turbulence problem has been found. The equations of motion have been analyzed into details, but it is still very difficult to perform any quantitative calculations without using empirical data. Statistical methods applied on the equations of motion always results in more unknowns than the number of available governing equations. This is called the *closure problem* in the theory of turbulence. Thus a number of assumptions have to be taken in order to balance the number of unknowns and available equations. This will be considered later.

A very common method of analysis is the so-called method of dimensional analysis. In many cases, it is possible to stipulate that the turbulence structure depends only on a few variables or parameters. If this is the case, the dimensional analysis commonly gives a relation between the dependent and the independent variables. This means a solution which is deterministic besides the numerical value of a constant has to be found. A good example of this is the so-called energy spectrum for the turbulent kinetic energy.

Another method is to attempt to find asymptotic properties for the turbulent flow. Turbulent flow is characterized by a high Reynolds number. It therefore seems reasonable to require that an assumed behavior of the turbulent field should develop in a certain way as the Reynolds number is increased to infinity. This approach often enables specific results. An example is the theory of turbulent boundary layers. Turbulent flow tends to be independent of the fluid viscosity except for the smallest scale motions. The asymptotic behavior results in Reynolds number similarity or asymptotic invariance.

Associated with but different from asymptotic invariance is the idea of local invariance or self-preservation. For simple flow geometries, it has been found that the properties of the turbulent motion at a certain point in the flow are mainly determined by the conditions in the immediate environment. It is then possible to assume that the turbulence is dynamically uniform everywhere if the local time and length scales are applied.

9.4 About length scales in a turbulent flow

In a turbulent flow, a broad spectrum of length scales exists. The biggest eddies are of the same size as the overall dimensions of the flow field. The size of the smallest eddies is determined by viscous effects. The smaller the eddy, the bigger is the velocity gradient inside it. Because the dissipation of the turbulent kinetic energy has been found to be proportional to the viscosity and the velocity gradient, it is reasonable to assume that the smallest eddies are responsible for the dissipation of turbulent kinetic energy. In the same way as various length scales exist, several time scales are present.

9.5   The origin of turbulence

A laminar flow becomes instable at high Reynolds number. For pipe or tube flow, this occurs at $Re = U_m D / \nu \approx 2300$.

The boundary layer along a flat plate becomes turbulent at

$$Re = \frac{U\delta^*}{\nu} \approx 600 (Re_{x,kr} = 3 \cdot 10^5 - 3 \cdot 10^6)$$

The turbulence can however not survive by its own, but energy must be transferred from the closest environment. A common source of energy for the turbulent velocity fluctuations is the existence of velocity gradients or shear in the mean flow field. This results in production or generation of turbulent kinetic energy.

Turbulence can also be created at the interface between two fluids at different velocities, e.g., jets and wakes.

9.6   Equations of motion for turbulent flow

In Fig. 9.1 is shown how the velocity u in the x-direction varies with respect to time at a certain point in the flow field.

The instantaneous velocity u can be interpreted as composed of a time-averaged value $\bar{u}$ and a randomly varying or fluctuating component u'. The velocities v and w in the y- and z-directions, respectively, can be splitted up similarly, and one writes

$$u = \bar{u} + u' \tag{9.1}$$

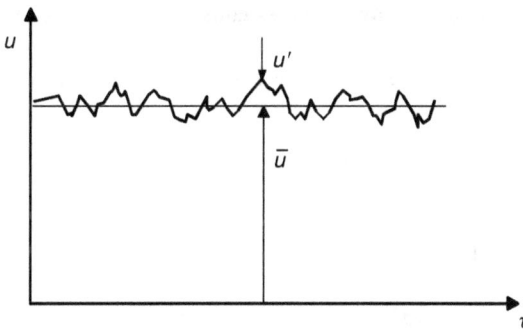

Figure 9.1: The velocity variation versus time for turbulent flow.

$$v = \bar{v} + v' \tag{9.2}$$

$$w = \bar{w} + w' \tag{9.3}$$

For the time-averaged value, the following definition holds

$$\bar{u} = \frac{1}{\tau_2 - \tau_1} \int_{\tau_1}^{\tau_2} u \, d\tau \tag{9.4}$$

From eqs. (9.1) and (9.4), it follows

$$\bar{u'} = \frac{1}{\tau_2 - \tau_1} \int_{\tau_1}^{\tau_2} u \, d\tau \equiv 0 \tag{9.5}$$

Completely analogous results appear in the y- and z-directions. The time interval $\tau_2 - \tau_1$ must be sufficiently big so that a true time-averaged value is obtained. Because the time-averaged values of the fluctuating components are zero, the strength of the turbulent motion or a specific fluctuating component is defined by its root mean square value, i.e., the RMS-value, e.g.,

$$u' \, (\text{RMS}) = \sqrt{\bar{u'^2}} \tag{9.6}$$

Figure 9.2 shows the variation of the RMS values for the fluctuating velocity components in a turbulent boundary layer along a flat plate.

In a similar way, scalar quantities like the pressure p is decomposed. One has

$$p = \bar{p} + p'$$

Note: The decomposition shown is commonly referred to as the Reynolds' decomposition, see, for example, Ref. [2].

9.6.1   Continuity equation or the mass conservation equation

For incompressible flow, the continuity equation reads (from eq. (6.4) with $p =$ constant)

$$\frac{\partial u}{\partial x} + \frac{\partial v}{\partial y} + \frac{\partial w}{\partial z} = 0 \tag{9.7}$$

Substituting eqs. (9.1)–(9.3) in eq. (9.7), one obtains

$$\frac{\partial \bar{u}}{\partial x} + \frac{\partial \bar{v}}{\partial y} + \frac{\partial \bar{w}}{\partial z} + \frac{\partial u'}{\partial x} + \frac{\partial v'}{\partial y} + \frac{\partial w'}{\partial z} = 0$$

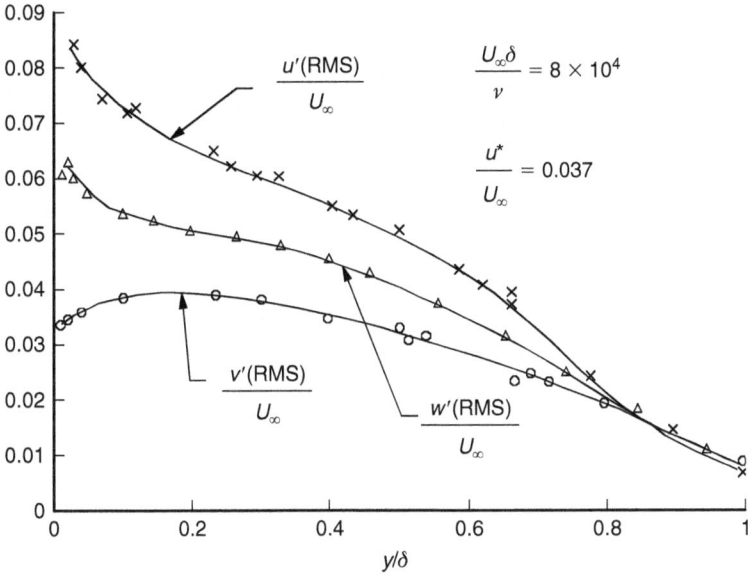

Figure 9.2: u' (RMS), v' (RMS), and w' (RMS) in a turbulent boundary layer along a flat plate. (Based on Klebanoff, see Hinze [1].)

If this equation is time-averaged, one has

$$\frac{\partial \bar{u}}{\partial x} + \frac{\partial \bar{v}}{\partial y} + \frac{\partial \bar{w}}{\partial z} = 0 \qquad (9.8)$$

Thus the continuity equation has the same form as in the laminar case. However, it is also easily shown that

$$\frac{\partial u'}{\partial x} + \frac{\partial v'}{\partial y} + \frac{\partial w'}{\partial z} = 0 \qquad (9.9)$$

9.6.2 Navier–Stokes' equations

Navier–Stokes' equations were given in Chapter 6 and these are also valid for the turbulent case if instantaneous values are used. For the incompressible case without body force, the equation in the x-direction reads (from eqs. (6.7), (6.12), (6.14), and (6.15))

$$\rho \left(\frac{\partial u}{\partial \tau} + u \frac{\partial u}{\partial x} + v \frac{\partial u}{\partial y} + w \frac{\partial u}{\partial z} \right) = -\frac{\partial p}{\partial x} + \mu \left(\frac{\partial^2 u}{\partial x^2} + \frac{\partial^2 u}{\partial y^2} + \frac{\partial^2 u}{\partial z^2} \right) \qquad (9.10)$$

Substituting eqs. (9.1)–(9.3) and (9.7) in eq. (9.10), one has

$$
\rho \left\{ \frac{\partial \bar{u}}{\partial \tau} + \frac{\partial u'}{\partial \tau} + \bar{u}\frac{\partial \bar{u}}{\partial x} + \bar{u}\frac{\partial u'}{\partial x} + u'\frac{\partial \bar{u}}{\partial x} + u'\frac{\partial u'}{\partial x} + \bar{v}\frac{\partial \bar{u}}{\partial y} + \bar{v}\frac{\partial u'}{\partial y} + v'\frac{\partial \bar{u}}{\partial y} + v'\frac{\partial u'}{\partial y} \right.
$$
$$
\left. + \bar{w}\frac{\partial \bar{u}}{\partial z} + \bar{w}\frac{\partial u'}{\partial z} + w'\frac{\partial \bar{u}}{\partial z} + w'\frac{\partial u'}{\partial z} \right\}
$$
$$
= -\frac{\partial p}{\partial x} - \frac{\partial p'}{\partial x} + \mu \left(\frac{\partial^2 \bar{u}}{\partial x^2} + \frac{\partial^2 \bar{u}}{\partial y^2} + \frac{\partial^2 \bar{u}}{\partial z^2} + \frac{\partial^2 u'}{\partial x^2} + \frac{\partial^2 u'}{\partial y^2} + \frac{\partial^2 u'}{\partial z^2} \right) \quad (9.11)
$$

To make this equation more handy, a time-averaging procedure is applied, and the result is

$$
\rho \left(\frac{\partial \bar{u}}{\partial \tau} + \bar{u}\frac{\partial \bar{u}}{\partial x} + \bar{v}\frac{\partial \bar{u}}{\partial y} + \bar{w}\frac{\partial \bar{u}}{\partial z} + \overline{u'\frac{\partial u'}{\partial x}} + \overline{v'\frac{\partial u'}{\partial y}} + \overline{w'\frac{\partial u'}{\partial z}} \right)
$$
$$
= -\frac{\partial \bar{p}}{\partial x} + \mu \left(\frac{\partial^2 \bar{u}}{\partial x^2} + \frac{\partial^2 \bar{u}}{\partial y^2} + \frac{\partial^2 \bar{u}}{\partial z^2} \right) \quad (9.12)
$$

With eq. (9.9), the last three terms within the parenthesis in the left-hand side of eq. (9.12) can be written

$$
\overline{u'\frac{\partial u'}{\partial x}} + \overline{v'\frac{\partial u'}{\partial y}} + \overline{w'\frac{\partial u'}{\partial z}} = \frac{\partial}{\partial x}\overline{u'^2} + \frac{\partial}{\partial y}\overline{u'v'} + \frac{\partial}{\partial z}\overline{u'w'} \quad (9.13)
$$

Eq. (9.13) in eq. (9.12) gives finally

$$
\frac{\partial \bar{u}}{\partial \tau} + \bar{u}\frac{\partial \bar{u}}{\partial x} + \bar{v}\frac{\partial \bar{u}}{\partial y} + \bar{w}\frac{\partial \bar{u}}{\partial z}
$$
$$
= -\frac{1}{\rho}\frac{\partial \bar{p}}{\partial x} + \frac{\mu}{\rho}\left(\frac{\partial^2 \bar{u}}{\partial x^2} + \frac{\partial^2 \bar{u}}{\partial y^2} + \frac{\partial^2 \bar{u}}{\partial z^2} \right) - \frac{\partial}{\partial x}\overline{u'^2} - \frac{\partial}{\partial y}\overline{u'v'} - \frac{\partial}{\partial z}\overline{u'w'} \quad (9.14)
$$

Equation (9.14) is the Navier–Stokes' equation in the x-direction for a turbulent flow. Compared to the laminar case, three additional terms have been achieved. The equation is thus more complicated than for the laminar case, and approximations have to be introduced to enable solutions. In the y-direction, one obtains similarly:

$$
\frac{\partial \bar{v}}{\partial \tau} + \bar{u}\frac{\partial \bar{v}}{\partial x} + \bar{v}\frac{\partial \bar{v}}{\partial y} + \bar{w}\frac{\partial \bar{v}}{\partial z}
$$
$$
= -\frac{1}{\rho}\frac{\partial \bar{p}}{\partial y} + \frac{\mu}{\rho}\left(\frac{\partial^2 \bar{v}}{\partial x^2} + \frac{\partial^2 \bar{v}}{\partial y^2} + \frac{\partial^2 \bar{v}}{\partial z^2} \right) - \frac{\partial}{\partial x}\overline{u'v'} - \frac{\partial}{\partial y}\overline{v'^2} - \frac{\partial}{\partial z}\overline{v'w'} \quad (9.15)
$$

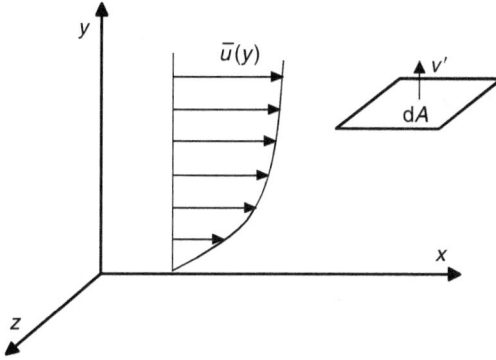

Figure 9.3: Illustration of turbulent shear stress.

The equation in the z-direction becomes

$$\frac{\partial \overline{w}}{\partial \tau} + \overline{u}\frac{\partial \overline{w}}{\partial x} + \overline{v}\frac{\partial \overline{w}}{\partial y} + \overline{w}\frac{\partial \overline{w}}{\partial z}$$

$$= -\frac{1}{\rho}\frac{\partial \overline{p}}{\partial z} + \frac{\mu}{\rho}\left(\frac{\partial^2 \overline{w}}{\partial x^2} + \frac{\partial^2 \overline{w}}{\partial y^2} + \frac{\partial^2 \overline{w}}{\partial z^2}\right) - \frac{\partial}{\partial x}\overline{u'w'} - \frac{\partial}{\partial y}\overline{v'w'} - \frac{\partial}{\partial z}\overline{w'^2} \quad (9.16)$$

9.6.3 Turbulent stresses

Consider the flow field in Fig. 9.3 where the time-averaged velocity is given by $u = u(y)$. Through the surface element dA in the xz-plane, a mass flow $\dot{m}_y$ is established due to the turbulent velocity fluctuation v':

$$\dot{m} = \rho v' \, dA \quad (9.17)$$

According to the momentum theorem, a related force is acting in the x-direction according to

$$F_x = \dot{m}_y u = \rho \, dA v'(\overline{u} + u') \quad (9.18)$$

Because the surface element is moving in the x-direction, the force on the element will be $-F_x$. Especially, the time-averaged value of this force is of interest:

$$\sigma \, dA = -\overline{F}_x$$

One has

$$\sigma = \rho \overline{v'(\overline{u} + u')} = -\rho \overline{u'v'} \quad (9.19)$$

Thus it has become evident that the turbulence contribution $-\rho \overline{u'v'}$ can be interpreted as a shear stress. This will be called a turbulent shear stress and is given

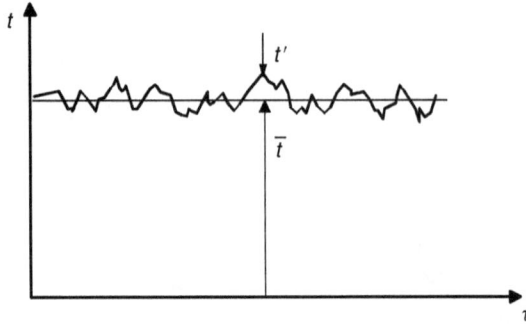

Figure 9.4: Temperature variation versus time for turbulent flow.

the notation σ_{turb}. In an analogous manner, the terms $-\rho\overline{u'u'}$ and $-\rho\overline{u'w'}$ can be interpreted as a normal stress and a shear stress, respectively. Also the turbulence terms in eqs. (9.15) and (9.16) can be interpreted as turbulent stresses.

For a laminar flow, $u = u(y)$, a shear stress due to molecular action is (see Chapter 6)

$$\sigma_{\text{lam}} = \mu \frac{\partial \overline{u}}{\partial y} \tag{9.20a}$$

For the case in Fig. 9.3 including both the laminar and turbulent contributions, one has

$$\sigma = \sigma_{\text{lam}} + \sigma_{\text{turb}} = \mu \frac{\partial \overline{u}}{\partial y} - \rho \overline{u'v'} \tag{9.20b}$$

(With the notations in Chapter 6, the shear stress σ corresponds to the stress σ_{yx}.)

9.7 Temperature field equation for turbulent flow

Figure 9.4 shows how the temperature t varies with time at a certain point in a turbulent flow field. In a similar manner as for the velocities and pressure, the temperature is decomposed into a time-averaged temperature and a fluctuating component. Thus one writes

$$t = \overline{t} + t' \tag{9.21}$$

The temperature field equation (6.36) with the time-dependent term added is valid for a turbulent flow if instantaneous values for the variables are used. Thus one has

$$\frac{\partial t}{\partial \tau} + u \frac{\partial t}{\partial x} + v \frac{\partial t}{\partial y} + w \frac{\partial t}{\partial z} = \frac{\lambda}{\rho c_{\text{p}}} \left(\frac{\partial^2 t}{\partial x^2} + \frac{\partial^2 t}{\partial y^2} + \frac{\partial^2 t}{\partial z^2} \right) \tag{9.22}$$

Equations (9.1)–(9.3) and (9.21) are inserted in eq. (9.22). If the time-averaging procedure is applied and the appearing turbulence terms are rewritten by using eq. (9.9), one finds after some algebraic operations

$$\frac{\partial \bar{t}}{\partial \tau} + \bar{u}\frac{\partial \bar{t}}{\partial x} + \bar{v}\frac{\partial \bar{t}}{\partial y} + \bar{w}\frac{\partial \bar{t}}{\partial z}$$

$$= \frac{\lambda}{\rho c_p}\left(\frac{\partial^2 \bar{t}}{\partial x^2} + \frac{\partial^2 \bar{t}}{\partial y^2} + \frac{\partial^2 \bar{t}}{\partial z^2}\right) - \frac{\partial}{\partial x}\overline{u't'} - \frac{\partial}{\partial y}\overline{v't'} - \frac{\partial}{\partial z}\overline{w't'} \quad (9.23)$$

9.7.1 Turbulent heat flux

Consider again the flow in Fig. 9.3. As an effect of the fluctuating velocity component v' in the y-direction, an energy transport across the surface element dA in the y-direction is established. This transport is

$$q(\tau) = \frac{d\dot{Q}}{dA} = \frac{d\dot{m}h}{dA} = \rho v' c_p(\bar{t} + t')$$

The time-average of this energy transport (or heat flux) is of particular interest. One finds

$$q_{turb} = \rho c_p \overline{v't'} \quad (9.24)$$

In an analogous manner, the terms $\rho c_p \overline{u't'}$ and $\rho c_p \overline{w't'}$ can be interpreted as heat fluxes due to the turbulent motion in the x- and z-directions, respectively.

For the molecular heat conduction in a fluid, one has (see Chapter 6)

$$q_y = -\lambda \frac{\partial \bar{t}}{\partial y} \quad (9.25)$$

For the case in Fig. 9.3 with both molecular heat conduction and turbulent heat transport, the total heat flux can be written as

$$q = q_{molecular} + q_{turb}$$

In the y-direction, eqs. (9.24) and (9.25) give

$$q = -\lambda \frac{\partial \bar{t}}{\partial y} + \rho c_p \overline{v't'} \quad (9.26)$$

9.8 Boundary layer equations for turbulent flow

For the two-dimensional case with steady mean motion and constant physical properties, it is possible to derive the so-called boundary layer equations for a turbulent flow similar to what was done for laminar flow. One finds

$$\bar{u}\frac{\partial \bar{u}}{\partial x} + \bar{v}\frac{\partial \bar{u}}{\partial y} = -\frac{1}{\rho}\frac{dp}{dx} + \frac{\mu}{\rho}\frac{\partial^2 \bar{u}}{\partial y^2} - \frac{\partial}{\partial y}\overline{u'v'} \tag{9.27}$$

$$\frac{\partial \bar{u}}{\partial x} + \frac{\partial \bar{v}}{\partial y} = 0 \tag{9.28}$$

$$\frac{dp}{dx} = -\rho U \frac{dU}{dx} \tag{9.29}$$

$$\bar{u}\frac{\partial \bar{t}}{\partial x} + \bar{v}\frac{\partial \bar{t}}{\partial y} = \frac{\lambda}{\rho c_p}\frac{\partial^2 \bar{t}}{\partial y^2} - \frac{\partial}{\partial y}\overline{v't'} \tag{9.30}$$

9.9 Turbulent viscosity and turbulent diffusivity

In eqs. (9.19) and (9.24), the turbulent shear stress and the turbulent heat flux were defined. To transfer the boundary layer equations (9.27)–(9.30) to a more handy form, it is common to introduce a turbulent viscosity $\rho \varepsilon_m$ in order to relate the turbulence terms to the mean motion. The definition is

$$\sigma_{turb} = -\rho\overline{u'v'} = \rho\varepsilon_m\frac{\partial \bar{u}}{\partial y} \tag{9.31}$$

Equation (9.31) is analogous to eq. (9.20a) for the laminar shear stress. The assumption in eq. (9.31) is sometimes referred to as the Boussinesq's approximation (1877). For the turbulent heat flux, a turbulent diffusivity ε_q is introduced as

$$q_{turb} = \rho c_p\overline{v't'} = -\rho c_p\varepsilon_q\frac{\partial \bar{t}}{\partial y} \tag{9.32}$$

By the definitions (9.31) and (9.32), the determination of the turbulence terms has been transferred to a determination of the variables ε_m and ε_q.

Commonly, also a turbulent Prandtl number Pr_t is introduced as

$$Pr_t = \frac{\varepsilon_m}{\varepsilon_q} \tag{9.33}$$

From eqs. (9.31)–(9.33), the total shear stress and turbulent heat flux can be written as

$$\sigma = \mu\frac{\partial \bar{u}}{\partial y} + \rho\varepsilon_m\frac{\partial \bar{u}}{\partial y} = \rho(\nu + \varepsilon_m)\frac{\partial \bar{u}}{\partial y} \tag{9.34}$$

$$q = -\lambda\frac{\partial \bar{t}}{\partial y} - \rho c_p\varepsilon_q\frac{\partial \bar{t}}{\partial y} = \rho c_p\left(\frac{\lambda}{\rho c_p} + \frac{\varepsilon_m}{Pr_t}\right)\frac{\partial \bar{t}}{\partial y} = -\rho c_p\left(\frac{\nu}{Pr} + \frac{\varepsilon_m}{Pr_t}\right)\frac{\partial \bar{t}}{\partial y} \tag{9.35}$$

The turbulent viscosity $\rho\varepsilon_m$ is nowadays often denoted as μ_{turb}. This is in particular true in the literature concerning turbulence modeling. The turbulent diffusivity $\rho\varepsilon_q$ is often written as μ_{turb}/Pr_t. The notations are then completely analogous to the laminar case.

9.10   Reynolds' analogy

Consider a two-dimensional boundary layer when the fluid has constant physical properties and the mean motion is steady. For the shear stress, eq. (9.34) is valid and for the heat flux, eq. (35) is valid. The ratio between these two equations is now established:

$$\frac{q}{\sigma} = \frac{-\rho c_p \left((\nu/Pr) + (\varepsilon_m/Pr_t)\right)(\partial \bar{t}/\partial y)}{\rho(\nu + \varepsilon_m)(\partial \bar{u}/\partial y)} = -\frac{c_p \left((\nu/Pr) + (\varepsilon_m/Pr_t)\right)(\partial \bar{t}/\partial y)}{(\nu + \varepsilon_m)(\partial \bar{u}/\partial y)}$$

It is now assumed that q/σ is constant and equal to q_w/σ_w. If in addition, the Prandtl numbers are assumed to be unity ($Pr = 1$ is a good approximation for gases and $Pr_t = 1$ gives the right order of magnitude for Pr_t, see Refs. [3, 4]):

$$\frac{q_w}{\sigma_w} = -c_p \frac{\partial \bar{t}/\partial y}{\partial \bar{u}/\partial y}$$

or

$$\frac{\partial \bar{t}}{\partial y} = -\frac{1}{c_p} \frac{q_w}{\sigma_w} \frac{\partial \bar{u}}{\partial y} \qquad (9.36)$$

Equation (9.36) is now integrated from $y = 0$ ($\bar{u} = 0$, $\bar{t} = \bar{t}_w$) to $y = \delta$ ($\bar{u} = U$, $\bar{t} = \bar{t}_\infty$)

$$\bar{t}_\infty - \bar{t}_w = -\frac{1}{c_p} \frac{q_w}{\sigma_w} U \qquad (9.37)$$

The heat flux q_w is written as

$$q_w = \alpha(\bar{t}_w - \bar{t}_\infty)$$

and the shear stress σ_w (compare with Chapter 7) is written as

$$\sigma_w = C_F \frac{\rho U^2}{2}$$

Inserting these expressions in eq. (9.37) gives

$$\bar{t}_\infty - \bar{t}_w = -\frac{1}{c_p} \frac{\alpha(\bar{t}_w - \bar{t}_\infty)}{C_F \rho U^2/2} U$$

or

$$\frac{\alpha}{\rho c_p U} = \frac{C_F}{2}. \qquad (9.38)$$

The relation (9.38) is sometimes called *Reynolds' analogy* and relates the heat transfer coefficient α to the shear stress coefficient C_F. Thus the problem to determine α for turbulent flow is transferred to determine C_F.

In Chapter 7, the Stanton number St was introduced according to eq. (7.42). Equation (9.38) can therefore be written as

$$St = \frac{Nu_x}{Re_x Pr} = \frac{C_F}{2} \tag{9.39}$$

If eq. (9.39) is compared with eq. (7.43), one finds that the Prandtl number dependence is not included in eq. (9.39). In the derivation of eq. (9.38) or (9.39), $Pr = 1$ was assumed. To eliminate this assumption, Colburn ([5]) suggested that eq. (9.39) should be written as eq. (7.43), i.e.,

$$St = \frac{Nu_x}{Re_x Pr} = \frac{C_F}{2} Pr^{-2/3} \tag{9.40}$$

Equation (9.40) is often called *Colburn's analogy*. However, one should note that eq. (9.40) for turbulent flows is only approximate while eq. (7.43) is exactly valid for laminar flow along a flat plate. Reynolds' analogy and Colburn's analogy are important as the heat transfer coefficient is determined for turbulent flows.

For pipe or tube flow, a corresponding derivation can be performed and one finds, see Özisik [6]

$$St = \frac{Nu_D}{Re_D Pr} = \frac{C_F}{2} Pr^{-2/3} \tag{9.41}$$

Equations (9.40) and (9.41) are often written in the form

$$j = St\, Pr^{2/3} = \frac{C_F}{2}$$

This is sometimes called the modified Reynolds' analogy or *Chilton–Colburn's analogy*. j is called the Colburn factor.

9.11 The velocity profile in a turbulent boundary layer and in pipe flow

To enable determination of the shear stress coefficient C_F, it is necessary to know the velocity profile. A turbulent boundary layer along a flat plate is now considered. The equation of motion, eq. (9.27), reads after inclusion of eq. (9.34)

$$\bar{u}\frac{\partial \bar{u}}{\partial x} + \bar{v}\frac{\partial \bar{u}}{\partial y} = \frac{1}{\rho}\frac{\partial \sigma}{\partial y} \tag{9.42}$$

Closest to the wall surface, i.e., as $y \to 0$, the velocities $\bar{u}$ and $\bar{v}$ are very small and then eq. (9.42) gives

$$0 = \frac{1}{\rho}\frac{\partial \sigma}{\partial y}$$

which implies

$$\sigma = \text{constant} = \sigma_w = \text{wall shear stress} \qquad (9.43)$$

Equations (9.43) and (9.34) now give

$$(\nu + \varepsilon_m)\frac{\partial \bar{u}}{\partial y} = \frac{\sigma_w}{\rho} \qquad (9.44)$$

In the limit $y \to 0$, one has $\varepsilon_m \to 0$ (because u' and v' approach zero). Equation (9.44) now gives

$$\nu\frac{\partial \bar{u}}{\partial y} = \frac{\sigma_w}{\rho}$$

Integration gives

$$\bar{u} = \frac{\sigma_w}{\mu}y + c_1$$

The constant $c_1 = 0$ because $\bar{u} = 0$ for $y = 0$. If $\sigma_w = \rho u_\tau^2$ is introduced, one can write ($u_\tau = \sqrt{\sigma_w/\rho}$ is the so-called friction velocity, which is also denoted by u^* in the literature)

$$\frac{\bar{u}}{u_\tau} = \frac{u_\tau y}{\nu} \qquad (9.45)$$

where the dimensionless coordinate $y^+ = u_\tau y/\nu$ has also been introduced.

Equation (9.45) gives the mean velocity distribution in the so-called viscous sublayer where the viscous forces dominate. It has been found that the velocity distribution equation (9.45) is valid when (see Hinze [1])

$$0 < y^+ < 5 \qquad (9.46)$$

Somewhat further away from the wall, where eq. (9.43) is still valid, one has $\varepsilon_m \gg \mu/\rho$. Equation (9.44) then gives

$$\varepsilon_m\frac{\partial \bar{u}}{\partial y} = \frac{\sigma_w}{\rho} \qquad (9.47)$$

In order to solve eq. (9.47), one has to know ε_m.

Prandtl in 1925 introduced the so-called mixing length concept. The mixing length is related to the exchange of momentum between various layers and was interpreted by Prandtl as the distance a macroscopically identified mass fraction on the average is moving by the turbulent motion before it loses its identity.

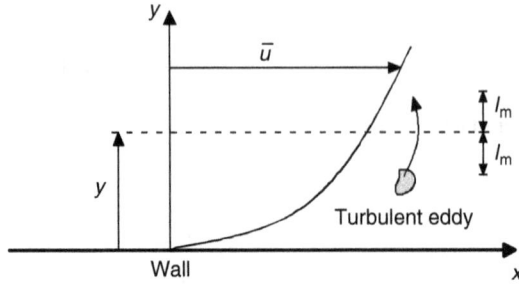

Figure 9.5: Turbulent stress and mixing length.

Consider now Fig. 9.5.

At the level $y + l_m$, the following relation holds

$$\bar{u}(y + l_m) = \bar{u}(y) + l_m \frac{\partial \bar{u}}{\partial y}$$

and at the level $y - l_m$, the following relation holds

$$\bar{u}(y - l_m) = \bar{u}(y) - l_m \frac{\partial \bar{u}}{\partial y}$$

Prandtl postulated that the turbulent velocity fluctuation u' is proportional to the velocity difference according to

$$u' \approx l_m \frac{\partial \bar{u}}{\partial y} \tag{9.48}$$

Prandtl assumed that the velocity v' is of the same order of magnitude as u'. The turbulent stress can then be written as

$$\sigma_{turb} = -\rho \overline{u'v'} = \rho l_m^2 \left(\frac{\partial \bar{u}}{\partial y} \right)^2 \tag{9.49}$$

A comparison with eq. (9.31) gives

$$\varepsilon_m = l_m^2 \frac{\partial \bar{u}}{\partial y} \tag{9.50}$$

The mixing length l_m varies considerably across the boundary layer. Several investigations have been performed over the years to consider this variation, e.g., Ref. [1]. Prandtl's hypothesis was that the mixing length close to the body surface (the wall) is proportional to the distance from the surface, i.e.,

$$l_m = \kappa y \tag{9.51}$$

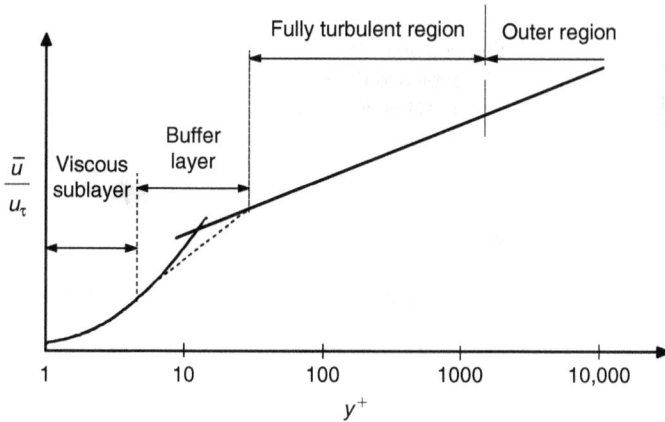

Figure 9.6: Sketch of the velocity distribution in a turbulent boundary layer.

Substituting eqs. (9.50) and (9.51) in eq. (9.47), one has

$$\kappa^2 y^2 \left(\frac{\partial \bar{u}}{\partial y} \right)^2 = \frac{\sigma_w}{\rho}$$

With $\sigma_w = \rho u_\tau^2$, one finds

$$\frac{\partial \bar{u}}{\partial y} = \frac{1}{\kappa} \frac{u_\tau}{y}$$

Integration gives

$$\bar{u} = \frac{u_\tau}{\kappa} \ln y + C$$

which can be written as

$$\frac{\bar{u}}{u_\tau} = \frac{1}{\kappa} \ln y^+ + A \tag{9.52}$$

The constant κ is called the von Karman constant. The velocity distribution (9.52) is valid in the so-called fully turbulent region ($\varepsilon_m \gg \nu$) or in the so-called logarithmic region. It has been found that it is valid for

$$y^+ > 30$$

In the interval $5 < y^+ < 30$, a transitional region between eqs. (9.45) and (9.52) exists. This is called the buffer layer, see Fig. 9.6.

The velocity distributions (9.45) and (9.52) cover together about 20% of the boundary layer, the so-called wall layer. As the derivation was carried out, the shear stress was assumed to be constant. Experiments show that this is valid. The wall layer is sometimes also called the constant stress layer.

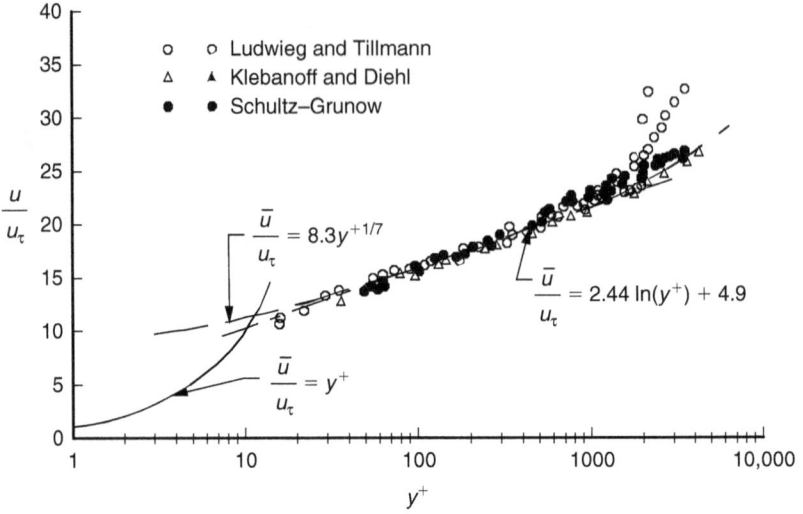

Figure 9.7: The velocity distribution $\bar{u}/u_\tau$ as function of $y^+ = u_\tau y/\nu$ in a turbulent boundary layer.

In Figs. 9.7 and 9.8, comparisons are presented between the theoretical velocity distributions ((9.45) and (9.52)) and experimental data. The values of the constants in eq. (9.52) are recommended as

$$\frac{1}{\kappa} = 2.44$$
$$A = 4.9 - 5.5 \tag{9.53}$$

The velocity distributions (9.45) and (9.52) are examples of the so-called Reynolds number similarity. For the outer region, $y/\delta > 0.2$, the velocity distribution is more commonly expressed as in Fig. 9.8, i.e., $(U - \bar{u})/u_\tau$ as a function of y/δ. It is found that $(U - \bar{u})/u_\tau = \text{function}(y/\delta)$ is a good approximation and this velocity distribution is an example of the so-called local invariance or self-preservation. For more details, see Refs. [1, 2]. The analysis above can also be carried out for turbulent pipe or tube flow. In Fig. 9.9, a comparison between experimental data and theoretical velocity distributions is shown.

Sometimes the turbulent velocity profile is approximated by the formula

$$\bar{u} = Cy^{1/7} \tag{9.54}$$

From Fig. 9.7, it can be concluded that eq. (9.54) gives a reasonable representation of the velocity profile except in the viscous sublayer.

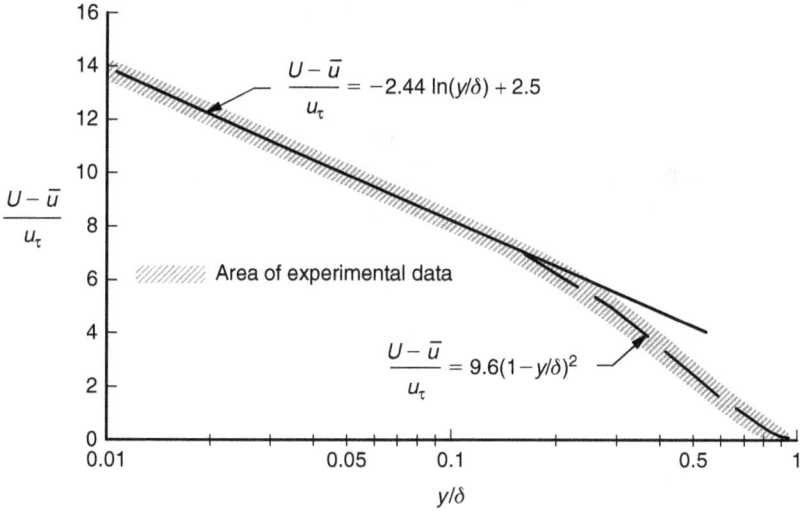

Figure 9.8: The velocity distribution $(U - \bar{u})/u_\tau$ as function of y/δ in a turbulent boundary layer.

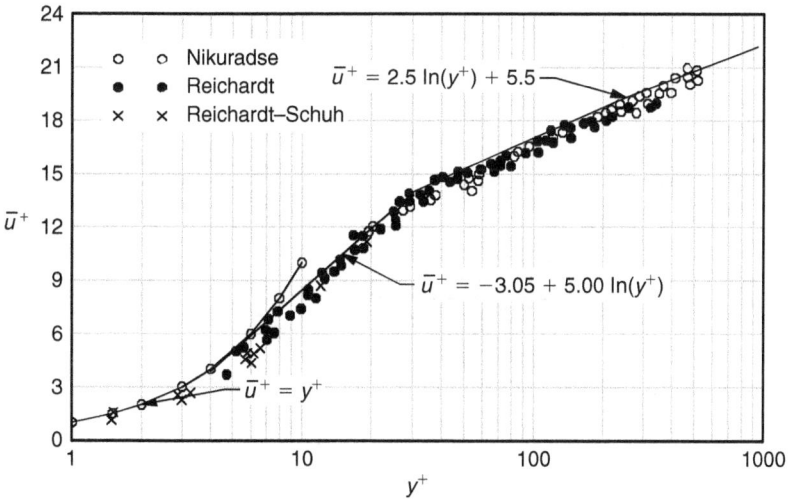

Figure 9.9: The velocity distribution in the near wall region for turbulent pipe flow.

9.12   Determination of the shear stress coefficient

9.12.1   Pipe flow

Consider a circular pipe or tube in Fig. 9.10.

The pressure drop in a pipe for a fully developed flow was in Chapter 8 can be written as

$$\Delta p = f \frac{L}{D} \frac{\rho u_m^2}{2} \qquad (9.55)$$

Equilibrium for a fluid element occupying the whole cross section and a length L gives

$$\Delta p \frac{\pi D^2}{4} = \sigma_w \pi D L$$

For σ_w, one obtains

$$\sigma_w = \frac{\Delta p D}{4L} = f \frac{\rho \bar{u}_m^2}{8} \qquad (9.56)$$

With the definition for C_F according to $\sigma_w = C_F \rho \bar{u}_m^2 / 2$, one obtains

$$C_F = \frac{f}{4} \qquad (9.57)$$

Equation (9.57) implies that Reynolds–Colburn's analogy for pipe flow is written as

$$St = \frac{Nu_D}{Re_D Pr} = \frac{f}{8} Pr^{-2/3}$$

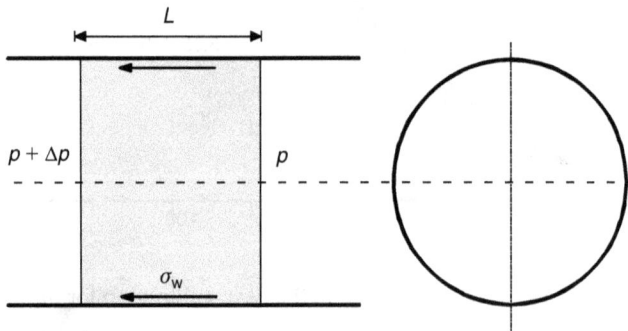

Figure 9.10: Circular pipe or tube.

To determine C_F (and hence the heat transfer coefficient), one has to determine the friction factor f. With $\sigma_w = \rho u_\tau^2$, eq. (9.56) gives

$$f = \frac{8}{(\bar{u}_m/u_\tau)^2} = \frac{8}{(\bar{u}_m^+)^2} \tag{9.58}$$

One now has to determine $\bar{u}_m^+ = \bar{u}_m/u_\tau$. For the mass flow $\dot{m}$, one has

$$\dot{m} = \rho \frac{\pi D^2}{4} \bar{u}_m = \int_0^R \rho \bar{u} 2\pi r \, dr$$

For $\bar{u}_m$, one finds

$$\bar{u}_m = \frac{8}{D^2} \int_0^R \bar{u} r \, dr \tag{9.59}$$

The following variables are now introduced:

$$\bar{u}^+ = \frac{\bar{u}}{u_\tau}; \quad y = R - r$$

$$R^+ = \frac{u_\tau R}{\nu}; \quad y^+ = \frac{u_\tau y}{\nu}$$

After some algebraic operation, eq. (9.59) can then be written as

$$\bar{u}_m^+ = \frac{2}{(R^+)^2} \int_0^{R^+} \bar{u}^+ (R^+ - y^+) dy^+ \tag{9.60}$$

From eq. (9.60) $\bar{u}_m^+$ can be determined if $\bar{u}^+(y^+)$ is known. The velocity profile (9.52) is now assumed to be valid over the whole cross section and then eq. (9.60) can be written as

$$\bar{u}_m^+ = \frac{2}{(R^+)^2} \int_0^{R^+} \left(\frac{1}{\kappa} \ln y^+ + A\right)(R^+ - y^+) dy^+ = \cdots = \frac{1}{\kappa} \ln R^+ - \frac{3}{2\kappa} + A$$

With $1/\kappa = 2.5$ and $A = 5.5$ (see Fig. 9.9), one finds

$$\bar{u}_m^+ = 2.5 \ln R^+ + 1.75 \tag{9.61}$$

Equations (9.58) and (9.61) give for the friction factor f

$$f = \frac{8}{(2.5 \ln R^+ + 1.75)^2}$$

However, R^+ can be written as

$$R^+ = \frac{u_\tau R}{v} = \frac{u_\tau}{\bar{u}_m} \frac{\bar{u}_m D}{2v} = \frac{Re_D}{2}\sqrt{\frac{f}{8}}$$

Thus, one obtains

$$f = \frac{8}{\left(2.5 \ln\left(\frac{Re_D}{2}\sqrt{\frac{f}{8}}\right) + 1.75\right)^2}$$

which can be written as

$$\frac{1}{\sqrt{f}} = 0.84 \ln(\sqrt{f}Re_D) - 0.91$$

Transfer to logarithms with the base 10 gives

$$\frac{1}{\sqrt{f}} = 2.03 \log(\sqrt{f}Re_D) - 0.91 \tag{9.62a}$$

It has been found that eq. (9.62a) agrees better with experimental data if the constants are chosen as 2.0 instead of 2.03 and 0.8 instead of 0.91, see, for example, Ref. [6]. One then writes

$$\frac{1}{\sqrt{f}} = 2.0 \log(\sqrt{f}Re_D) - 0.8 \tag{9.62b}$$

The expression (9.62b) is the so-called *Prandtl's friction law* and it enables that f can be calculated if Re_D is known. As f has been determined, C_F can be determined from eq. (9.57) and St from eq. (9.41). A simpler expression for f can be obtained if the velocity distribution in eq. (9.54) is used. An analysis similar to the one above then gives (see Eckert [7])

$$f = \frac{0.3164}{(Re_D)^{0.25}} \tag{9.63}$$

Equation (9.63) is commonly called *Blasius' relation* and is valid for smooth pipes when $4000 < Re_D < 10^5$.

Equations (9.41), (9.57), and (9.63) now gives

$$St = \frac{Nu_D}{Re_D Pr} = \frac{1}{8}\frac{0.3164}{(Re_D)^{0.25}}Pr^{-2/3}$$

or

$$Nu_D = 0.0396Re_D^{3/4}Pr^{1/3} \tag{9.64}$$

In Fig. 9.11, the friction factors f calculated from eqs. (9.62b) and (9.63) are compared to experimental results. For rough or nonsmooth surfaces, similar expressions

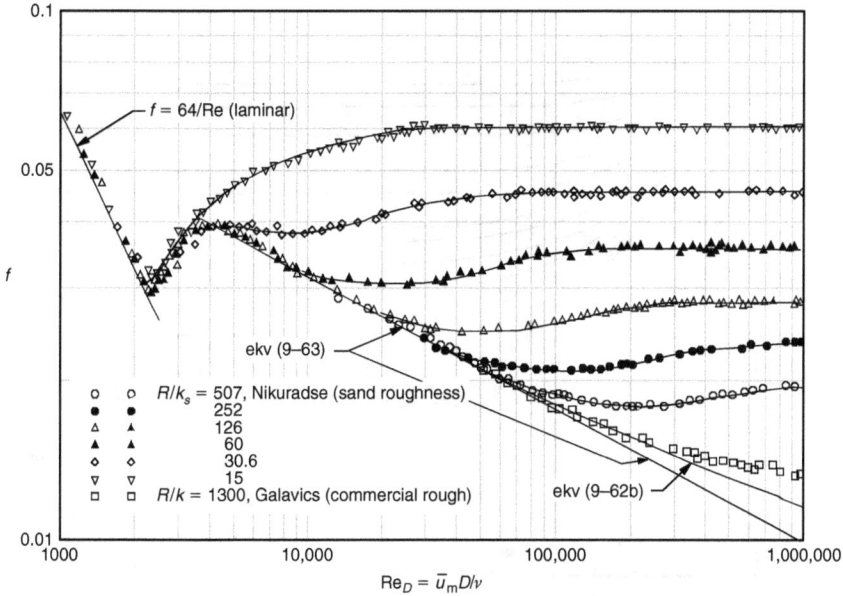

Figure 9.11: The friction factor f for smooth and rough surfaces.

have been developed for f, see, for example, Ref. [8]. Figure 9.11 also presents the friction factor f for rough surfaces and the roughness parameter R/k_s. R is the pipe radius and k_s is a measure of the surface roughness. The results are based on measurements by Nikuradse (see Ref. [8]) where the rough surface was created by gluing sand on the pipe surface.

Commercially available pipes or tubes usually have some natural roughness. The friction factor for such pipes behaves somewhat differently compared to the curves according to Nikuradse. In Fig. 9.12, results from Moody [9] are presented. The diagram is commonly called the Moody chart and is presented in most textbooks, e.g., Ref. [13]. The influence of the roughness is determined by the parameter ε/D. ε is a measure of the roughness which is commonly given by the manufacturer or dealer. D is the pipe or tube diameter.

If the roughness of the pipe surface and the Reynolds number are known, the friction factor can be determined from Figs. 9.11 and 9.12, and then the heat transfer coefficient can be determined with eqs. (9.57) and (9.41).

9.12.2   Turbulent flow along flat plates

Similar to pipe flows, it is possible to find expressions for the shear stress coefficient C_F for flat plates by using the velocity distributions (9.52) and (9.54). With eq. (9.54), it is possible to derive the expression below

$$C_{F,x} = 0.0592 \mathrm{Re}_x^{-1/5} \qquad (9.65)$$

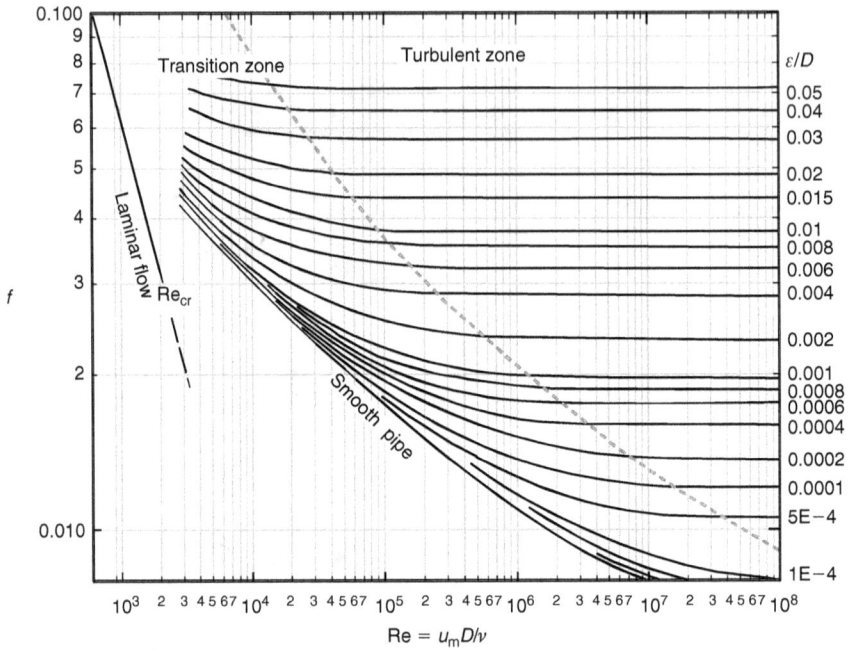

Figure 9.12: The friction factor f for pipe flow according to Moody [9].

The formula (9.65) is recommended in the Reynolds number interval $\mathrm{Re}_x = 5 \cdot 10^5 - 10^7$.

In the Reynolds number range $10^7 - 10^9$, an equation called Schultz–Grunow's formula is recommended

$$C_{F,x} = \frac{0.370}{(\log \mathrm{Re}_x)^{2.584}} \qquad (9.66)$$

Combining eqs. (9.65) and (9.40) gives the well-known formula

$$\mathrm{St} = \frac{\mathrm{Nu}_x}{\mathrm{Re}_x \mathrm{Pr}} = 0.0296 \mathrm{Re}_x^{-1/5} \mathrm{Pr}^{-2/3}$$

or

$$\mathrm{Nu}_x = 0.0296 \mathrm{Re}_x^{4/5} \mathrm{Pr}^{1/3} \qquad (9.67)$$

In Chapter 7, in the discussion of the superposition principle, it was mentioned that Nu_0 is for the case $t_w = $ constant and turbulent flow. Eq. (9.67) is the requested equation.

9.13   Improvements of Reynolds' analogy

In the derivation of Reynolds' analogy, it was assumed that $\text{Pr} = 1$ and $\text{Pr}_t = 1$. The obtained differential equation was integrated directly from $y = 0$ to $y = \delta$ (or R for a pipe). In this section, a few procedures improving Reynolds' analogy will be presented. Prandtl and Taylor (see Refs. [6, 10]) assumed that the flow field consisted of two layers, a viscous sublayer, where the molecular diffusivities ($\nu = \mu/\rho$, $a = \lambda/\rho c_p$) are dominating, and a turbulent core, where the turbulent diffusivities (ε_m, ε_q) are dominating. Without derivation, the result for pipe flow is given

$$\text{St} = \frac{f/8}{1 + 7.54\text{Pr}^{-1/6}\sqrt{f/8}(\text{Pr} - 1)} \tag{9.68}$$

Equation (9.68) is in principle *Prandtl–Taylor's analogy*, but a modification according to Hoffman has been introduced, see Ref. [10].

Von Kármán further developed the Prandtl–Taylor's analogy by dividing the flow field in three different layers, namely a viscous sublayer, a buffer layer, and a turbulent core. The assumptions for the viscous sublayer and the turbulent core were the same as those of Prandtl and Taylor. For the buffer layer, von Kármán assumed that the molecular and turbulent diffusivities are of the same order of magnitude.

$$St = \frac{f/8}{1 + 5\sqrt{f/8}\{(\text{Pr} - 1) + \ln[(5\text{Pr} + 1)/6]\}} \tag{9.69}$$

Equation (9.69) is valid also for a flat plate if $f/8$ is exchanged by $C_F/2$. Especially if $Pr = 1$, eq. (9.69) will be identical to Reynolds' analogy, eq. (9.39).

9.14   Formulas for determination of the heat transfer coefficient for turbulent flow

9.14.1   Pipe or tube flow, smooth surfaces

For the boundary condition *constant heat flux* q_w at the tube wall, eq. (9.70) is recommended, see Ref. [3].

$$\text{St} = \frac{\text{Nu}_D}{\text{Re}_D \text{Pr}} = \frac{f/8}{1.07 + 12.7\sqrt{f/8}(\text{Pr}^{2/3} - 1)} \tag{9.70}$$

The friction factor f must be calculated from Prandtl's friction law (9.62b). Equation (9.70) was developed by the Russian researchers Petukhov and Popov, and has been shown to agree reasonably well with experimental data (within $\pm 2\%$) for Pr numbers in the interval 0.7–50.

Equation (9.70) requires fully developed thermal conditions. Gnielinski [11] modified eq. (9.70) so that it becomes more accurate at low Reynolds numbers.

The modified equation reads

$$Nu_D = \frac{(f/8)(Re_D - 1000)Pr}{1.0 + 12.7\sqrt{f/8}(Pr^{2/3} - 1)}$$

The friction factor f must be determined from

$$f = (0.79 \ln Re_D - 1.64)^{-2}$$

The *Gnielinski formula* is valid for $0.5 < Pr < 2000$, $2300 < Re_D < 5 \cdot 10^6$. It has also been found that eq. (9.70) and the Gnielinski formula can be used for $t_w = $ constant and $q_w = $ constant.

For the case with *constant wall temperature* ($t_w = $ constant), eq. (9.64) could also be used. However, it has been found that eq. (9.64) overpredicts the heat transfer coefficient compared to experimental data. For smooth pipes or tubes, correlations based on experimental data are recommended. So, for instance, Colburns equation reads

$$\overline{St} = \frac{Nu_D}{Re_D Pr} = 0.023 Re_D^{-0.2} Pr^{-2/3} \tag{9.71}$$

It is valid if $Re_D > 10^5$, $0.7 < Pr < 160$, $L/D > 60$.

A common version of eq. (9.71) is the so-called Dittus–Boelter equation (for smooth pipes or tubes and $t_w = $ constant)

$$\overline{Nu}_D = 0.023 Re_D^{0.8} Pr^n \tag{9.72}$$

where

$$n = 0.4 \quad \text{if } t_w > t_B$$
$$n = 0.3 \quad \text{if } t_w < t_B$$

Equation (9.72) is valid if

$$Re_D > 10^4, \quad 0.7 < Pr < 160, \quad L/D > 60$$

Equations (9.71) and (9.72) are valid if both the flow and temperature fields are fully developed. Nusselt studied available experimental data in the range $L/D = 10 - 100$ and found that the influence of the entrance length could be accounted for by using the formula below:

$$\overline{Nu}_D = 0.036 Re_D^{0.8} Pr^{1/3} (D/L)^{0.055} \tag{9.73}$$

if $10 < L/D < 400$.

For eqs. (9.72) and (9.73), it is recommended that the fluid properties are determined at the mean bulk temperature. For eq. (9.71), the St number should be calculated with the fluid properties at the mean bulk temperature, while the Reynolds and Prandtl numbers are calculated with the fluid properties evaluated at the mean value of the wall and bulk temperatures.

At this point, it is worthwhile to mention that the so-called entrance length in turbulent pipe flow is short, typically $10 < L_{\text{entrance}}/D < 60$.

9.14.2   Flat plate

For a flat plate, eq. (9.67) is reasonably accurate in the range $5 \cdot 10^5 < \text{Re}_x < 10^7$. For a tangentially approached flat plate, the boundary layer is laminar up to $x = x_{\text{critical}}$, and then a turbulent boundary layer is established. For engineering calculations, often an average value of the heat transfer coefficient is needed. If the wall temperature (plate surface temperature) is constant, the following formula is valid:

$$\bar{\alpha} = \frac{\int\limits_0^{x_{\text{critical}}} \alpha_{\text{lam}}\, dx + \int\limits_{x_{\text{critical}}}^{L} \alpha_{\text{turb}}\, dx}{L} \tag{9.74}$$

With eqs. (9.67), (7.41a), and (9.71) one finds for $\overline{\text{Nu}}$:

$$\overline{\text{Nu}_L} = 0.037\text{Pr}^{1/3}(\text{Re}_L^{4/5} - 23{,}550) \tag{9.75}$$

if $\text{Re}_{x,\text{critical}} = 5 \cdot 10^5$.

9.15   Stanton number for pipe flow

Consider a circular pipe with a constant wall temperature t_w according to Fig. 9.13.

The Stanton number can be determined directly if the inlet and outlet bulk temperatures are known. This is a reason why it is common to present the heat

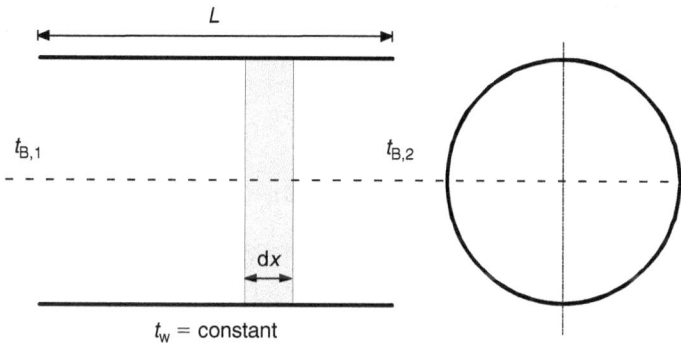

Figure 9.13: Circular pipe with constant wall temperature.

transfer coefficient as $\overline{St} = Nu_D/Re_DPr = \bar{\alpha}/\rho c_p \bar{u}_m$. An energy balance for the element dx in Fig. 9.13 gives

$$d\dot{Q} = \dot{m}c_p \frac{dt_B}{dx}dx = \bar{\alpha}\pi D(t_w - t_B)dx$$

$$\frac{dt_B}{dx} - \frac{\bar{\alpha}\pi D}{\rho(\pi D^2/4)\bar{u}_m c_p}(t_w - t_B) = 0$$

Introducing $\vartheta = t_B - t_w$ gives

$$\frac{d\vartheta}{dx} + \frac{4}{D}St\vartheta = 0$$

and

$$\vartheta = c_1 e^{-(4/D)Stx}$$

For $x = 0$, one has $\vartheta = \vartheta_1 = t_{B_1} - t_w$, and for $x = L$ one has $\vartheta = \vartheta_2 = t_{B_2} - t_w$.
  The Stanton number can now be determined as

$$St = \frac{D}{4L} \ln \frac{t_{B_1} - t_w}{t_{B_2} - t_w} \tag{9.76}$$

Thus if the bulk temperatures at the inlet and outlet are measured, the St number can be calculated if the wall temperature and the pipe dimensions are known. If the mass flow rate is also measured, correlations of the type (9.71) can be established.

9.15.1   Noncircular pipes

For moderate temperature differences, the formulas for pipe flow can also be applied to noncircular tubes if the hydraulic diameter $D_h = 4A/P$ (4×(cross sectional area)/(perimeter)) is used.
  In most cases, the application of the hydraulic diameter concept gives satisfactory results for the friction factor and heat transfer coefficient, but there are cases for which this is not the case, see Irvine [12].

9.16   Dimensionless or universal temperature profile

Consider now the temperature field equation for a turbulent boundary layer. Equations (9.30), (9.32), and (9.33) give

$$\bar{u}\frac{\partial \bar{t}}{\partial x} + \bar{v}\frac{\partial \bar{t}}{\partial y} = \frac{\partial}{\partial y}\left[\left(\frac{\nu}{Pr} + \frac{\varepsilon_m}{Pr_t}\right)\frac{\partial \bar{t}}{\partial y}\right] = -\frac{1}{\rho c_p}\frac{\partial}{\partial y}(q) \tag{9.77}$$

Close to the solid surface, i.e., $y \to 0$, one has $y \to 0: \bar{u} \to 0, \bar{v} \to 0$. Equation (9.77) then gives

$$\frac{\partial}{\partial y}(q) \to 0 \tag{9.78}$$

From eq. (9.78), one concludes that for small y, one has

$$q = \text{constant} = q_w \tag{9.79}$$

Equations (9.77) and (9.79) now give

$$q_w = -\rho c_p \left(\frac{\nu}{\text{Pr}} + \frac{\varepsilon_m}{\text{Pr}_t} \right) \frac{\partial \bar{t}}{\partial y}$$

or

$$\frac{\partial \bar{t}}{\partial y} = -\frac{q_w/\rho c_p}{(\nu/\text{Pr}) + (\varepsilon_m/\text{Pr}_t)} \tag{9.80}$$

Introduce $y^+ = (u_\tau y)/\nu$ and $T^+ = (\bar{t}_w - \bar{t})\rho c_p u_\tau / q_w$, and then eq. (9.80) can be written as

$$\frac{\partial T^+}{\partial y^+} = \frac{1}{(1/\text{Pr}) + (\varepsilon_m/\nu)/\text{Pr}_t} \tag{9.81}$$

Integration of eq. (9.81) gives the dimensionless or universal temperature profile T^+:

$$T^+ = \int_0^{y^+} \frac{dy^+}{(1/\text{Pr}) + (\varepsilon_m/\nu)\text{Pr}_t} \tag{9.82}$$

Equation (9.82) is the so-called law of the wall for the temperature field. If Pr and Pr_t are constants and if a correlation (based on experiments) for ε_m/ν is used, eq. (9.82) can be integrated numerically. Deissler [13] has performed such an integration for $\text{Pr}_t = 1$, and the results are presented in Fig. 9.14. For ε_m/ν, eq. (9.83) was used:

$$\frac{\varepsilon_m}{\nu} = a\bar{u}^+ y^+ (1 - e^{-a\bar{u}^+ y^+}) \tag{9.83}$$

with $a = 0.125$.

Closest to the wall surface, the temperature distribution is

$$T^+ = \text{Pr}\, y^+ \tag{9.84}$$

while at greater distances a logarithmic region is obtained. In this region, the following formula is a good approximation:

$$T^+ = \frac{\text{Pr}_t}{\kappa} \ln y^+ + A_t(\text{Pr}) \tag{9.85}$$

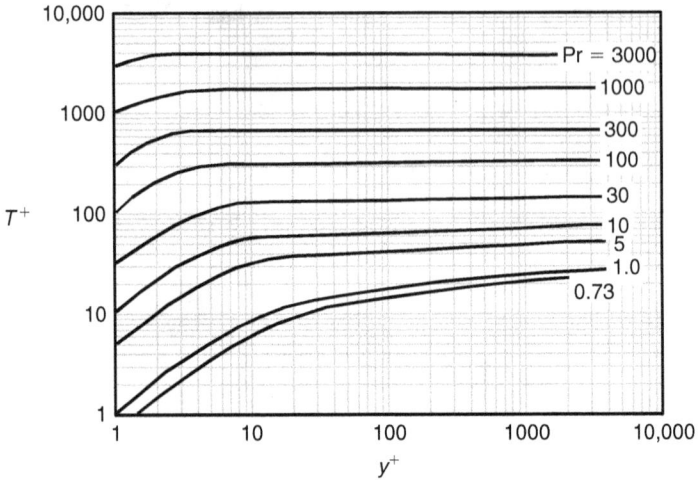

Figure 9.14:　Dimensionless temperature profile. (From Deissler [13].)

As is evident from Fig. 9.14, $A_t(\mathrm{Pr})$ is varying considerably with the Pr number. Two empirical expressions for $A_t(\mathrm{Pr})$ are given below:

$$A_t(\mathrm{Pr}) = 13\mathrm{Pr}^{2/7} - 7 \tag{9.86}$$

$$A_t(\mathrm{Pr}) = (3.85\mathrm{Pr}^{1/3} - 1.3)^2 + 2.12\ln\mathrm{Pr} \tag{9.87}$$

Equation (9.86) is reasonably accurate for $\mathrm{Pr} > 0.7$, while eq. (9.87) can be used for Pr values as low as $\mathrm{Pr} = 0.03$. For $\mathrm{Pr} = 1$ and $\mathrm{Pr}_t = 1$, it is found that the temperature distribution T^+ is identical to the velocity distribution $u^+ = \bar{u}_m/u_\tau$ according to eq. (9.52).

The group $q_w/(\rho c_p u_\tau)$ is sometimes denoted as T^* and corresponds to the friction velocity $u_\tau (u^*)$. The temperature distribution (9.85) has been found to be rather independent of the pressure gradient in the flow but is greatly affected by injection and suction through the wall surface. For more details, see White [14].

9.17　Turbulence modeling

As was pointed out in the analysis of the governing equations, one always obtain more unknowns than equations as Reynolds decomposition is introduced. In order to enable a mathematical solution, models for the turbulent motion and the corresponding momentum and heat exchanges. In this chapter, the Prandtl mixing length concept was introduced. The international research frontier is characterized by huge activities in terms of turbulence modeling. Numerical solution of partial differential equations by powerful computers is common. A detailed investigation of various turbulence models is beyond the scope of this book. Instead Refs. [15–17] are recommended for further studies.

References

[1] J.O. Hinze, Turbulence, 2nd ed., McGraw-Hill, New York (1975).

[2] H. Tennekes and J.L. Lumley, A First Course in Turbulence, MIT Press, MA (1972).

[3] E.R.G. Eckert and R.M. Drake Jr., Analysis of Heat and Mass Transfer, McGraw-Hill, Tokyo (1972).

[4] A.J. Reynolds, The prediction of turbulent Prandtl and Schmidt numbers, Int. J. Heat and Mass Transfer, 18, 1055–1069 (1975).

[5] A.P. Colburn, A method of correlating forced convection heat transfer data and a comparison with liquid frictions, Trans. AIChE, 29, 174–210 (1933).

[6] M.N. Özisik, Basic Heat Transfer, McGraw-Hill, Tokyo (1977).

[7] E.R.G. Eckert, Introduction to Heat and Mass Transfer, McGraw-Hill, New York (1963).

[8] H. Schlichting, Boundary Layer Theory, 7th ed., McGraw-Hill, New York (1979).

[9] L.F. Moody, Friction factor for pipe flow, Trans. ASME, 66, 671–684 (1944).

[10] W.M. Rohsenow, Heat transfer in turbulent flow, Proceedings of the Advanced Study Institute on Turbulent Forced Convection in Channels and Rod Bundles, Istanbul, Turkey (1978).

[11] V. Gnielinski, New equations for heat and mass transfer in turbulent pipe and channel flow, Int. Chem. Eng., 16, 359–368 (1976).

[12] T.R. Irvine, Noncircular convective heat transfer, in Modern Developments in Heat Transfer, Academic Press, New York (1963).

[13] R.G. Deissler, Convective heat transfer and friction in flow of liquids, High Speed Aerodyn. Jet Propulsion, 5, 288 ff (1959).

[14] F.M. White, Viscous Fluid Flow, 2nd ed., McGraw-Hill, New York (1991).

[15] B.E. Launder, On the computation of convective heat transfer in complex turbulent flows, ASME J. Heat Transfer, 110, 1112–1128 (1988).

[16] D.C. Wilcox, Turbulence Modeling for CFD, DCW Industries Inc., La Canada, CA (1993).

[17] H.K. Versteeg and W. Malalasekera, An Introduction to Computational Fluid Dynamics: The Finite Volume Method, 2nd ed., Pearson-Prentice Hall, Glasgow (2007).

10   Natural convection

10.1   Introduction

The body forces, which are created by the density variation in a fluid, might be of great importance for the fluid movement and thus affect the heat transfer significantly. How important this effect will be depends on the ratio between these body forces and other forces affecting the fluid motion. As will be shown in due course, the ratio between these body forces (created by the density variation) and the inertia forces is governed by the Grashof number (Gr) divided by the square of the Reynolds number (Re). The convective heat transfer can then be classified according to

$$\frac{Gr}{Re^2} \ll 1 \quad \text{forced convection} \tag{10.1}$$

$$\frac{Gr}{Re^2} \gg 1 \quad \text{natural convection} \tag{10.2}$$

$$\frac{Gr}{Re^2} \approx 1 \quad \text{mixed convection or combined forced and natural convection} \tag{10.3}$$

where

$$Gr = \frac{g\beta(t_w - t_\infty)L^3}{\nu^2} \tag{10.4}$$

$$Re = \frac{U_\infty L}{\nu} \tag{10.5}$$

In this chapter, the natural convection will be studied in detail. Natural convection is also called free convection.

10.2   Natural convection along vertical surfaces

An important but relatively simple problem of natural convection is the case with a heated or cooled vertical wall. The case with a heated wall is shown in Fig. 10.1. The origin of the coordinate system is placed at the bottom edge of the wall, and the x-axis is parallel to the wall while the y-axis is perpendicular to the wall.

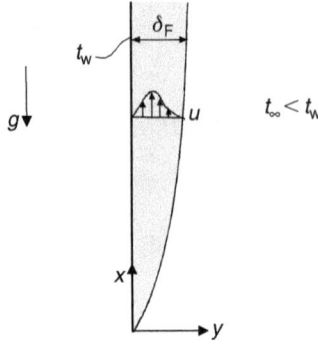

Figure 10.1: Boundary layer (laminar) along a heated vertical wall for natural convection.

Due to the temperature difference between the wall and the fluid, the fluid density will vary. The fluid closest to the wall will be heated and its density will be decreased and the fluid will move upward. At large distances from the wall, the fluid has a uniform temperature and will remain at rest. At the wall surface, the fluid velocity must be zero due to the frictional forces. Thus the velocity has a maximum at a certain distance from the wall. If the Grashof number is sufficiently large ($Gr > 10^4$), it has been found that the heat transfer mainly occurs in a thin layer, the boundary layer, close to the wall. Similar to the forced convection case, the boundary layer flow might be either laminar or turbulent.

The laminar boundary layer will now be analyzed. From Chapter 6, the basic equations for convective heat transfer are given by

$$\rho \left(u \frac{\partial u}{\partial x} + v \frac{\partial u}{\partial y} \right) = -\frac{dp}{dx} - \rho g + \mu \frac{\partial^2 u}{\partial y^2} \tag{10.6a}$$

$$\frac{\partial u}{\partial x} + \frac{\partial v}{\partial y} = 0 \tag{10.7}$$

$$\rho c_p \left(u \frac{\partial t}{\partial x} + v \frac{\partial t}{\partial y} \right) = \lambda \frac{\partial^2 t}{\partial y^2} \tag{10.8}$$

with the boundary conditions

$$y = 0: u = v = 0, t = t_w \tag{10.9}$$

$$y = \delta_F: u = 0, t = t_\infty \tag{10.10}$$

Equation (10.6a) is applied outside the boundary layer. It follows

$$0 = -\frac{dp}{dx} - \rho_\infty g \tag{10.11}$$

Equation (10.6a) can now be written as

$$\rho\left(u\frac{\partial u}{\partial x}+v\frac{\partial u}{\partial y}\right)=(\rho_\infty-\rho)g+\mu\frac{\partial^2 u}{\partial y^2} \qquad (10.6b)$$

The thermal expansion coefficient β is now introduced according to

$$V=V_\infty(1+\beta(t-t_\infty)) \qquad (10.12)$$

For a mass element, one has

$$\rho V=\rho_\infty V_\infty$$

Then one finds

$$\frac{\rho_\infty}{\rho}=1+\beta(t-t_\infty) \qquad (10.13)$$

The relation (10.13) is called the Boussinesq approximation.
Equation (10.13) combined with eq. (10.6b) gives

$$\rho\left(u\frac{\partial u}{\partial x}+v\frac{\partial u}{\partial y}\right)=\rho g\beta(t-t_\infty)+\mu\frac{\partial^2 u}{\partial y^2} \qquad (10.6c)$$

Equations (10.6c), (10.7), and (10.8) with the boundary conditions (10.9) and (10.10) are the basic equations here.

10.2.1   The ratio Gr/Re²

Assume that a characteristic velocity U_∞ and a characteristic length L exist.
The inertia forces in eq. (10.6c) can then be estimated as

$$u\frac{\partial u}{\partial x}+v\frac{\partial u}{\partial y}\sim\frac{U_\infty^2}{L} \qquad (10.14)$$

The body forces due to the density variation can be estimated as

$$g\beta(t-t_\infty)\sim g\beta(t_w-t_\infty) \qquad (10.15)$$

The ratio between eqs. (10.15) and (10.14) becomes

$$\frac{g\beta(t-t_\infty)}{u(\partial u/\partial x)+v(\partial u/\partial y)}\sim\frac{g\beta(t_w-t_\infty)}{U_\infty^2/L}=\frac{g\beta(t_w-t_\infty)L^3}{\nu^2}\cdot\frac{1}{U_\infty^2 L^2/\nu^2}=\frac{Gr}{Re^2}$$

$$(10.16)$$

Equation (10.16) shows that Gr/Re² can be interpreted as the ratio between the body forces due to the density variations and the inertia forces.

10.2.2   The thermal expansion coefficient β

For an ideal gas one has

$$p = \rho R T \tag{10.17}$$

where T is the temperature in Kelvin. Outside the boundary layer, the relation below is valid:

$$p_\infty = \rho_\infty R T_\infty \tag{10.18}$$

Because the pressure is constant across the boundary layer, i.e., $p = p_\infty$ and then eqs. (10.17) and (10.18) give

$$\frac{\rho_\infty}{\rho} = \frac{T}{T_\infty}$$

which can be written as

$$\frac{\rho_\infty}{\rho} = 1 + \frac{T - T_\infty}{T_\infty} = 1 + \frac{t - t_\infty}{T_\infty} \tag{10.19}$$

A comparison with eq. (10.13) gives

$$\beta = \frac{1}{T_\infty} \tag{10.20}$$

One observes that eq. (10.20) is valid only for ideal gases. For common gases like air, eq. (10.20) is reasonably accurate. For liquids, tables or handbooks must be used.

The thermal expansion coefficient β is defined in thermodynamics (see, for example, [1]) as

$$\beta = \frac{1}{v}\left(\frac{\partial v}{\partial t}\right)_p = -\frac{1}{\rho}\left(\frac{\partial \rho}{\partial t}\right)_p$$

For small variations, eq. (10.12) is obtained. If the ideal gas law is applied at $p = $ constant, eq. (10.20) is obtained.

10.2.3   Solution of the basic equations

In Figs. 10.2 and 10.3, experimental results for natural convection along a vertical wall are shown. Figure 10.2 shows the velocity distribution, and Fig. 10.3 shows the temperature distribution at some positions along the wall.

From these figures, it is obvious that the velocity profiles at different positions and the temperature profiles at different positions are similar. This indicates that similarity solutions should exist, i.e., the basic equations (10.6c), (10.7), and (10.8) can be transformed to ordinary differential equations. Pohlhausen [4] showed that by introducing the stream function ψ, according to

$$u = \frac{\partial \psi}{\partial y} \quad \text{and} \quad v = -\frac{\partial \psi}{\partial x} \tag{10.21}$$

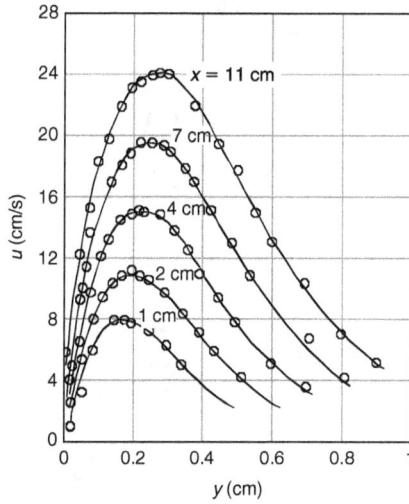

Figure 10.2: Velocity distribution in the boundary layer along a vertical plate for natural convection. (Based on Schmidt and Beckman [2, 3].)

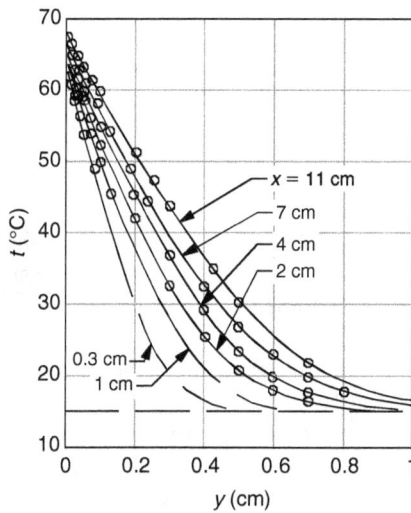

Figure 10.3: Temperature distribution in the boundary layer along a vertical wall for natural convection. (Based on Schmidt and Beckman [2, 3].)

and the nondimensional variables below

$$\theta = \frac{t - t_\infty}{t_w - t_\infty} \tag{10.22}$$

$$\eta = \frac{y}{x}\left(\frac{Gr_x}{4}\right)^{1/4} \tag{10.23}$$

$$\psi^* = \frac{\psi}{4\nu}\left(\frac{Gr_x}{4}\right)^{-1/4} \tag{10.24}$$

Eqs. (10.6c), (10.7), and (10.8) can be transformed to

$$\frac{d^3\psi^*}{d\eta^3} + 3\psi^*\frac{d^2\psi^*}{d\eta^2} - 2\left(\frac{d\psi^*}{d\eta}\right)^2 + \theta = 0 \tag{10.25}$$

$$\frac{d^2\theta}{d\eta^2} + 3Pr\psi^*\frac{d\theta}{d\eta} = 0 \tag{10.26}$$

Two coupled ordinary differential equations have now been achieved. In eq. (10.25) the first term represents the friction forces, the second and third terms correspond to the inertia forces, while the forth one corresponds to the body force due to the temperature dependence of the density. In eq. (10.26) the first term represents the molecular diffusion and the second term the macroscopic energy transport (convection).

The boundary conditions of eqs. (10.25) and (10.26) are

$$\eta = 0: \psi^* = \frac{d\psi^*}{d\eta} = 0, \ \theta = 1 \tag{10.27}$$

$$\eta \to \infty: \frac{d\psi^*}{d\eta} \to 0, \ \theta \to 0 \tag{10.28}$$

Pohlhausen [4] solved eqs. (10.25) and (10.26) subject to the boundary conditions (10.27) and (10.28) for air with $Pr = 0.733$. The results are shown in Figs. 10.4 and 10.5. In these figures, the experimental results by Schmidt and Beckman are also included. As is found, similarity solutions exist.

Equations (10.25) and (10.26) can be solved for arbitrary Pr numbers and results are shown in Figs. 10.6 and 10.7.

From these figures, the big difference between oils (high Pr) and liquid metals (low Pr) can be noted. For engineering calculations, the convective heat transfer coefficient is of great interest. As mentioned earlier, the heat transfer coefficient is defined according to

$$q_w = -\lambda\left(\frac{\partial t}{\partial y}\right)_{y=0} = \alpha(t_w - t_\infty) \tag{10.29}$$

With eqs. (10.22) and (10.23), one finds

$$\frac{\partial t}{\partial y} = \frac{\partial\theta}{\partial\eta}(t_w - t_\infty)\frac{\partial\eta}{\partial y} = \frac{(t_w - t_\infty)}{x}\left(\frac{Gr_x}{4}\right)^{1/4}\frac{\partial\theta}{\partial\eta} \tag{10.30}$$

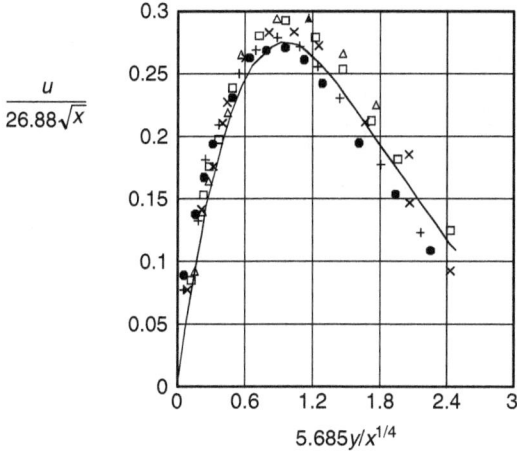

Figure 10.4: Dimensionless velocity distribution.

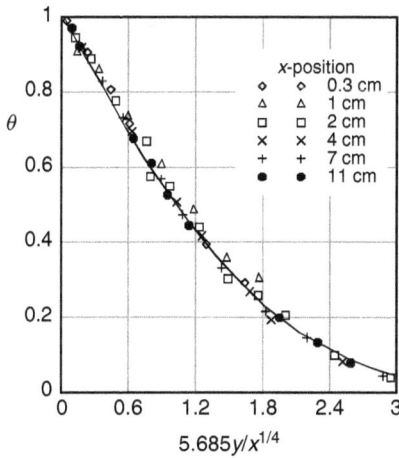

Figure 10.5: Dimensionless temperature distribution.

Equations (10.30) and (10.29) result in

$$\alpha = -\frac{\lambda}{x}\left(\frac{\text{Gr}_x}{4}\right)^{1/4}\left(\frac{\partial\theta}{\partial\eta}\right)_{\eta=0} \qquad (10.31)$$

or

$$\text{Nu}_x = \frac{\alpha x}{\lambda} = \left(\frac{\text{Gr}_x}{4}\right)^{1/4}\left(\frac{\partial\theta}{\partial\eta}\right)_{\eta=0}$$

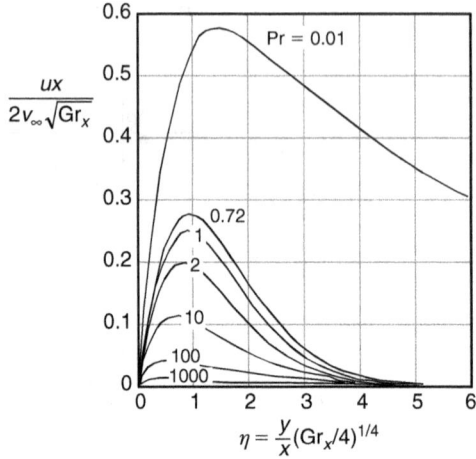

Figure 10.6: Velocity distribution for various Pr. (Based on Ostrach [5].)

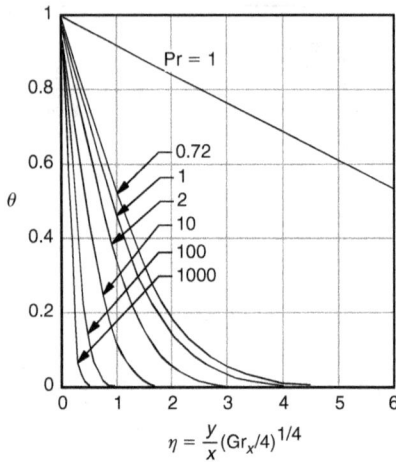

Figure 10.7: Temperature distribution for various Pr. (Based on Ostrach [5].)

which can be written as

$$\text{Nu}_x = \left(\frac{\text{Gr}_x}{4}\right)^{1/4} \phi_1(\text{Pr}) \tag{10.32}$$

The function ϕ_1 (Pr) is tabulated in Table 10.1.

Most often, the mean value of Nu_x over the entire plate or wall length L is of interest. A simple calculation gives

$$\overline{\text{Nu}} = \frac{4}{3}\text{Nu}_{x=L} \tag{10.33}$$

Table 10.1: ϕ_1 (Pr) [5].

Pr	0.01	0.72	0.733	1	2	10	100	1000
ϕ_1 (Pr)	0.0812	0.5046	0.5080	0.5671	0.7165	1.1694	2.191	3.966

The question is now, when are the above formulas valid?

At the beginning it was assumed that the boundary layer was laminar. This assumption has been found to be valid if

$$\text{Gr} \cdot \text{Pr} < 10^9 \tag{10.34}$$

The product Gr · Pr is often called the *Rayleigh* number and is denoted by Ra.

10.2.4   Integrated form of the basic equations

For turbulent flow, a similar analysis is much more difficult as is evident from Chapter 9. Sometimes it is successful to use the integrated form of the basic equations. In this section, this form will be derived. Note that u, v, and t here represent time-averaged quantities.

One has

$$\frac{\partial u}{\partial x} + \frac{\partial v}{\partial y} = 0 \tag{10.35}$$

$$u\frac{\partial u}{\partial x} + v\frac{\partial u}{\partial y} = g\beta(t - t_\infty) + \frac{\partial}{\partial y}\left((v + \varepsilon_m)\frac{\partial u}{\partial y}\right) \tag{10.36}$$

$$u\frac{\partial t}{\partial x} + v\frac{\partial t}{\partial y} = \frac{\partial}{\partial y}\left((a + \varepsilon_q)\frac{\partial t}{\partial y}\right) \tag{10.37}$$

Taking advantage of eq. (10.35), the left-hand side of eqs. (10.36) and (10.37) can be written as

$$u\frac{\partial u}{\partial x} + v\frac{\partial u}{\partial y} = \frac{\partial(u^2)}{\partial x} + \frac{\partial(uv)}{\partial y} \tag{10.38}$$

$$u\frac{\partial t}{\partial x} + v\frac{\partial t}{\partial y} = \frac{\partial(ut)}{\partial x} + \frac{\partial(vt)}{\partial y} \tag{10.39}$$

Then the basic equations become

$$\frac{\partial(u^2)}{\partial x} + \frac{\partial(uv)}{\partial y} = g\beta(t - t_\infty) + \frac{\partial}{\partial y}\left((v + \varepsilon_m)\frac{\partial u}{\partial y}\right) \tag{10.40}$$

$$\frac{\partial(ut)}{\partial x} + \frac{\partial(vt)}{\partial y} = \frac{\partial}{\partial y}\left((a + \varepsilon_q)\frac{\partial t}{\partial y}\right) \tag{10.41}$$

Equations (10.40) and (10.41) can now be integrated across the boundary layer, i.e., from $y = 0$ to $y = \delta_F$, with the boundary conditions: (a) $y = 0$: $u = v = 0$, $\varepsilon_m = \varepsilon_q = 0$, (b) $y = \delta_F$: $u = v = 0$, $\partial u/\partial y = 0$, $\partial t/\partial y = 0$, $\varepsilon_m = \varepsilon_q = 0$

$$\int_0^{\delta_F} \frac{\partial (u^2)}{\partial x} dy = \int_0^{\delta_F} g\beta(t - t_\infty) dy - v\left(\frac{\partial u}{\partial y}\right)_{y=0} \tag{10.42a}$$

$$\int_0^{\delta_F} \frac{\partial (ut)}{\partial x} dy = -a\left(\frac{\partial t}{\partial y}\right)_{y=0} \tag{10.43a}$$

The following relation is now applied

$$\tau_w = \mu\left(\frac{\partial u}{\partial y}\right)_{y=0} = \rho v\left(\frac{\partial u}{\partial y}\right)_{y=0} \quad \text{and}$$

$$q_w = -\lambda\left(\frac{\partial t}{\partial y}\right)_{y=0} = -a\rho c_p\left(\frac{\partial t}{\partial y}\right)_{y=0}$$

Then one obtains

$$\frac{\partial}{\partial x}\int_0^{\delta_F} (u^2) dy = \int_0^{\delta_F} g\beta(t - t_\infty) dy - \frac{\tau_w}{\rho} \tag{10.42b}$$

$$\frac{\partial}{\partial x}\int_0^{\delta_F} (ut) dy = \frac{q_w}{\rho c_p} \tag{10.43b}$$

Equations (10.42b) and (10.43b) express the so-called integrated conditions for natural convection along a vertical surface. They are valid for both laminar and turbulent cases because ε_m and ε_q do not appear in the final equations. By using eqs. (10.42b) and (10.43b), approximative solutions can be found.

10.2.5   The turbulent case

If $\mathrm{Gr} \cdot \mathrm{Pr} > 10^9$ is part of the boundary layer, flow will be turbulent. The velocity and temperature profiles will then be changed and the relation between the Nusselt number and the Grashof and Prandtl numbers will be different from eq. (10.32). Exact analytical methods to find this relation are not available. Approximative results can be achieved by using the integral conditions (10.42b) and (10.43b). The details will not be given here but only a few formulas will be provided.

For the local heat transfer coefficient, the following formula has been found

$$\mathrm{Nu}_x = 0.0295\mathrm{Gr}_x^{2/5}\mathrm{Pr}^{7/15}(1 + 0.494\mathrm{Pr}^{2/3})^{-2/5} \tag{10.44}$$

$$\mathrm{Gr}_x \cdot \mathrm{Pr} > 10^9$$

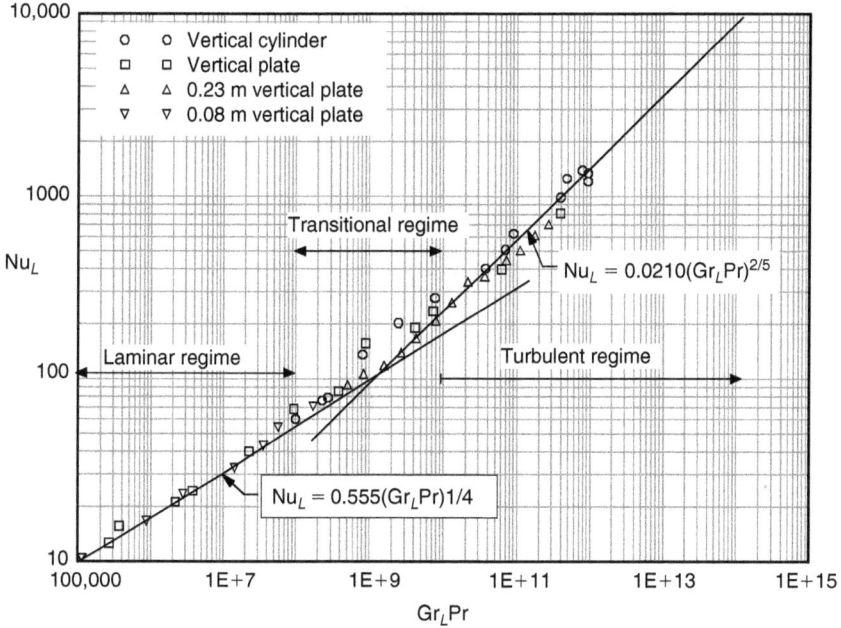

Figure 10.8: Comparison between experimental data and empirical formulas. (Based on Eckert and Jackson [8].)

For the average heat transfer coefficient, the formula below is commonly applied

$$\overline{Nu} = 0.0210(Gr\,Pr)^{2/5} \tag{10.45}$$

$$Gr_L \cdot Pr > 10^9$$

In the derivation of the above formulas, it was assumed that the whole boundary layer is turbulent. Another formula applicable for all $Gr \cdot Pr$ has been suggested by Churchill and Chu, see Refs. [6, 7]. This is based on a large number of experimental data.

$$\overline{Nu} = \left[0.825 + \frac{0.387(Gr\,Pr)^{1/6}}{(1 + (0.492/Pr)^{9/16})^{8/27}} \right]^2$$

In Fig. 10.8, it is shown how eq. (10.45) and simplified forms of eqs. (10.32) and (10.33) agree with experimental data. As is evident the agreement is good. In Fig. 10.8, experimental data for vertical cylinders are also provided. These data also agree well with the formulas. Generally speaking, as long as the boundary layer is thin, the results for vertical plane surfaces and vertical cylinders agree well.

10.3   Horizontal circular cylinders

For a horizontal circular cylinder as shown in Fig. 10.9, most of the surface is not parallel with the gravity vector. The equations for a vertical surface are not directly applicable.

An analysis of a cross section of the cylinder is given in Fig. 10.10.

The main difference compared to the vertical surface is that the gravitational force is varying along the surface.

In the stagnation region ($\theta = 0$ in Fig. 10.10), the partial differential equations can be reduced to ordinary differential equations. Along the cylinder surface, the complete boundary layer equations must be solved by some suitable method. In Fig. 10.11, the distribution of the local heat transfer coefficient, $(Nu_D/Gr_D)^{1/4}$, for ($Pr = 0.714$) is given based on calculations by Hermann [9]. In this figure, calculations and experiments by Eckert and Soehngen [10] are also provided. It should be noted that for horizontal cylinders, the characteristic length is the cylinder diameter D.

The average Nusselt number over the whole cylinder surface is of great interest in engineering calculations. Hyman et al. [11] have found that the following equation gives a reasonable accurate $\overline{Nu}_D$ for several fluids:

$$\overline{Nu}_D = 0.53\left[\frac{Pr}{Pr + 0.952}Gr_D Pr\right]^{1/4} \qquad (10.46a)$$

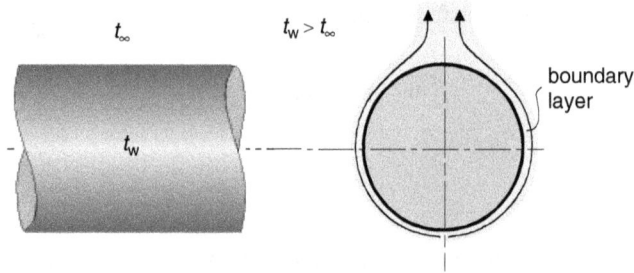

Figure 10.9: Natural convection around a horizontal circular cylinder.

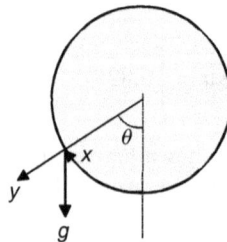

Figure 10.10: Cross section of the horizontal cylinder. Natural convection.

If $Pr \ll 1$ (liquid metals), one finds

$$\overline{Nu}_D = 0.53\left[Gr_D Pr^2\right]^{1/4} \tag{10.46b}$$

If $Pr \gg 1$, one obtains

$$\overline{Nu}_D = 0.53[Gr_D Pr]^{1/4} \tag{10.46c}$$

Equation (10.46c) is identical to a formula proposed by McAdams [12] to be used in the range of $10^4 < Gr_D Pr < 10^9$. Equation (10.46c) can therefore be used even if Pr is not much greater than unity.

For $Gr_D Pr < 10^4$ the area in which the density variations are important becomes of the same order of magnitude as the cylinder diameter. The boundary layer equations are then not applicable and the complete equations of motion and the temperature field equation must be solved.

If $Gr_D Pr > 10^9$ transition to turbulent flow in the boundary layer will occur. The following formula can then be used if $10^9 < Gr_D Pr < 10^{12}$ (see McAdams [12])

$$\overline{Nu}_D = 0.126(Gr_D Pr)^{1/3} \tag{10.47}$$

Churchill and Chu [13] have recommended the following formula

$$\overline{Nu} = \left[0.60 + \frac{0.387(Gr\,Pr)^{1/6}}{(1 + (0.559/Pr)^{9/16})^{8/27}}\right]^2 \tag{10.48}$$

It is valid in the range of $10^{-5} < Gr_D Pr < 10^{12}$.

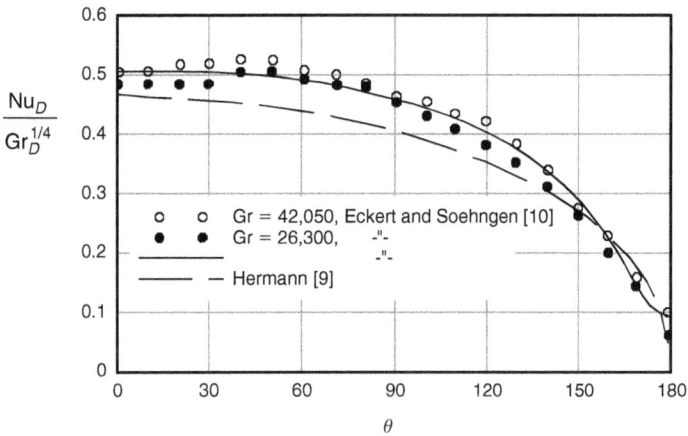

Figure 10.11: Local heat transfer coefficient along a horizontal circular cylinder for natural convection.

10.4   Natural convection for vertical surfaces when the wall heat flux is prescribed

For the cases considered so far, it was assumed that the surface or wall temperature was constant and uniform, t_w. As electric heating of walls occurs, the wall boundary conditions is instead a specified heat flux, q_w. This case has also received a lot of attention in the literature, both experimentally [14–16] and theoretically [17]. The following formulas are recommended

$$\text{Nu}_x = 0.60(\text{Gr}_x^*\text{Pr})^{1/5} \tag{10.49}$$

$$10^5 < \text{Gr}_x^*\text{Pr} < 10^{11}$$

$$\text{Nu}_x = 0.568(\text{Gr}_x^*\text{Pr})^{0.22} \tag{10.50}$$

$$2 \cdot 10^{13} < \text{Gr}_x^*\text{Pr} < 10^{16}$$

Gr_x^* is a modified Grashof number defined according to

$$\text{Gr}_x^* = \frac{g\beta q_w x^4}{\lambda \nu^2} \tag{10.51}$$

10.5   Reference temperature

In the given formulas for calculation of Nu, Gr, etc. some thermophysical properties of the fluid are included. Commonly these properties are temperature dependent. To take this into account, these properties are evaluated at a reference temperature t_{ref} given by

$$t_{\text{ref}} = \frac{t_w + t_\infty}{2} \tag{10.52}$$

For gases, it is however recommended that the coefficient of thermal expansion β is calculated as

$$\beta = \frac{1}{T_\infty}$$

10.6   Natural convection in enclosures

10.6.1   Vertical layers

Consider the fluid in the space between two parallel plates as shown in Fig. 10.12.
Let the left plate surface have the temperature t_1 and the right one the temperature t_2. If $t_1 > t_2$, heat will be transferred from the left to the right plate. If the temperature

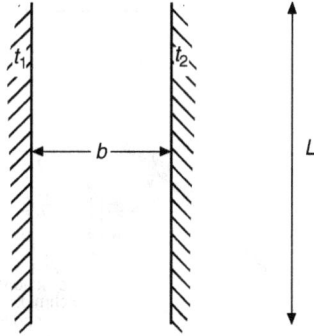

Figure 10.12:   Fluid in the space between two vertical plates.

difference is small, the density variations will be negligible and the heat transfer will be due to heat conduction only. One then has

$$\dot{Q} = \lambda \frac{A}{b}(t_1 - t_2) \qquad (10.53)$$

Equation (10.53) is valid for $Gr_b \, Pr \leq 2 \cdot 10^3$, where

$$Gr_b = \frac{g\beta(t_w - t_\infty)b^3}{\nu^2}$$

If the temperature difference becomes larger and the density variation becomes large accordingly, the fluid will raise along the hot wall and fall down along the cold one. The heat transfer will now be affected by this motion. The heat transfer rate is written as

$$\dot{Q} = \alpha A(t_1 - t_2) = Nu_b \lambda \frac{A}{b}(t_1 - t_2)$$

An apparent thermal conductivity λ_a is introduced, and the heat transfer rate is written as (compare with eq. (10.53))

$$\dot{Q} = \lambda_a \frac{A}{b}(t_1 - t_2) \qquad (10.54)$$

If the last two expressions are compared, one finds

$$Nu_b \equiv \frac{\lambda_a}{\lambda} \qquad (10.55)$$

The apparent thermal conductivity λ_a and thus the Nu_b depends on $Gr_b \, Pr$ and the ratio b/L. In Fig. 10.13, experimental results for λ_a are presented.

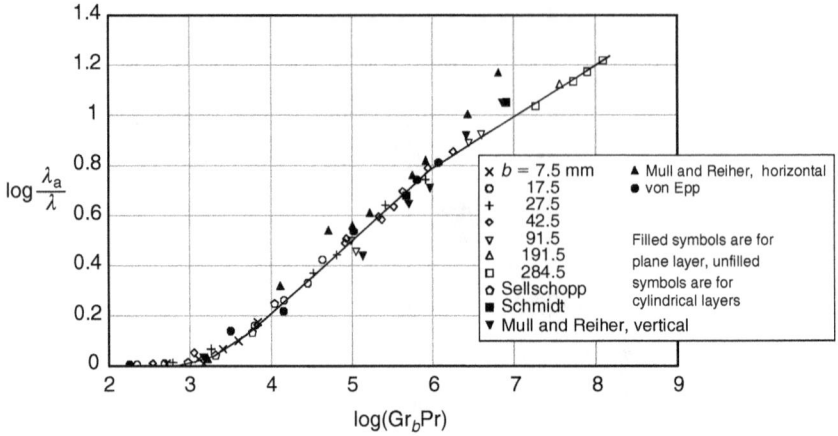

Figure 10.13: Graph for determination of λ_a for vertical layers.

Also formulas of the type $\mathrm{Nu} = \text{function}(\mathrm{Gr}_b, \mathrm{Pr}, b/L)$ are available in the literature. Jakob [18] provides the following formulas for gases:

$$\mathrm{Nu}_b = 0.197(\mathrm{Gr}_b\mathrm{Pr})^{1/4}\left(\frac{b}{L}\right)^{1/9} \tag{10.56}$$

$$6 \cdot 10^3 < \mathrm{Gr}_b\mathrm{Pr} < 2 \cdot 10^5$$

$$3.1 < \frac{L}{b} < 42.2$$

and

$$\mathrm{Nu}_b = 0.073(\mathrm{Gr}_b\mathrm{Pr})^{1/3}\left(\frac{b}{L}\right)^{1/9} \tag{10.57}$$

$$2 \cdot 10^5 < \mathrm{Gr}_b\mathrm{Pr} < 1.1 \cdot 10^7$$

$$3.1 < \frac{L}{b} < 42.2$$

For liquids, Emery and Chu [19] suggest

$$\mathrm{Nu}_b = 0.280(\mathrm{Gr}_b\mathrm{Pr})^{1/4}\left(\frac{b}{L}\right)^{1/4} \tag{10.58}$$

$$10^3 < \mathrm{Gr}_b\mathrm{Pr} < 10^7$$

If $L/b < 3$, then equations (10.32), (10.33), and (10.45) for a single plate can be used.

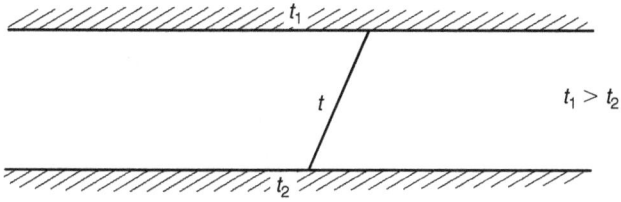

Figure 10.14: Horizontal plates. Hot plate at the top.

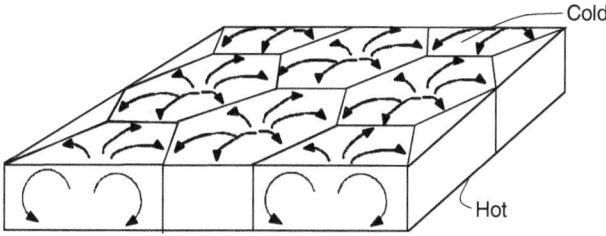

Figure 10.15: Flow structure for natural convection between two parallel plates. Hot plate at the bottom and cold plate at the top.

10.6.2 Horizontal layers

Consider now two parallel horizontal plates with temperatures t_1 and t_2, respectively, and $t_1 > t_2$. If, as in Fig. 10.14, the hot plate is placed at the top, hot fluid will always be above cold. A stable stratification is achieved and the heat transfer is due to heat conduction only. Equation (10.53) is then valid.

Consider now the case with the hot plate at the bottom. The fluid closest to the hot plate will be heated up and will start to raise toward the cold plate. If the temperature difference $t_1 - t_2$ is sufficiently large, the heated fluid will reach the cold plate where it is cold down and then it starts to fall down toward the hot lower plate. It has been found that in the range $1700 < \mathrm{Gr}_b \mathrm{Pr} < 50{,}000$ a hexagonal flow structure is obtained as shown in Fig. 10.15.

The flow structure is called a Benard cell, named after the person who first observed the phenomenon (Benard, 1900). If $\mathrm{Gr}_b \mathrm{Pr} > 50{,}000$, the flow will become turbulent and the motion or flow structure will be irregular and more random. The heat transfer rate is calculated similar to the vertical layers, i.e., an apparent thermal conductivity λ_a according to eq. (10.54) is introduced. The apparent thermal conductivity can be determined by another diagram provided in Fig. 10.16.

Holman [20] suggests that for a fluid between isothermal horizontal plates, the apparent thermal conductivity λ_a can be calculated from

$$\frac{\lambda_a}{\lambda} = C(\mathrm{Gr}_b \mathrm{Pr})^n \qquad (10.59)$$

where C and n depend on $\mathrm{Gr}_b \mathrm{Pr}$ according to Table 10.2.

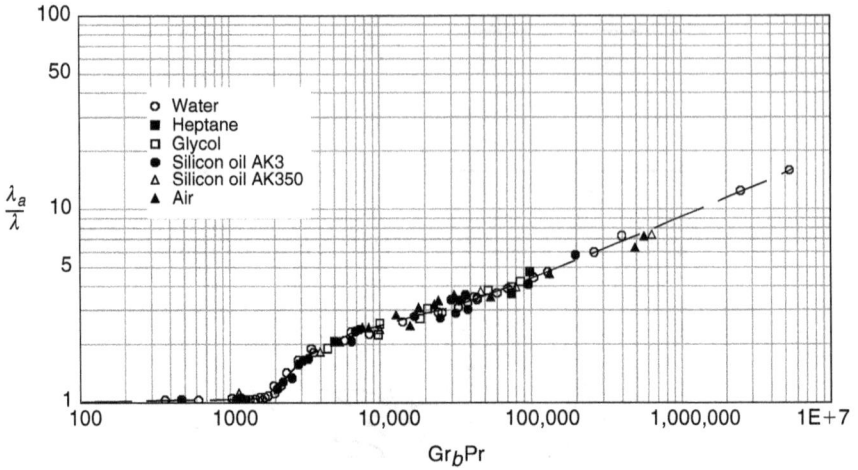

Figure 10.16: Diagram for determination of the apparent thermal conductivity for horizontal layers.

Table 10.2: C and n for calculation of the apparent thermal conductivity for the space between horizontal plates with the hot plate at the bottom [20].

	$Gr_b\,Pr$	C	n	Comments
Gases	<1700	1	0	Pure heat conduction
	1700–7000	0.059	0.4	
	7000–$3.2\cdot10^5$	0.212	0.25	
	>$3.2\cdot10^5$	0.061	1/3	
Liquids	<1700	1	0	Pure heat conduction
	1700–6000	0.012	0.6	
	6000–$3.7\cdot10^4$	0.375	0.2	
	$3.7\cdot10^4$–10^8	0.13	0.3	
	>10^8	0.057	1/3	

10.7 More than two layers

For natural convection between two parallel plates, placed either horizontally or vertically, the heat transfer rate was written as

$$\dot{Q} = \lambda_a \frac{A}{b}(t_1 - t_2) \qquad (10.54)$$

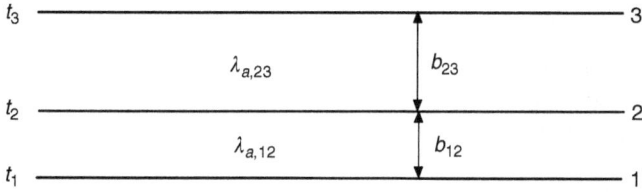

Figure 10.17: Three parallel and horizontal plates.

This equation can be written in a thermal resistance manner similar to what was done in Chapter 3, i.e.,

$$t_1 - t_2 = \dot{Q} \cdot \frac{b}{\lambda_a A} \tag{10.60}$$

Consider now three plates with two layers as shown in Fig. 10.17. The temperature difference $t_1 - t_3$ can be written as

$$t_1 - t_3 = (t_1 - t_2) + (t_2 - t_3) \tag{10.61}$$

If the layer between plates 1 and 2 is called 12 and the layer between plates 2 and 3 is called 23, eq. (10.61) can, by using eq. (10.60), be written as

$$t_1 - t_3 = \dot{Q} \left(\frac{b_{12}}{\lambda_{a_{12}} A} + \frac{b_{23}}{\lambda_{a_{23}} A} \right)$$

or

$$\dot{Q} = \frac{t_1 - t_3}{(b_{12}/\lambda_{a_{12}} A) + (b_{23}/\lambda_{a_{23}} A)} \tag{10.62}$$

To determine $\lambda_{a_{12}}$ and $\lambda_{a_{23}}$, the temperature t_2 is needed. This temperature is not known a priori so some form of iterative calculation procedure has to be applied (condition: $\dot{Q}_{12} = \dot{Q}_{23} = \dot{Q}$).

Equation (10.62) can be generalized to include heat conduction in the plates as well as convective cooling or heating at the outer plate surfaces.

For the heat transmission across a window with an arbitrary number of glass sheets, one finds that if thermal radiation is neglected

$$\dot{Q} = \frac{t_{room} - t_{surroundings}}{(1/\alpha_{inside} A) + \sum (b_{glass}/\lambda_{glass} A) + \sum (b_i/\lambda_{a_i} A) + (1/\alpha_{outside} A)} \tag{10.63}$$

10.8   Summary

In this chapter, natural convection for a number of important cases were considered. Formulas for engineering calculations were presented. Additional cases can be found in, for example, Wong [21].

References

[1] Y.A. Cengel and M.A. Boles, Thermodynamics: An Engineering Approach, 7th ed., McGraw-Hill, New York (2011).

[2] E. Schmidt, Heat transfer by natural convection, Paper Presented at 1961 International Heat Transfer Conference, Boulder, CO (1961).

[3] E. Schmidt and W. Beckman, Das Temperatur- und Geschwindigkeitsfeld von einer varme abgebenden senkrechten Platte bei naturlicher Konvektion, Tech. Mechanik Thermodynamik, 1, 341–390 (1930).

[4] E. Pohlhausen, Der Wärmeaustausch zwischen festen Körper und Flüssigkeiten mit kleiner Reibung und kleiner Wärmeleitung, Z. Angew. Math. Mech., 1, 115–121 (1921).

[5] S. Ostrach, An analysis of laminar free convection flow and heat transfer about a flat plate parallel to the direction of the generating body force, NACA Rep. 1111 (1953).

[6] S.W. Churchill and H.H.S. Chu, Correlating equations for laminar and turbulent free convection from a vertical plate, Int. J. Heat Mass Transfer, 18, 1323–1328 (1975).

[7] F.P. Incropera, D.P. DeWitt, T. Bergman and A. Lavine, Introduction to Heat Transfer, 5th ed., John Wiley & Sons, New York (2007).

[8] E.R.G. Eckert and J.W. Jackson, Analysis of turbulent free convection boundary layer on a flat plate, NACA Rep. 1015 (1951).

[9] R. Hermann, Wärmeübergang bei freier Stromung am waagerechten Zylinder in zweiatomigen Gasen, VDI-Forschungsheft 379, Berlin (1936).

[10] E.R.G. Eckert and E. Soehngen, Studies on heat transfer in laminar free convection with the Zehnder-Mach Interferometer, USAF Tech. Rep. 5747 (1948).

[11] S.C. Hyman, C.F. Bonilla and S.W. Ehrlich, Natural convection transfer processes. 1. Heat transfer to liquid metals and non-metals at horizontal cylinders, Chem. Eng. Prog. Symp. Ser. 49(5), 21–33 (1953).

[12] W.H. McAdams, Heat Transmission, 3rd ed., McGraw-Hill, New York (1954).

[13] S.W. Churchill and H.H.S. Chu, Correlating equations for laminar and turbulent free convection from a horizontal cylinder, Int. J. Heat Mass Transfer, 18, 1049–1053 (1975).

[14] E.M. Sparrow and J.L. Gregg, Laminar free convection from a vertical plate, Trans. ASME, 78, 435–440 (1956).

[15] G.C. Vliet, Natural convection local heat transfer on constant heat-flux inclined surfaces, J. Heat Transfer, 91C, 511–516 (1969).

[16] G.C. Vliet and C.K. Liu, An experimental study of natural convection boundary layers, J. Heat Transfer, 91C, 517–531 (1969).

[17] O.G. Martynenko and Y.A. Sokovishin, Heat transfer in buildings with panel heating, Proceedings of the International Seminar on Heat Transfer in Buildings, Dubrovnik (1977).

[18] M. Jakob, Heat Transfer, vol. I, John Wiley & Sons, New York (1959).

[19] A. Emery and N.C. Chu, Heat transfer across vertical layers, Trans. ASME J. Heat Transfer, 87, 110–116 (1965).

[20] J.P. Holman, Heat Transfer, 10th ed., McGraw-Hill, Tokyo (2009).

[21] H.Y. Wong, Handbook of Essential Formulae and Data on Heat Transfer for Engineers, Longman, London (1977).

11 Forced convective heat transfer for bodies in external flow

11.1 Introduction

In many engineering applications, heat transfer across immersed bodies like tubes or circular cylinders occurs. For instance, the so-called shell-and-tube heat exchangers are common equipment in the process industries. The flow structure and the convective heat transfer process are very much depending on the flow velocity (Reynolds number) and to enable understanding of the complexity, it is essential to know the flow phenomena being present. Spherical objects appear in some heat transfer apparatus, especially in the chemical process industry. Heat transfer from spheres will be considered in this chapter as well.

11.2 Flow field around a circular cylinder (tube) in cross flow

The Reynolds number, $Re_D = U_\infty D/\nu$, can physically be interpreted as the ratio between the inertia forces and the viscous forces in the fluid. The magnitude of the Reynolds number is therefore important for the kind of flow structure being established. Figure 11.1 shows how the flow characteristics are changed as the Reynolds number is increased from about 5 to about 10^7, see Ref. [1].

For $Re_D < 5$, the friction forces are dominating and of importance both close to the cylinder surface and far away. The flow is attached across the cylinder surface, i.e., no separation occurs.

In the range of $5 < Re_D < 47$, a closed separation region (with counter-rotating vortices) is formed at the rear part of the cylinder. The fluid is not able to follow the cylinder surface anymore. The length of the separated region depends on the Reynolds number, see Fig. 11.2. The friction forces are still important in the whole flow domain.

At Reynolds number >47–50, the separated region becomes instable and the counter-rotating vortices will separate from the cylinder surface in an alternating manner and a regular vortex street is established (sometimes called Karman vortex street).

The region where the viscous forces are of importance will be diminished as the Reynolds number is increased. At Reynolds number $>10^3$, the viscous forces will be of significance only in a thin layer on the upstream side of the cylinder,

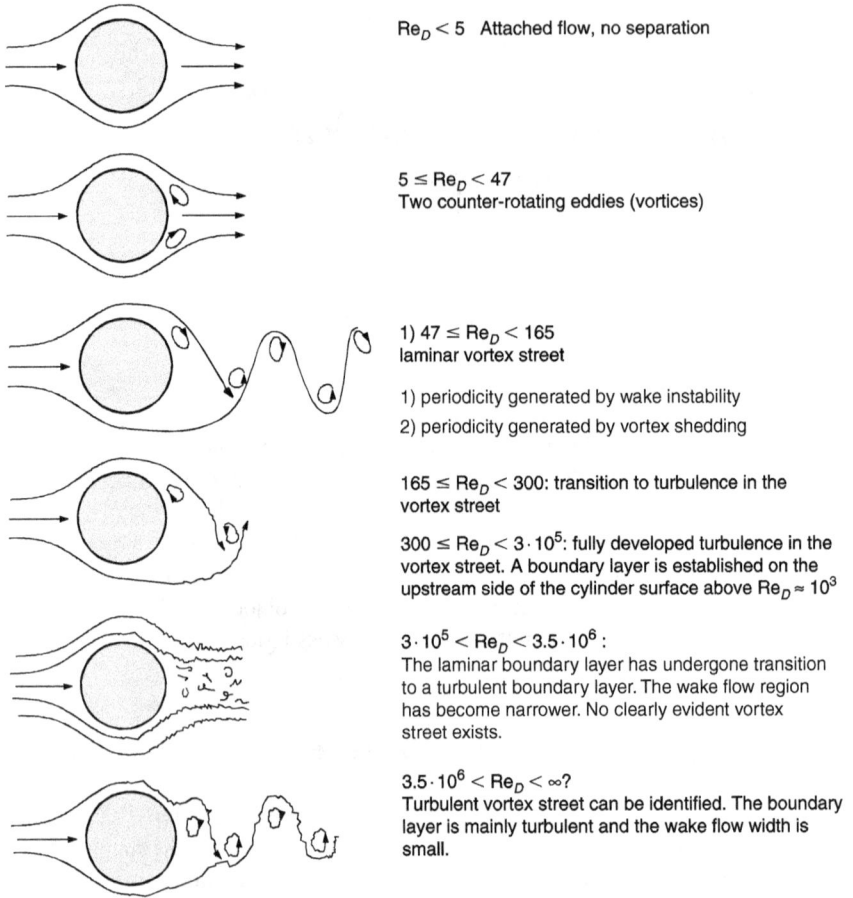

Figure 11.1: Flow characteristics across a circular cylinder in cross flow for various Reynolds numbers.

i.e., in the boundary layer. In the range of $10^3 - 3 \cdot 10^5$, the boundary layer is laminar. This boundary layer separates from the cylinder surface at a position about $80°$ downstream the front stagnation point. The separation from the cylinder surface occurs in an alternating manner from the upper and lower cylinder sides. The vortex motion on the rear side will gradually be more intensive and a transition to turbulent flow with three-dimensional effects will occur.

For Reynolds number $>3 \cdot 10^5$, the boundary layer on the upstream side will become more and more turbulent. As an effect, the separation point will move downstream to a position about 110–$120°$ downstream the front stagnation point. The extent of the separated region on the rear side will become smaller as shown in Fig. 11.1. Up to about $\mathrm{Re}_D = 3.5 \cdot 10^6$, no evident vortex street appears. At even higher Reynolds numbers, a vortex street might be reestablished. A more detailed

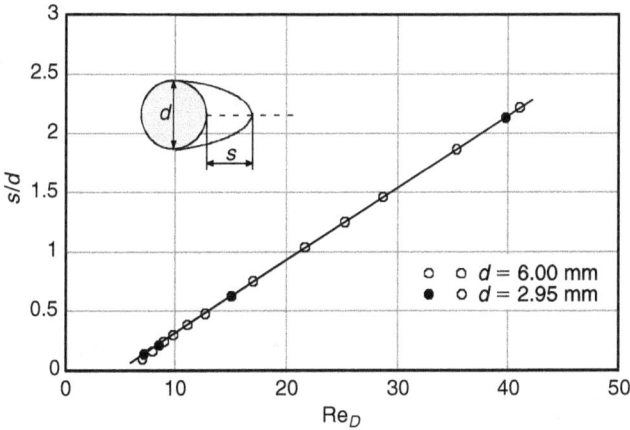

Figure 11.2: Length of the separated region versus the Reynolds number, $Re_D = U_\infty D/\nu$. (Based on Ref. [2].)

description of the flow structure is given in Fig. 11.1 and in Refs. [1, 3–7]. The precise values of the Reynolds number for appearance of the various flow structures are not exact, and in the literature different values might be given. In Williamson [8], more details are given.

11.2.1   Pressure distribution along the cylinder surface (circumferential)

Figures 11.3 and 11.4 show characteristic pressure distributions on the tube or cylinder surface for different Reynolds numbers.

In Fig. 11.3, the dashed line represents the pressure distribution which is valid for inviscid flow. From the front stagnation point, $\beta = 0°$, the pressure is decreased and reaches a minimum at $\beta = 90°$. For $\beta > 90°$, the pressure increases again and at $\beta = 180°$ the same value as at $\beta = 0°$ is achieved. As friction forces are acting, a boundary layer is formed on the upstream side of the cylinder (tube). The fluid outside the boundary layer moves against the increasing pressure (on the downstream side), and kinetic energy (maximum at the pressure minimum location) is converted to pressure energy. For the fluid inside the boundary layer, part of energy is consumed to beat the friction forces. Thus the fluid in the boundary layer can only move a limited distance in the direction of the increasing pressure until the kinetic energy has been consumed. The flow is reverted and separation from the surface occurs. Due to the flow separation, the pressure distribution on the rear side is significantly changed. How far the fluid will move along the cylinder surface, i.e., where separation takes place, depends on the amount of available kinetic energy. This energy content is higher for a turbulent boundary layer than for a laminar one and as an effect separation will be delayed. For a turbulent boundary layer, separation occurs at $\beta \approx 110°$–$120°$ (Fig. 11.3, $Re_D = 6.7 \cdot 10^5$) and at $\beta \approx 80°$ for a laminar boundary layer (Fig. 11.3, $Re_D = 1.1 \cdot 10^5$).

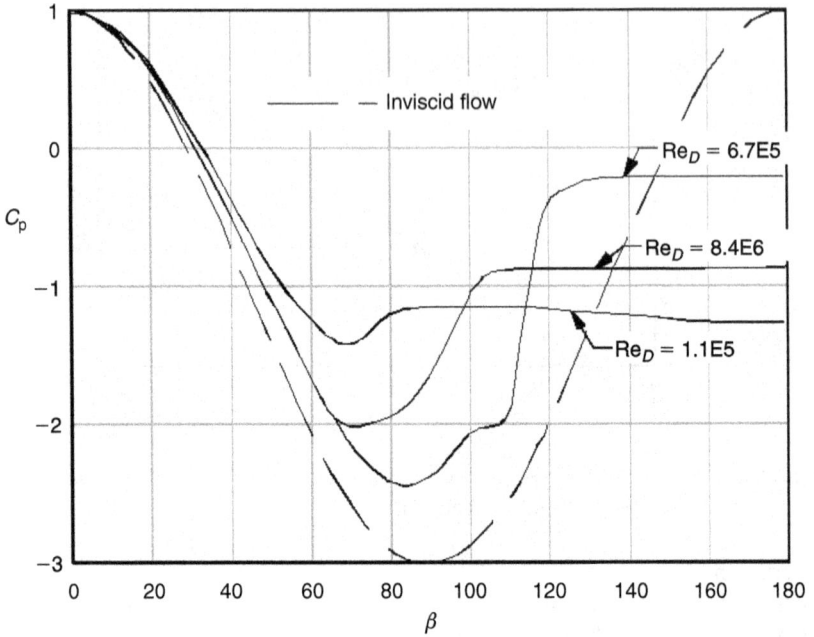

Figure 11.3: Pressure distribution along a circular cylinder (tube) at high Reynolds number. $C_p = (p - p_\infty)/(\rho U_\infty^2/2)$. (From Ref. [2].)

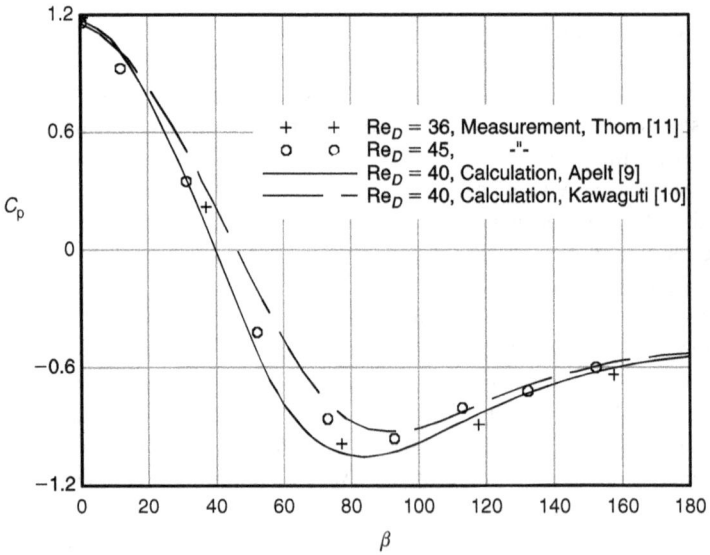

Figure 11.4: Pressure distribution on a circular cylinder (tube) at low Reynolds number. $C_p = (p - p_\infty)/(\rho U_\infty^2/2)$.

Figure 11.4 shows experimental and theoretical pressure distributions on the cylinder surface at low Reynolds number. For these cases, the viscosity and the viscous forces play a significant role for the whole flow field not only in a thin boundary layer. The pressure distributions are quite different from those in Fig. 11.3.

From Figs. 11.3 and 11.4, it is evident that the pressure on the rear side is lower than that at the front side. This means that a force in the flow direction is set up and commonly this is called the form drag. In addition, friction or viscous forces act on the cylinder surface. These give rise to the friction drag. The form and friction drag components together constitute the total drag or resistance and commonly an overall drag coefficient C_D is introduced. This is defined according to

$$F = C_D \cdot L \cdot D \cdot \frac{\rho U_\infty^2}{2} \tag{11.1}$$

where

$F =$ total force (N)
$L =$ cylinder or tube length (m)
$D =$ cylinder diameter (m)
$\rho =$ density of the fluid (kg/m^3)
$U_\infty =$ approach flow velocity (m/s)

Figure 11.5 shows the drag coefficient as a function of Reynolds number. At low Reynolds number, no separation occurs, see Fig. 11.1, and the overall resistance or force is dominated by the viscous or friction forces.

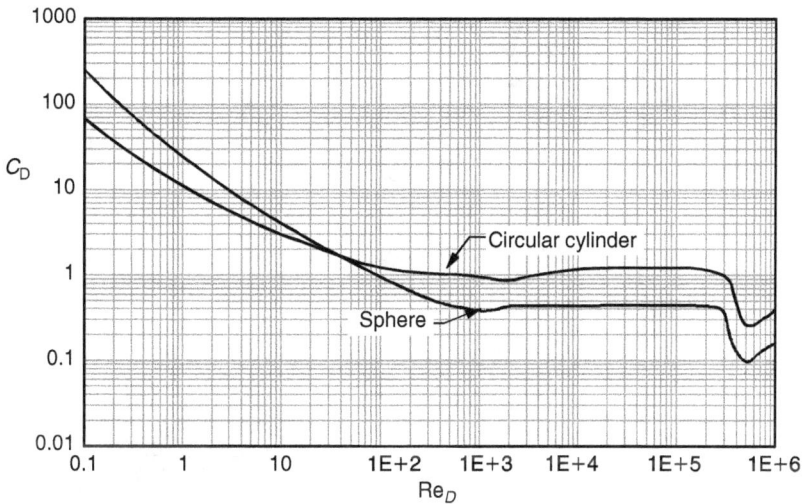

Figure 11.5: Drag coefficient C_D for a circular cylinder and a sphere as function of Reynolds number.

From $Re_D > 50$, vortex streets are formed with alternating shedding from the upper and lower sides of the cylinder. The frequency for the vortex shedding is given by the so-called Strouhal number Sr according to

$$\text{Sr} = \frac{f_s D}{U_\infty} \tag{11.2}$$

In the Reynolds number interval, $10^3 < Re_D < 10^5$, the Strouhal number is $\text{Sr} \approx 0.2$. For more details, see Ref. [1].

The variation in C_D with increasing Reynolds number in Fig. 11.5 is coupled to the changes in the flow structure being described in Fig. 11.1.

11.3   Convective heat transfer from a circular cylinder (tube)

The different flow processes which have been discussed are very important for heat transfer from the cylinder or tube surface. In Fig. 11.6, measured values for the local heat transfer coefficient ($V_D = \alpha D/\lambda_f$) are shown for some Reynolds number.

For the Reynolds number $Re_D = 70,800$ and $101,300$, the minimum of the heat transfer coefficient occurs approximately at the position $\beta = 80°$, where the boundary layer separates. On the rear side of the cylinder, the heat transfer coefficient is increased due to the intensive vortex motion which is often turbulent. For the

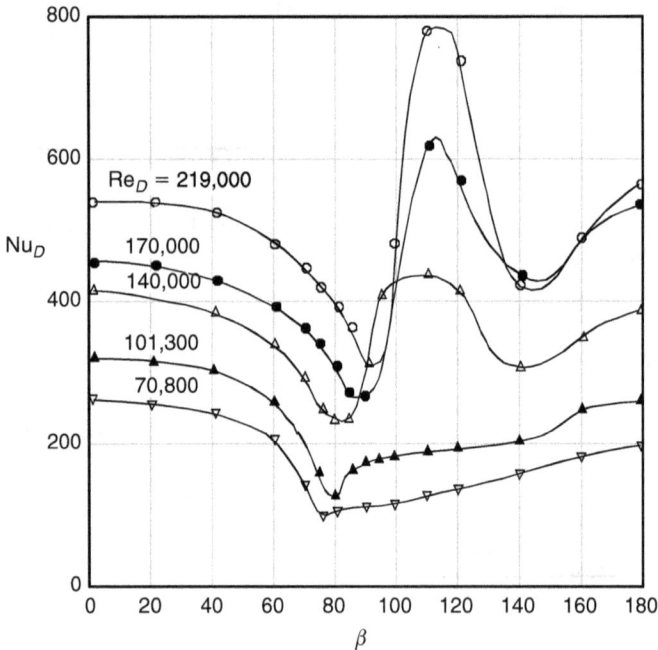

Figure 11.6: Local heat transfer coefficient (Nu_D) along the cylinder surface for air flow at high Reynolds number. β is the angle from the front stagnation point. (Based on Refs. [1, 2]).)

higher Reynolds numbers, two minima appear. The first one is associated with transition from laminar to turbulent flow in the boundary layer, while the second one is assumed to coincide with the separation point for the turbulent boundary layer. The region between the first minimum and the following maximum is the transition region. The first minimum may also be interpreted as the location for separation of the laminar boundary layer. The following maximum can then be interpreted as reattachment of the flow and start of the development of the turbulent boundary layer. As is evident, the influence of the Reynolds number is strong and as transition occurs, the heat transfer coefficient is increased considerably.

Figure 11.7 shows distributions of the heat transfer coefficient at low Reynolds numbers.

Compared to the results in Fig. 11.6, the heat transfer coefficients are low and the upstream side of the cylinder dominates the heat transfer. The low heat transfer on the rear side is explained by the low intensity of the vortex motion or separated flow.

As shown in Chapters 7–10, the Pr number (i.e., the fluid) also affects the heat transfer coefficient. The results in Figs. 11.6 and 11.7 are for air.

For engineering calculations, usually an average or mean value of the heat transfer coefficient is needed. Figure 11.8 shows $\overline{Nu_D} = \overline{\alpha}D/\lambda_f$ as function of the Reynolds number for air, see Refs. [14, 15].

Many formulas, based on experimental data, are available in the international literature. A few formulas are given below. A common one is eq. (11.3), see Refs. [14, 16, 17].

$$\overline{Nu_D} = \frac{\overline{\alpha}D}{\lambda_f} = CRe_D^n Pr^{1/3} \qquad (11.3)$$

where C and n are given in Table 11.1.

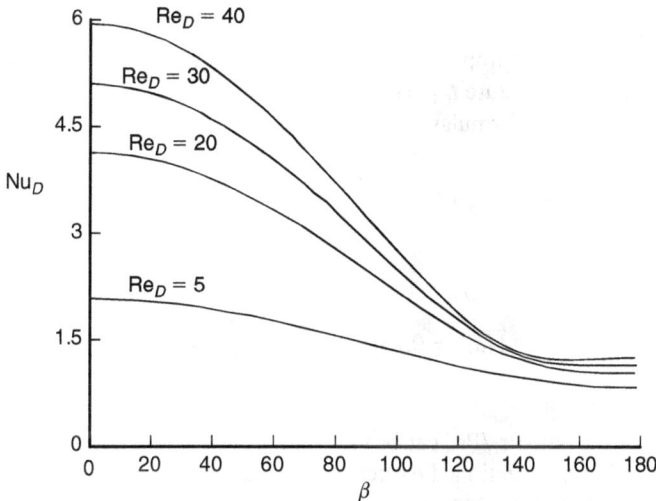

Figure 11.7: Local heat transfer coefficient (Nu_D) along the cylinder surface for air flow at low Reynolds number. β is the angle from the front stagnation point. (From Ref. [13]).

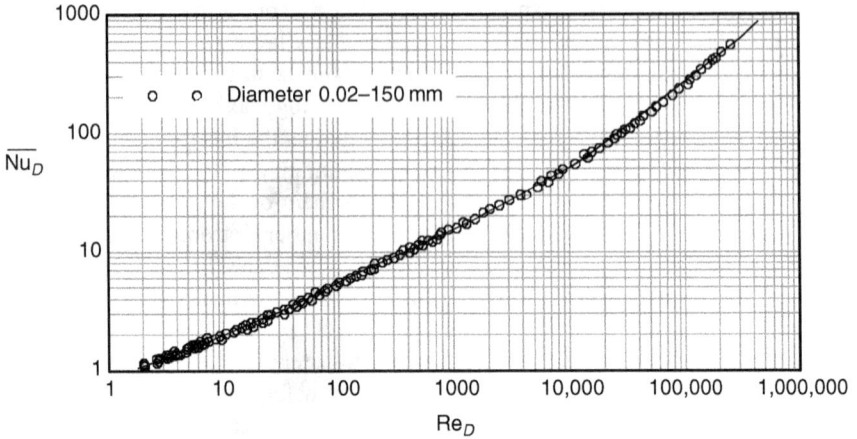

Figure 11.8: Average heat transfer coefficient ($\overline{Nu_D}$) versus Reynolds number.

Table 11.1: Constants C and n in eq. (11.3).

Re_D	C	n
0.4–4	0.989	0.330
4–40	0.911	0.385
40–4000	0.683	0.466
4000–40,000	0.193	0.618
40,000–400,000	0.0266	0.805

The thermophysical properties in eq. (11.3), i.e., λ_f, ν, Pr, should be determined at the reference temperature $t_{ref} = (t_w + t_\infty)/2$.

More complicated formulas are given below, see Refs. [14, 17].

$$\overline{Nu_D} = (0.43 + 0.50 Re_D^{0.5}) Pr^{0.38} \left(\frac{Pr_f}{Pr_w} \right)^{0.25} \tag{11.4}$$

if $1 < Re_D < 10^3$

$$\overline{Nu_D} = 0.25 Re_D^{0.6} V^{0.38} \left(\frac{Pr_f}{Pr_w} \right)^{0.25} \tag{11.5}$$

if $10^3 < Re_D < 2 \cdot 10^5$

For gases, the ratio Pr_f/Pr_w can be set to unity, and the thermophysical properties are calculated at t_{ref} mentioned earlier. For liquids, the thermophysical properties are calculated at the temperature of the approaching flow. Pr_w is determined by t_w.

Formulas for the average heat transfer coefficient are also available for cylinders of noncircular form, see, for example, Ref. [18].

11.4   Tube bundles

A tube bundle is an arrangement of parallel tubes being used to heat or cool a fluid which approaches a tube bundle perpendicular to the tube axes. (The heat transfer inside the tubes is treated by the methods introduced in Chapters 8 and 9). Shell-and-tube heat exchangers are common applications, see Chapter 15.

In Fig. 11.9, two base configurations are shown. In Fig. 11.9a, the tubes are placed in a so-called in-line arrangement, while in Fig. 11.9b the tubes are in a so-called staggered arrangement.

The flow field becomes much more complicated than for a single tube and is affected by the distance or pitch between the tubes. A detailed analysis of the flow and temperature fields is omitted here, but a method to find the heat transfer coefficient and the pressure drop is presented instead, see Refs. [19, 20]. The heat

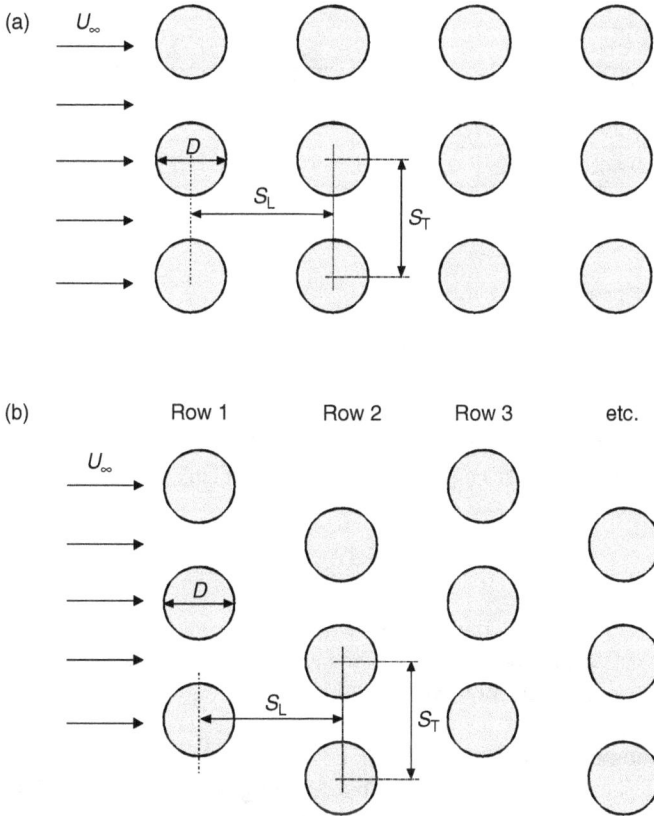

Figure 11.9: Base configurations for tube bundles. (a) in-line arrangement, (b) staggered arrangement.

Table 11.2: Constants C and n for tube bundles.

	S_T/D							
	1.25		1.5		2.0		3.0	
S_L/D	C	n	C	n	C	n	C	n
In-line								
1.25	0.386	0.592	0.305	0.608	0.111	0.704	0.0703	0.752
1.50	0.407	0.586	0.278	0.620	0.112	0.702	0.0753	0.744
2.00	0.464	0.570	0.332	0.602	0.254	0.632	0.220	0.648
3.00	0.322	0.601	0.396	0.584	0.415	0.581	0.317	0.608
Staggered								
0.600	–	–	–	–	–	–	0.236	0.636
0.900	–	–	–	–	0.495	0.571	0.445	0.581
1.000	–	–	0.552	0.558	–	–	–	–
1.125	–	–	–	–	0.531	0.565	0.575	0.560
1.250	0.575	0.556	0.561	0.554	0.576	0.556	0.579	0.562
1.500	0.501	0.568	0.511	0.562	0.502	0.568	0.542	0.568
2.000	0.448	0.572	0.462	0.568	0.535	0.556	0.498	0.570
3.000	0.344	0.592	0.395	0.580	0.488	0.562	0.467	0.574

transfer coefficient is determined by

$$\overline{Nu}_D = \frac{\overline{\alpha}D}{\lambda_f} = C(Re_{D,max})^n Pr^{1/3} \tag{11.6}$$

where C and n depend on the configuration (S_T/D, S_L/D, in-line, staggered). Table 11.2 provides values of C and n. $Re_{D,max}$ is calculated according to

$$Re_{D,max} = \frac{u_{max}D}{v_f}$$

where the velocity u_{max} is the average velocity which appears in the minimum cross flow area. This area depends on the tube layout configuration.

The thermophysical properties of the fluid in eq. (11.6) are determined at $t_{ref} = (t_w + t_\infty)/2$.

If the number of tubes in the main flow direction is less than 10 ($N < 10$), the average heat transfer coefficient $\overline{\alpha}$ according to eq. (11.6) and Table 11.2 is somewhat too big. A correction factor c_1 according to Table 11.3 must then be used so that

$$\overline{\alpha}_{N<10} = c_1\overline{\alpha} \tag{11.7}$$

Table 11.3: Correction factor c_1 in eq. (11.7).

N	1	2	3	4	5	6	7	8	9	10
In-line	0.64	0.80	0.87	0.90	0.92	0.94	0.96	0.98	0.99	1.0
Staggered	0.68	0.75	0.83	0.89	0.92	0.95	0.97	0.98	0.99	1.0

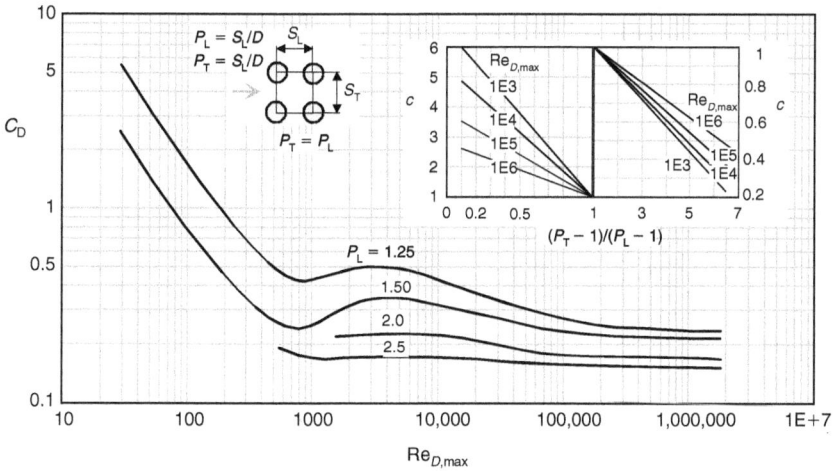

Figure 11.10: The coefficient C_D for tubes in in-line arrangement.

11.5   Pressure drop calculations for a tube bundle in cross flow

The pressure drop Δp across a tube bundle is calculated from

$$\Delta p = NC_D \rho u_{max}^2 / 2 \qquad (11.8)$$

where

N = number of tube rows in the main flow direction
ρ = fluid density
u_{max} = average flow velocity in the minimum flow area
C_D = drag coefficient for the tube bundle

The coefficient C_D can be found from Figs. 11.10 and 11.11 for tubes in in-line arrangement and staggered arrangement, respectively.

The presented curves are for a square and equally sized triangular tube arrangement, respectively. Deviations from these basic configurations can be handled by the inserted correction graphs. The read C_D is then multiplied by the correction factor c.

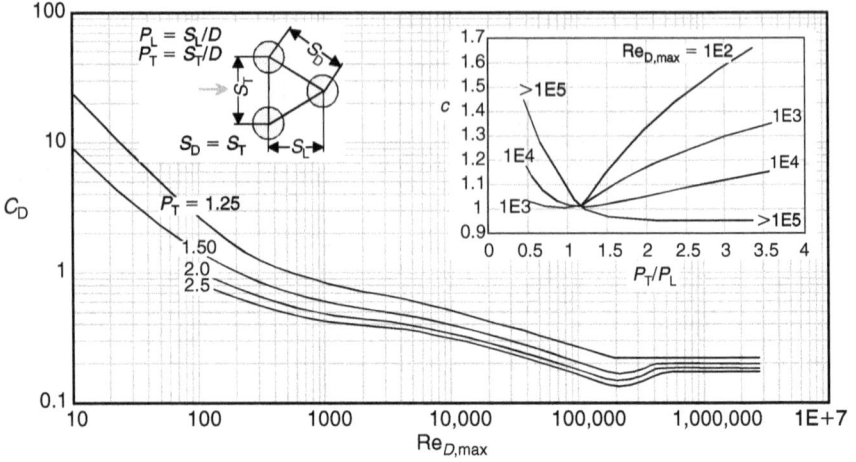

Figure 11.11: The coefficient C_D for tubes in staggered arrangement.

Additional information concerning flow and heat transfer for tube bundles can be found in Refs. [20, 21].

11.6   Heat transfer from spheres

The flow process and the convective heat transfer from spheres are very similar to those of tubes or circular cylinders. A great number of investigations have been performed, but here only some applicable formulas for the heat transfer coefficient are given.

For convective heat transfer from a sphere to a gas stream, a recommended formula is (see McAdams [22])

$$\overline{Nu_D} = \frac{\overline{\alpha}D}{\lambda_f} = 0.37 \mathrm{Re}_D^{0.6} \qquad (11.9)$$

The Reynolds number should be in the range of

$$17 < \mathrm{Re}_D = \frac{U_\infty D}{v_f} < 70{,}000 \quad \text{and} \quad t_{\mathrm{ref}} = \frac{(t_w + t_\infty)}{2}$$

Another formula (from Ref. [23]) can be used for both gases and liquids:

$$\overline{Nu_D} = 2 + (0.4 \mathrm{Re}_D^{0.5} + 0.06 \mathrm{Re}_D^{2/3})\mathrm{Pr}^{0.4} \left(\frac{\mu_\infty}{\mu_w}\right)^{1/4} \qquad (11.10)$$

This is valid for

$$3.5 < \mathrm{Re}_D < 7.6 \cdot 10^4, \quad 0.71 < \mathrm{Pr} < 380, \quad 1 < \frac{\mu_\infty}{\mu_w} < 3.2$$

The thermophysical properties in eq. (11.10) should be determined at t_∞. μ_w is however the dynamic viscosity at t_w.

A special case of heat and mass transfer from spheres is freely falling drops. Commonly, a formula established by Ranz and Marshall [24] is used.

$$\overline{Nu}_D = 2 + 0.6Re_D^{0.5}Pr^{1/3} \qquad (11.11)$$

Corrections to eq. (11.11) have been suggested, see Ref. [25]. These take oscillation of the drops as well as distortion of the drop shape into account.

The thermophysical properties should be determined at t_∞.

At this point, it should be mentioned that eq. (11.11) with the constant 0.6 replaced by 0.55 was already obtained by Frössling [26] in a study of evaporation of some specific liquid drops.

From eqs. (11.10) and (11.11), it is obvious that $Nu_D \rightarrow 2$ as $Re_D \rightarrow 0$. This corresponds to heat conduction from a spherical surface to a resting fluid. This result can be easily derived from eqs. (3.14) and (3.15).

References

[1] J.H. Lienhard, A Heat Transfer Textbook, 2nd ed., Prentice-Hall Inc., Englewood Cliffs, NJ (1987).

[2] G.K. Batchelor, An Introduction to Fluid Dynamics, Cambridge University Press, Cambridge, (1970).

[3] M.V. Morkovin, Flow around circular cylinder—a kaleiodoscope of challenging fluid phenomena, ASME-Symposium on Fully Separated Flows, Philadelphia, PA102–118 (1964).

[4] A. Roshko, On the drag and shedding frequency of two-dimensional bluff bodies, NACA TN 3169 (1964).

[5] A. Roshko, Experiments on the flow past a circular cylinder at very high Reynolds number, J. Fluid Mech., 10, 345–356 (1961).

[6] M.M. Zdravkovich, Discussion on "Flow around fixed circular cylinders: fluctuating loads", J. Engn. Mech. Div., 108, 567–568 (1982).

[7] C. Norberg, Reynolds number and freestream turbulence effects on the flow and fluid forces for a circular cylinder in cross flow, PhD thesis, CTH, Göteborg (1987).

[8] C.H.K. Williamson, Vortex dynamics in the cylinder wake, Ann. Rev. Fluid Mech., 28, 477–539, 1996.

[9] C.J. Apelt, The steady flow of a viscous fluid past a circular cylinder at Reynolds numbers 40 and 44, Aero. Res. Coun., Rep. Mem. 3175 (1958).

[10] M. Kawaguti, Numerical solution of the Navier–Stokes' equations for the flow around a circular cylinder at Reynolds number 40, J. Phys. Soc. Jpn, 8, 747–757 (1953).

[11] A. Thom, The flow past cylinders at low speeds, Proc. Roy. Soc. A141, 651–669 (1933).

[12] W.H. Giedt, Investigation of variation of point unit-heat-transfer coefficient around a cylinder normal to an air stream, Trans. ASME, 71, 375–381 (1949).

[13] B. Sunden, Conjugated heat transfer at low Reynolds number flow around a circular cylinder, Publ. 78/5, Inst. för Tillämpad Termodynamik och Strömningslära, CTH, Göteborg (1978).

[14] R. Hilpert, Wärmeabgabe von geheizen Drahten und Rohren in Luftstrom, Forsch. Geb. Ingenieurw., 4, 215–224 (1933).

[15] E.R.G. Eckert and R.M. Drake Jr., Analysis of Heat and Mass Transfer, McGraw-Hill, New York (1972).

[16] J.D. Knudsen and D.L. Katz, Fluid Dynamics and Heat Transfer, McGraw-Hill, New York (1958).

[17] J.P. Holman, Heat Transfer, 10th ed., McGraw-Hill, New York (2009).

[18] H.Y. Wong, Handbook of Essential Formulae and Data on Heat Transfer for Engineers, Longman, London (1977).

[19] E.D. Grimison, Correlation and utilization of new data on flow resistance and heat transfer for cross flow of gases over tube banks, Trans. ASME, 59, 583–594 (1937).

[20] Heat Exchanger Design Handbook, Hemisphere Publ. Corp., Washington, DC (1983).

[21] VDI—Wärmeatlas, 4th ed., VDI-Verlag, Dusseldorf (1984).

[22] W.H. McAdams, Heat Transmission, 3rd ed., McGraw-Hill, New York (1954).

[23] S. Whitaker, Forced convection heat transfer correlations for flow in pipes, past flat plates, single cylinders, single spheres and flow in packed beds and tube bundles, AIChE J., 18, 361–371 (1972).

[24] W. Ranz and W. Marshall, Evaporation from drops, Chem. Eng. Progr., 48, 141–146 (1952).

[25] F.P. Incropera, D.P. Dewitt, T. bergman and A. Lavine, Fundamentals of heat and mass transfer, 6th ed., John Wiley and Sons, New York (2010).

[26] N. Frössling, Uber die Verdunstung fallender Tropfen, Gerlands Beitrag zur Geophysik, Bd. 52, 170–216 (1938).

12 Thermal radiation

12.1 Introduction

The preceding chapters have shown how the heat transfer due to conduction and convection may be determined by mathematical analysis and empirical data. In this chapter, the heat transfer due to thermal radiation will be presented. Thermal radiation is the emission of electromagnetic radiation by a body or a substance due to its temperature. The nature, characteristics, and the common properties of thermal radiation will be described first. Then influences of material properties and the configuration of the involved geometries on the total radiant energy exchange will be considered.

Note that the notations $\rho, \alpha, \tau, \lambda, \varepsilon$, and ν in this chapter have different meanings when compared to that given in other chapters.

12.2 Physical mechanism

There are several types of electromagnetic radiation, the thermal is only one. Independent of the radiation type, the propagation speed is that of light, namely $3 \cdot 10^8$ m/s. This velocity can be written as

$$c = \lambda \nu \qquad (12.1)$$

where

$c =$ speed of light (m/s)
$\lambda =$ wavelength of the radiation (m)
$\nu =$ frequency of the radiation (1/s).

Part of the electromagnetic spectrum is shown in Fig. 12.1.

The thermal radiation is in the wavelength range from 0.1 to 100 µm, while the visible part of the spectrum is in the narrow range of 0.39–0.78 µm $(1 \, \mu m = 10^{-6} \, m)$.

The propagation of the thermal radiation takes place in form of discrete quanta, where each quantum has energy as

$$E_{\text{quanta}} = h\nu \qquad (12.2)$$

where h is Planck's constant with the value $h = 6.625 \cdot 10^{-34} \, J \cdot s$.

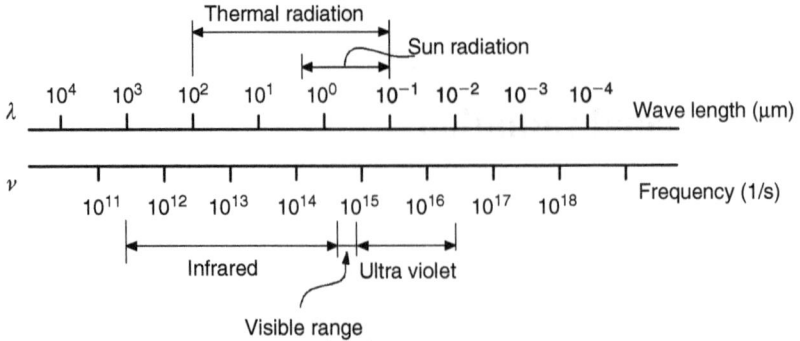

Figure 12.1: Spectrum of the electromagnetic radiation.

A very rough physical picture of the propagation process of the radiation may be achieved by considering every quantum as a particle with mass, energy, and momentum similar to how the molecules of a gas are considered. The radiation can then be considered as a "photon gas" which flows from one place to another.

A few notations are now introduced. The total emitted energy from a surface is given the notation E. It is commonly called the emissive power. This depends on the surface temperature and the surface properties. So for instance, some surfaces are emitting more than others despite the temperature is the same. The unit for E is W/m². The total incident radiation on a surface is denoted by G and has the unit W/m². The property G is also called irradiance.

12.3   Properties of thermal radiation

As radiant energy is approaching a surface, part of it will be reflected, another part will be absorbed while the remaining part will be transmitted (see the illustration in Fig. 12.2).

The following properties are now introduced:

reflectivity, ρ—the reflected part of the incident radiation
absorptivity, α—the absorbed part of the incident radiation
transmittivity, τ—the transmitted part of the incident radiation.

The following relation is valid:

$$\rho + \alpha + \tau = 1 \tag{12.3}$$

Most solid bodies do not transmit any part of the incident radiation, i.e., $\tau = 0$. They are said to be opaque. From eq. (12.3), one then has

$$\rho + \alpha = 1 \tag{12.4}$$

The reflection of the incident radiation may occur in two different ways. The refection might be regular which means that the angle of incidence is equal to

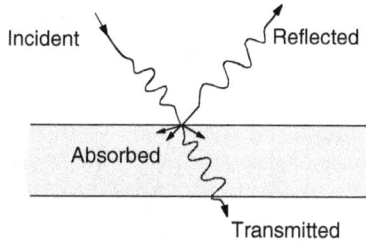

Figure 12.2: Reflection, absorption, and transmission of incident radiation.

the angle of reflection (mirror reflection). The other way is the so-called diffuse reflection where an incident beam is reflected uniformly in all directions.

The radiant energy from an opaque surface ($\tau = 0$) has two components, namely (a) the emitted energy and (b) the reflected part of the incident radiation.

This radiant energy is called the radiosity and is given the notation J. One then has

$$J = E + \rho G \tag{12.5}$$

An opaque surface ($\tau = 0$) which does not reflect any radiation ($\rho = 0$) is called a *blackbody* because it absorbs all incident radiation and it will be interpreted as black by the human eye.

12.4 Blackbody radiation

The total emitted radiant power (emissive power) E (W/m^2) from a surface is a total value in the sense that it represents the sum of the radiation for all wavelengths and all directions within a hemispherical space surrounding the surface. Sometimes it is necessary to consider the radiation within a certain wavelength range dλ. The monochromatic emitted radiant power (monochromatic emissive power) E_λ is then defined as the radiant emissive power per unit wavelength (W/m^3) within the wavelength range $[\lambda, \lambda + d\lambda]$.

The following relation is valid

$$E = \int_0^\infty E_\lambda \, d\lambda \tag{12.6}$$

A perfect emitter of radiant energy is a surface which emits maximum radiative energy in all wavelength intervals at a given temperature. From eq. (12.6) it is evident that such a surface also emits maximum total radiant energy at a given temperature. It can be relatively, easily shown that a perfect absorber, i.e., blackbody $\alpha = 1$, also is a perfect emitter and thus a perfect emitter is also called a blackbody. In the remaining part of this chapter, all properties related to blackbodies are given as subscript B (blackbody).

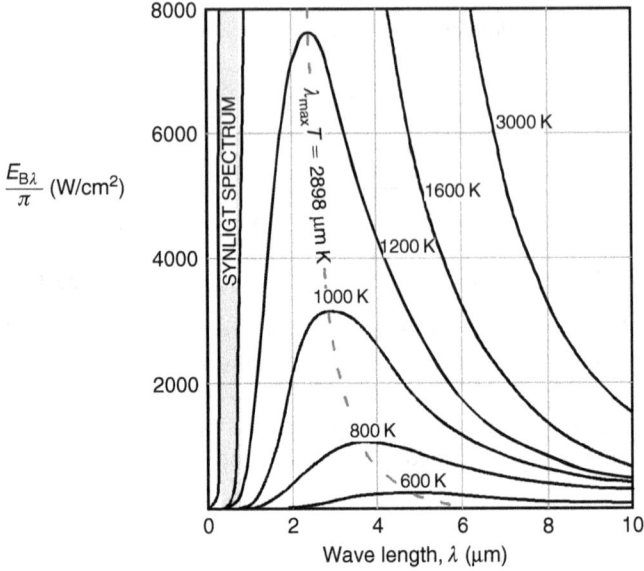

Figure 12.3: Spectral distribution of the blackbody radiation.

The blackbody radiation will now be described in details. It was found earlier that the laws of classical or conventional physics were not able to provide an analytical expression for the blackbody radiation which was in agreement with experiments. To enable such an expression, Max Planck (see Ref. [1]) in the year 1900 introduced the quantum theory which later become the origin of modern physics. From the quantum theory, Planck obtained

$$E_{B\lambda} = C_1 \frac{\lambda^{-5}}{e^{C_2/\lambda T} - 1} \tag{12.7}$$

where $C_1 = 3.742 \times 10^{-16} \, \text{W m}^2$ and $C_2 = 1.439 \times 10^{-2} \, \text{mK}$.

Figure 12.3 shows the spectral distribution of the blackbody radiation at a few different temperatures.

Equation (12.7) can be written as

$$\frac{E_{B\lambda}}{T^5} = \frac{C_1}{(\lambda T)^5 \{e^{C_2/\lambda T} - 1\}} = f(\lambda T) \tag{12.8}$$

This equation is commonly called the Wien's distribution law and is shown in Fig. 12.4.

An important result from this law is the maximum value of $E_{B\lambda}/T^5$ which is found at $\lambda T = 0.2898 \, \text{cm K}$. Thus, for a blackbody at a certain temperature, the

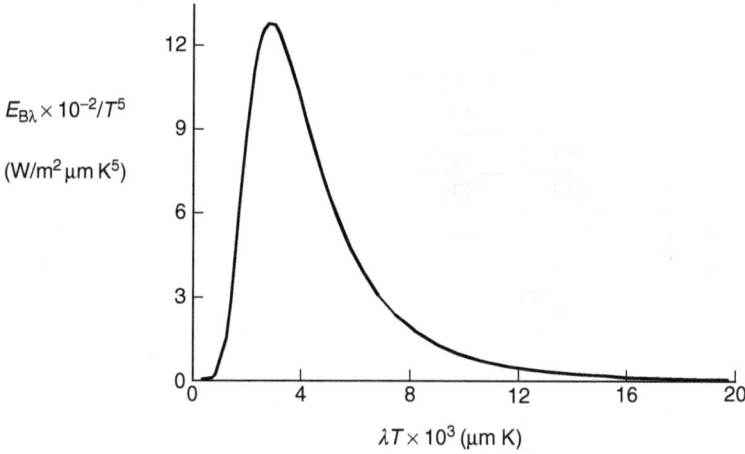

Figure 12.4: Wien's distribution law.

monochromatic emissive power is maximum at a wavelength λ_{max} which is given by

$$\lambda_{max} = \frac{0.002898}{T} \text{ [m]} \tag{12.9}$$

This relation is called Wien's displacement law. From Fig. 12.3, it is evident that as the temperature is increased, the radiation is displaced to shorter wavelengths. In general one can state that warm surfaces emit radiation with short wavelengths while cold surfaces emit radiation with long wavelengths.

An expression for E_B can now be found from eqs. (12.6) and (12.7). One finds

$$E_B = \int_0^\infty \frac{C_1 \lambda^{-5}}{e^{C_2/\lambda T} - 1} \, d\lambda = T^5 \int_0^\infty \frac{C_1 \, d\lambda}{(\lambda T)^5 \{e^{C_2/\lambda T} - 1\}}$$

$$= \left[\text{Introduce } x = \lambda T \Rightarrow \ dx = T \, d\lambda \right] = T^5 \int_0^\infty \frac{C_1 \, dx/T}{x^5 \{e^{C_2/x} - 1\}}$$

From this it follows

$$E_B = \sigma T^4 \tag{12.10}$$

Equation (12.10) is called Stefan–Boltzmann's law after the two persons who first derived it but without knowing Planck's equation (12.7). Stefan [2] (1879) suggested eq. (12.10) based on studies of experimental data by J. Tyndall. Boltzmann [3] derived eq. (12.10) using classical thermodynamics in the year 1884. The constant in eq. (12.10) has the value $\sigma = 5.67 \cdot 10^{-8}$ W/m^2 K^4.

Sometimes it is necessary to determine how much of the emitted radiation belongs to a certain wavelength interval, say $\lambda_1 - \lambda_2$. By using Table 12.1 it is

Table 12.1: Part of the emitted radiation in the interval $0-\lambda$.

$\lambda T\,(\mu m\,K)$	$E_{B,0-\lambda T}/\sigma T^4$	$\lambda T\,(\mu m\,K)$	$E_{B,0-\lambda T}/\sigma T^4$	$\lambda T\,(\mu m\,K)$	$E_{B,0-\lambda T}/\sigma T^4$
1000	0.00032	4100	0.49872	7200	0.81918
1100	0.00091	4200	0.51599	7300	0.82443
1200	0.00213	4300	0.53267	7400	0.82949
1300	0.00432	4400	0.54877	7500	0.83436
1400	0.00779	4500	0.56429	7600	0.83906
1500	0.01285	4600	0.57925	7700	0.84359
1600	0.01972	4700	0.59366	7800	0.84796
1700	0.02853	4800	0.60753	7900	0.85218
1800	0.03934	4900	0.62088	8000	0.85625
1900	0.05210	5000	0.63372	8200	0.86396
2000	0.06672	5100	0.64606	8400	0.87115
2100	0.08305	5200	0.65794	8600	0.87786
2200	0.10088	5300	0.66935	8800	0.88413
2300	0.12002	5400	0.68033	9000	0.88999
2400	0.14025	5500	0.69087	9400	0.90060
2500	0.16135	5600	0.70101	9800	0.90992
2600	0.18311	5700	0.71076	10200	0.91813
2700	0.20535	5800	0.72012	10600	0.92540
2800	0.22788	5900	0.72913	11000	0.93184
2900	0.25055	6000	0.73778	12000	0.94505
3000	0.27322	6100	0.74610	13000	0.95509
3100	0.29576	6200	0.75410	14000	0.96285
3200	0.31809	6300	0.76180	15000	0.96893
3300	0.34009	6400	0.76920	16000	0.97377
3400	0.36172	6500	0.77631	17000	0.97765
3500	0.38290	6600	0.78316	18000	0.98081
3600	0.40359	6700	0.78975	19000	0.98340
3700	0.42375	6800	0.79609	20000	0.98555
3800	0.44336	6900	0.80219	30000	0.99529
3900	0.46240	7000	0.80807	40000	0.99792
4000	0.48085	7100	0.81373	50000	0.99890

possible, at a certain temperature, to determine the fraction of the blackbody radiation in the intervals $0-\lambda_1$ and $0-\lambda_2$, respectively. The difference between these gives the fraction of the emitted radiation in the wavelength interval $\lambda_1 - \lambda_2$.

12.5   Radiation from nonblackbodies

The Stefan–Boltzmann's law (12.10) describes the radiation from a blackbody. Most bodies or surfaces in practical situations are nonblack. To enable description of the radiation from nonblack surfaces, the monochromatic emissivity ε_λ is introduced. It is defined as the ratio between the monochromatic emitted radiation from the non-black surface and the corresponding blackbody radiation at the same temperature and wavelength, i.e.,

$$\varepsilon_\lambda = \frac{E_\lambda}{E_{B\lambda}} \tag{12.11}$$

or

$$E_\lambda = \varepsilon_\lambda E_{B\lambda} \tag{12.12}$$

Figures 12.5 and 12.6 show some typical distributions of $\varepsilon_\lambda = $ function (λ).

A body or surface having ε_λ constant, i.e., independent of wavelength, is said to be a gray body. With eq. (12.12) one may write

$$E_\lambda = \varepsilon E_{B\lambda}$$

Integration over the wavelength interval $0 - \infty$ gives with eq. (12.10)

$$E_g = \varepsilon \sigma T^4 \tag{12.13}$$

where index g means gray body.

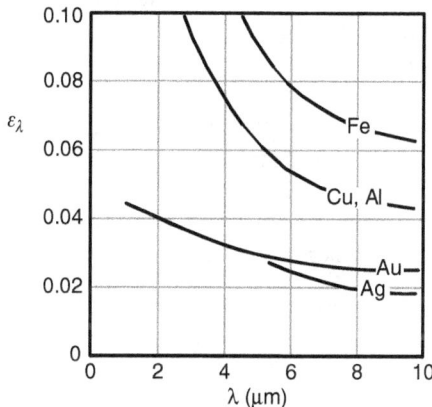

Figure 12.5: ε_λ as a function of λ for some metals (polished surfaces).

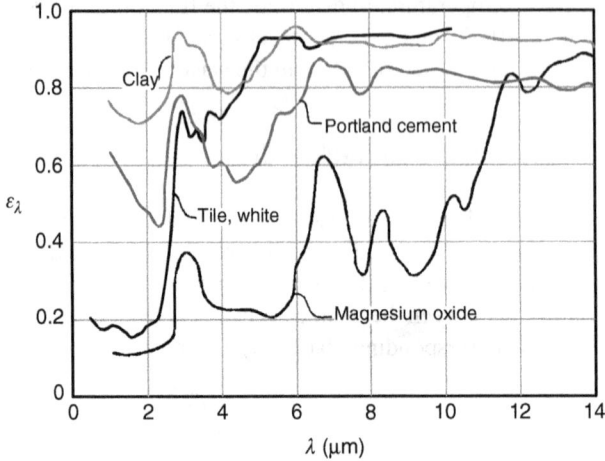

Figure 12.6: ε_λ as a function of λ for some materials being nonconductors.

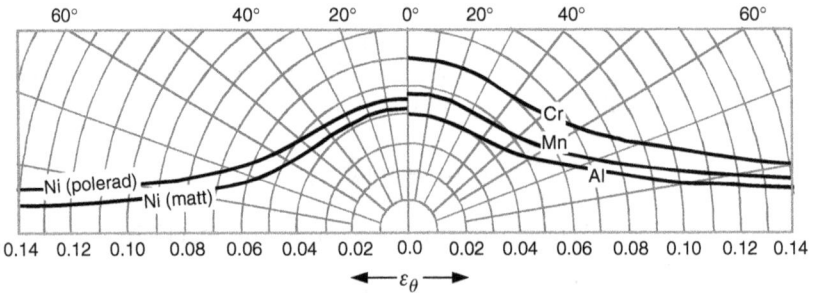

Figure 12.7: Emissivity for some electrical conductors. (Based on Ref. [7].)

Real surfaces do not emit diffusely but rather the emissivity is dependent on the direction being considered. Figure 12.7 shows the directional dependence of the emissivity for metals being electrical conductors while Fig. 12.8 shows corresponding values for nonconducting materials. From these figures, the characteristic difference between electrical conductors and nonconductors is evident. This difference can be explained by the electromagnetic wave theory, see, for example, Siegel and Howell [4].

Two important relations are now given without any proofs. For a body exchanging heat by radiation with other bodies at the same temperature, i.e., at thermal equilibrium, one has

$$\alpha = \varepsilon \qquad (12.14)$$

Equation (12.14) is one of the Kirchhoff's laws and it is valid at thermal equilibrium. For an arbitrary surface, even if no thermal equilibrium prevails, the monochromatic

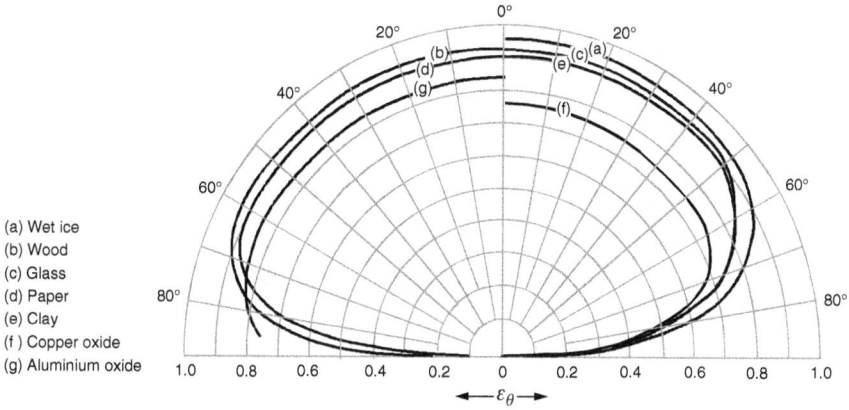

(a) Wet ice
(b) Wood
(c) Glass
(d) Paper
(e) Clay
(f) Copper oxide
(g) Aluminium oxide

Figure 12.8: Emissivity for some nonconducting materials. (Based on Ref. [7].)

absorptivity and the monochromatic emissivity are equal, i.e.,

$$\alpha_\lambda = \varepsilon_\lambda \qquad (12.15)$$

12.6   Radiation intensity

To enable calculations of the radiative energy from a surface in a certain direction, the radiation intensity, I, needs to be introduced. Consider the radiation from the surface element dA in Fig. 12.9. This surface is surrounded by a hemisphere with radius r, and all radiation from the surface dA must reach the surface of the hemisphere. Every surface element dA' on the hemisphere surface is occupying a solid angle $d\omega$ seen from the element dA. One has

$$d\omega = \frac{dA'}{r^2} \qquad (12.16)$$

The radiation intensity is defined as the radiative energy per unit area projected perpendicular to a given direction and per unit solid angle seen from the radiating surface. In Fig. 12.9, the radiative energy leaving surface dA and hitting surface dA' is $I\,dA\cos\theta\,d\omega$. This can be interpreted as an infinitely small fraction of the total radiation from surface dA. Thus one has

$$d(E\,dA) = I\,dA\cos\theta\,d\omega = I\,dA\cos\theta\frac{dA'}{r^2} \qquad (12.17)$$

One also has

$$E\,dA = \int d(E\,dA)$$

By using Fig. 12.10 it will be possible to determine $d\omega$.

One has $dA' = r \, d\theta \, r \sin\theta \, d\phi$. Thus one finds

$$d\omega = \sin\theta \, d\theta \, d\phi \tag{12.18}$$

Using eq. (12.17) it is found that

$$d(E \, dA) = I \, dA \cos\theta \sin\theta \, d\theta \, d\phi \tag{12.19}$$

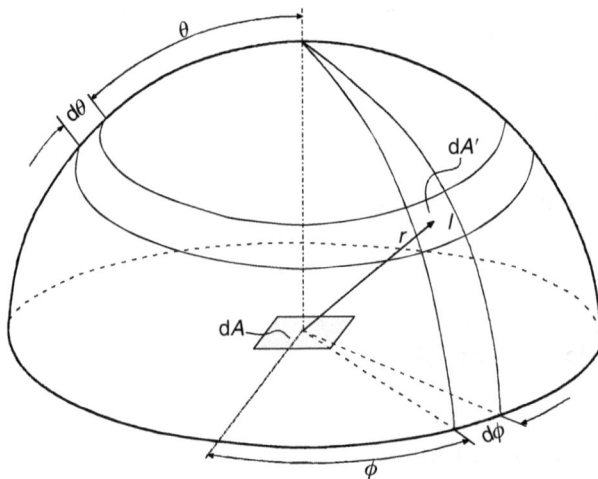

Figure 12.9: Emission from a plane surface.

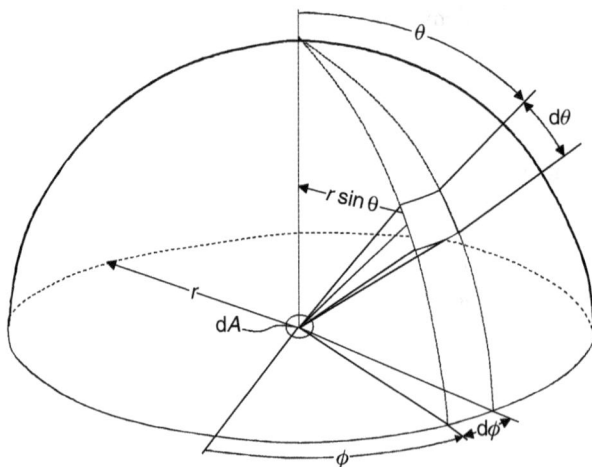

Figure 12.10: Sketch of geometry to determine $d\omega$.

Now a diffusely radiating surface is considered. Then, the intensity I is independent of direction. Integration of equation (12.19) gives

$$E \, dA = I \, dA \int\limits_0^{2\pi} d\phi \int\limits_0^{\pi/2} \cos\theta \sin\theta \, d\theta = I \, dA\pi$$

Thus one has

$$E = \pi I \qquad\qquad (12.20)$$

12.7   Angle factor, view factor, or shape factor

In many applications involving thermal radiation, it is convenient to introduce a so-called *angle factor, view factor, or shape factor*. This is defined as the fraction of the diffuse radiation from a surface which will reach another surface. Consider now the radiative exchange between two arbitrary surfaces as shown in Fig. 12.11. Consider the surface increments dA_1 and dA_2 on surfaces 1 and 2, respectively. The radiation leaving dA_1 and reaching dA_2 is traveling on the connecting line between theses increments. The radiation $d\Phi$ from dA_1 in this direction and per unit solid angle is

$$\frac{d\Phi}{d\omega_1} = I_1 \, dA_1 \cos\theta_1 \qquad\qquad (12.21)$$

The surface element dA_2 occupies the solid angle $d\omega_1$ (seen from dA_1)

$$d\omega_1 = \frac{dA_2 \cos\theta_2}{r^2} \qquad\qquad (12.22)$$

The radiation leaving dA_1 and reaching dA_2 is then

$$d\Phi_{dA_1 - dA_2} = I_1 \frac{\cos\theta_1 \cos\theta_2 \, dA_1 dA_2}{r^2} \qquad\qquad (23a)$$

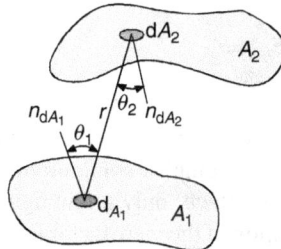

Figure 12.11: Sketch for determination of the view factor.

By integrating over the surfaces A_1 and A_2, the total radiation leaving surface 1 and reaching surface 2 becomes

$$\Phi_{A_1-A_2} = \int_{A_1} \int_{A_2} I_1 \frac{\cos\theta_1 \cos\theta_2}{r^2} dA_1\, dA_2 \tag{12.23b}$$

The total radiation from surface A_1 is

$$\Phi_{A_1} = E_1 A_1 = \pi I_1 A_1 \tag{12.24}$$

The view factor F_{12} is defined as

$$\Phi_{A_1-A_2} = F_{12}\Phi_{A_1} \tag{12.25}$$

With eqs. (12.23b) and (12.24), one finds

$$F_{12} = \frac{1}{A_1} \int_{A_1} \int_{A_2} \frac{\cos\theta_1 \cos\theta_2}{\pi r^2} dA_1\, dA_2 \tag{12.26}$$

If now instead the radiation from surface 2 which reaches surface 1 is considered a view factor F_{21} is found as

$$F_{21} = \frac{1}{A_2} \int_{A_1} \int_{A_2} \frac{\cos\theta_1 \cos\theta_2}{\pi r^2} dA_1\, dA_2 \tag{12.27}$$

By comparing eqs. (12.26) and (12.27) one finds

$$A_1 F_{12} = A_2 F_{21} \tag{12.28}$$

Equation (12.28) is called the *reciprocal identity*.

   In general, an arbitrary surface i is exchanging radiative heat with several other surfaces, say n. If the n other surfaces are enclosing the surface i, then from the definition of the view factor one finds

$$\sum_{j=1}^{j=n} F_{ij} = F_{i1} + F_{i2} + \cdots + F_{in} = 1 \tag{12.29}$$

The integration of eq. (12.26) can be carried out analytically for a number of simple configurations. For more complex configurations, the integration has to be performed numerically. At this stage, only a few figures (Figs. 12.12–12.15) are provided to enable determination of the view factor for some simple configurations. Additional cases and analytical solutions can be found in Howell [6], Wong [7], and Modest [8].

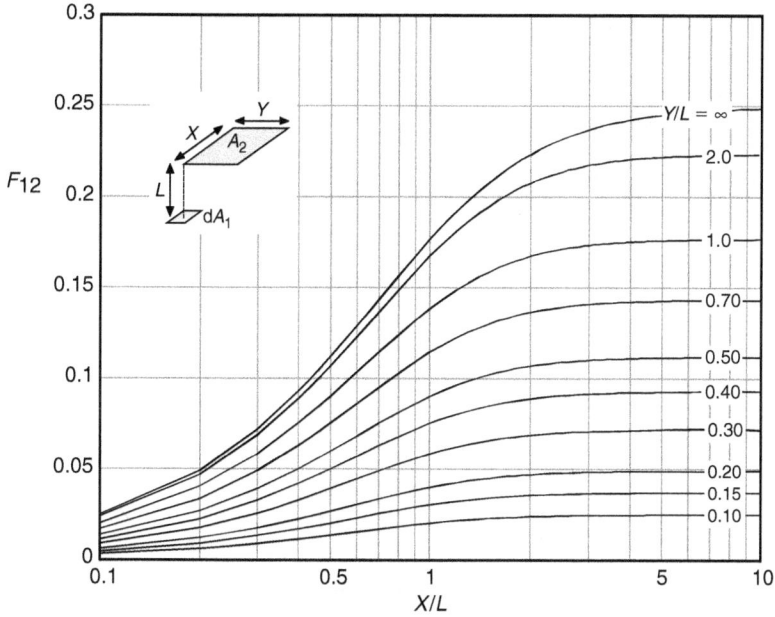

Figure 12.12: View factor F_{12} for a small surface increment dA_1 and a parallelly placed rectangle A_2.

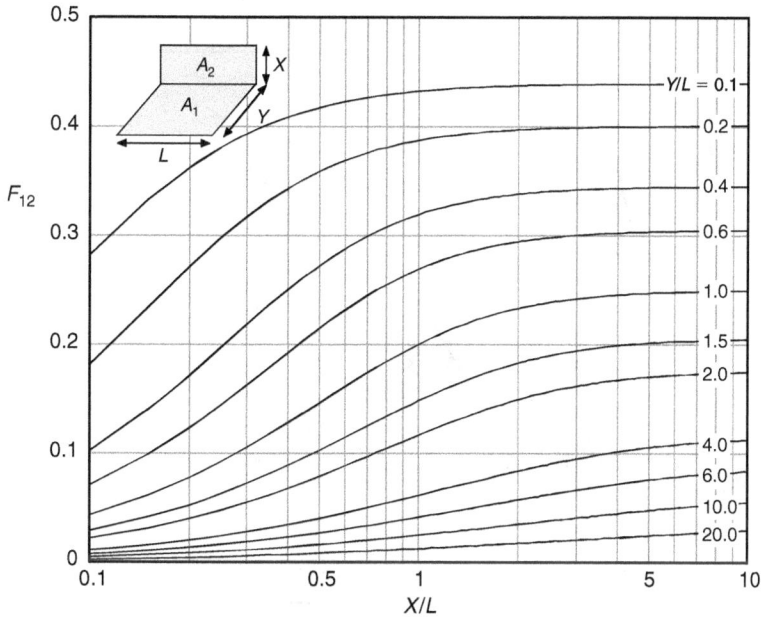

Figure 12.13: View factor F_{12} for two perpendicular rectangles with a common edge.

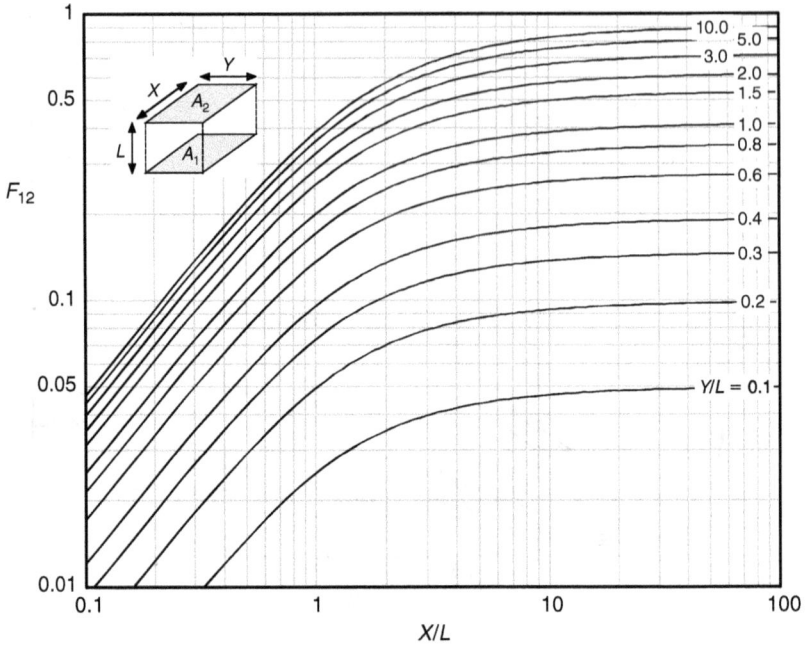

Figure 12.14: View factor F_{12} between two parallel rectangles.

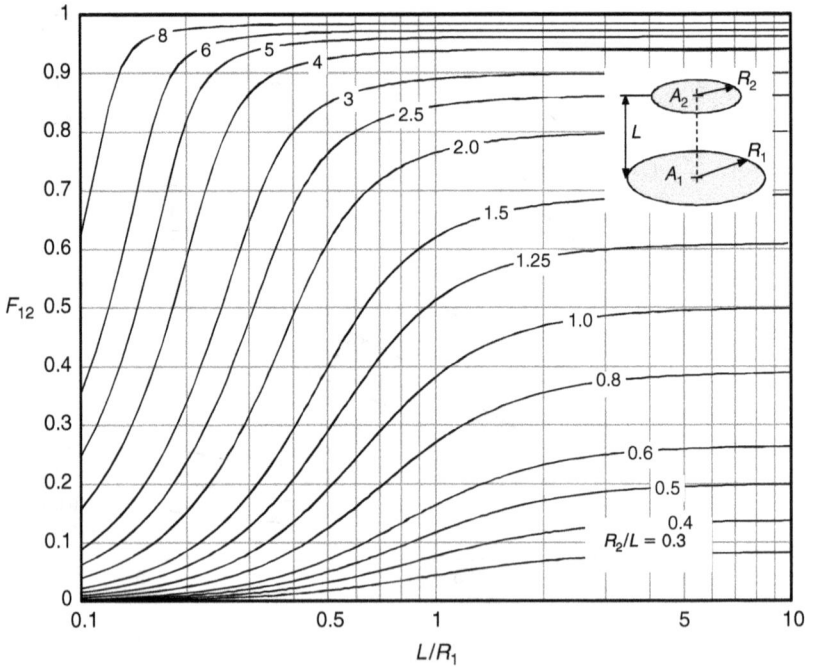

Figure 12.15: View factor F_{12} between two parallel co-axial disks.

12.8   Radiative exchange between blackbodies

Consider a room as shown in Fig. 12.16.

All surfaces are assumed black, i.e., all incident radiation is absorbed. This means $\alpha = \varepsilon = 1$. For an arbitrary surface i, one has

$$\dot{Q}_i = A_i\,(J_i - G_i) \tag{12.30}$$

$$J_i = E_{\mathrm{B},i} \tag{12.31}$$

For the incident radiation, one has

$$A_i G_i = \sum_k A_k J_k F_{ki} = \sum_k A_i J_k F_{ik} \tag{12.32}$$

Inserting eq. (12.32) in eq. (12.30) gives

$$\dot{Q}_i = A_i\left(J_i - \sum_k F_{ik}J_k\right) \tag{12.33}$$

If all surfaces k together enclose surface i, the following relation is valid.

$$\sum_k F_{ik} = 1$$

Equation (12.33) can then be written as

$$\dot{Q}_i = A_i\sum_k F_{ik}(J_i - J_k) = A_i\sum_k F_{ik}(E_{\mathrm{B},i} - E_{\mathrm{B},k}) \tag{12.34}$$

Equation (12.34) expresses the radiative exchange for surface i with all the other surfaces. Especially the exchange between surfaces i and k is

$$\dot{Q}_{i-k} = A_i F_{ik}(E_{\mathrm{B},i} - E_{\mathrm{B},k}) \tag{12.35}$$

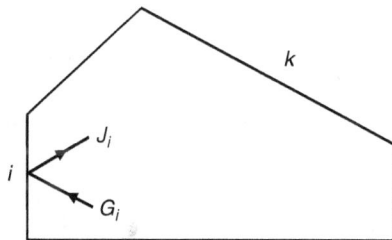

Figure 12.16: Room with black walls.

12.9   Radiative exchange between nonblackbodies

Consider again the room in Fig. 12.16 but let the surfaces be diffuse, i.e., the intensity is independent of direction. One now has $\alpha = \varepsilon \neq 1$. For an arbitrary surface i, the following relations hold:

$$\dot{Q}_i = A_i (J_i - G_i) \tag{12.36}$$

$$J_i = \varepsilon_i E_{B,i} + \rho_i G_i \tag{12.37}$$

The walls can often be considered as strongly absorbing (opaque), i.e., $\tau_i = 0$. By using eqs. (12.3) and (12.14), one finds

$$\rho_i = 1 - \varepsilon_i \tag{12.38}$$

From eqs. (12.37) and (12.38), G_i can be written as

$$G_i = \frac{1}{1 - \varepsilon_i} J_i - \frac{\varepsilon_i}{1 - \varepsilon_i} E_{B,i} \tag{12.39}$$

Substituting (12.39) in (12.36) gives

$$\dot{Q}_i = A_i \frac{\varepsilon_i}{1 - \varepsilon_i} (E_{B,i} - J_i) \tag{12.40}$$

G_i can also be written as

$$A_i G_i = \sum_k A_k J_k F_{ki} = \sum_k A_i F_{ik} J_k \tag{12.41}$$

If all surfaces k together enclose surface i, one has $\sum_k F_{ik} = 1$, and then $\dot{Q}_i$ can be written as

$$\dot{Q}_i = A_i \sum_k F_{ik}(J_i - J_k) \tag{12.42}$$

It should be noted that for black surfaces only one equation (eq. (12.34)) was obtained. For nonblack surfaces, two equations ((12.40) and (12.42)) for every surface are needed. As the radiosities in eqs. (12.40) and (12.42) are not known a priori, these equations need to be solved as a coupled system of equations.

   Comments: In the analysis of the radiative exchange, total values of the emissivity ε were used. This assumption means that the results are for so-called gray bodies, i.e., ε is independent of wavelength. This assumption can easily be dropped by considering monochromatic radiation and then eqs. (12.40) and (12.42) should have an index λ. The total exchange is then obtained by integration over the wavelength interval.

If the surfaces are reflecting the incident radiation regularly (according to the mirror laws), the equations above also have to be modified. For details, see Eckert and Drake [5], Holman [9], and Modest [8].

Note: For a surface with $A_i \rightarrow \infty$ one finds $J_i \rightarrow E_{B,i}$, i.e., a very big surface can be treated as a blackbody.

For a thermally insulated surface, $\dot{Q}_i = 0$, one also finds $J_i = E_{B,i}$.

12.10   Simple example

Consider two parallel planes with infinite extension as shown in Fig. 12.17.

The surfaces are assumed to be diffuse. Because the planes have infinite extension, they will enclose each other. Thus, one has $F_{12} = F_{21}$. Equations (12.40) and (12.42) give

$$\dot{Q}_1 = A_1 \frac{\varepsilon_1}{1 - \varepsilon_1}(E_{B,1} - J_1) \tag{12.43}$$

$$\dot{Q}_2 = A_2 \frac{\varepsilon_2}{1 - \varepsilon_2}(E_{B,2} - J_2) \tag{12.44}$$

$$\dot{Q}_1 = A_1(J_1 - J_2) \tag{12.45}$$

$$\dot{Q}_2 = A_2(J_2 - J_1) \tag{12.46}$$

Because $A_1 = A_2$, eqs. (12.45) and (12.46) give $\dot{Q}_2 = -\dot{Q}_1$ as expected. From eqs. (12.43) and (12.44), J_1 and J_2, respectively, can be found. These are substituted into eq. (12.45) and after some calculations one finds

$$\dot{Q}_1 = A_1 \frac{E_{B,1} - E_{B,2}}{(1/\varepsilon_1) + (1/\varepsilon_2) - 1} \tag{12.47}$$

In this case, the calculations and the final results were relatively simple. For cases involving more than two surfaces and of finite size, the solution procedure becomes more advanced.

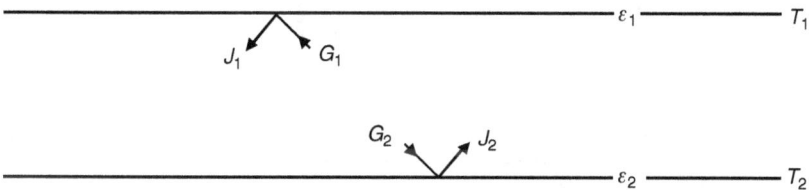

Figure 12.17: Radiative heat exchange between two parallel planes with infinite extension.

12.11   Gas radiation

The radiative exchange between a gas and a solid surface is much more complicated than was described in the preceding sections. In contrast to solid bodies, gases may transmit a considerable amount of the incident radiation. When a gas is emitting or absorbing radiation, the process commonly appears only in narrow wavelength intervals. In addition, elementary gases (molecules are of one kind only) such as H_2, O_2, and N_2 do not emit any thermal radiation and are transparent ($\tau = 1$) for incident radiation. In engineering applications, CO_2 and water vapor are most important as they are good emitters and also appear in high concentrations. Also CO, SO_2, and CH_4 are good emitters but usually they only appear in small concentrations and thus are less important.

Figures 12.18 and 12.19 show the monochromatic absorptivity for carbon dioxide and water vapor, respectively.

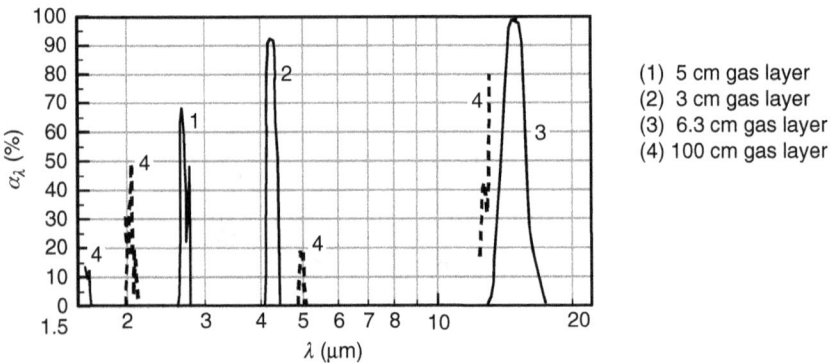

(1) 5 cm gas layer
(2) 3 cm gas layer
(3) 6.3 cm gas layer
(4) 100 cm gas layer

Figure 12.18: Absorptivity for CO_2.

(a) 127°C, gas layer 109 cm   (b) 127°C, gas layer 104 cm   (c) 127°C, gas layer 32.4 cm
(d) 81°C, gas layer 32.4 cm   (e) room temperature, gas layer 220 cm

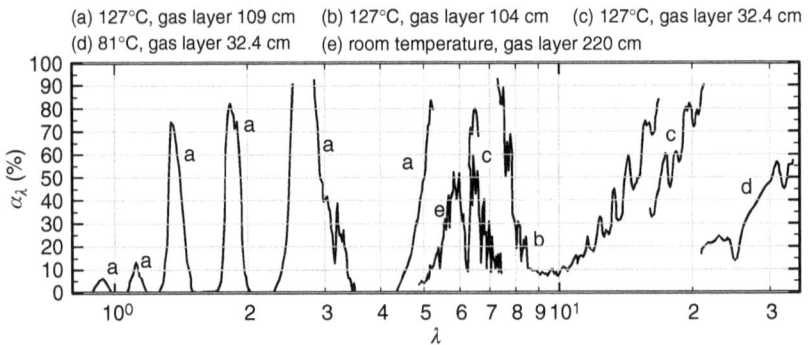

Figure 12.19: Absorptivity for water vapor.

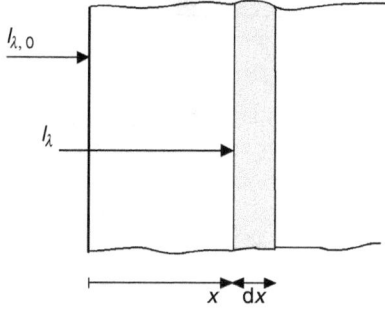

Figure 12.20: Absorption in a gas layer.

Consider now the absorption process in a gas layer as shown in Fig. 12.20.

A monochromatic beam or ray with intensity I_λ approaches a gas layer with thickness dx. The intensity is reduced because the gas layer absorbs part of the incident radiation. The absorption is assumed to be proportional to the intensity and the thickness of the gas layer. One then writes

$$dI_\lambda = -a_\lambda I_\lambda \, dx \qquad (12.48)$$

where the constant of proportionality a_λ is called the *monochromatic absorption coefficient*. Integration of eq. (12.48) gives

$$\int_{I_{\lambda,0}}^{I_\lambda} \frac{dI_\lambda}{I_\lambda} = -\int_0^x a_\lambda \, dx \qquad (12.49)$$

In general, a_λ depends on pressure and temperature of the gas. If the pressure and temperature are assumed to be uniform across the whole gas layer, eq. (12.49) can be integrated and one finds

$$I_\lambda = I_{\lambda,0} \, e^{-a_\lambda x} \qquad (12.50)$$

This equation is called Beer's law.

With the previously given definitions of transmittivity and absorptivity, one obtains

$$\tau_\lambda = e^{-a_\lambda x} \qquad (12.51)$$

If the gas is not reflecting any part of the radiation ($\rho_\lambda = 0$), one finds

$$\alpha_\lambda = 1 - e^{-a_\lambda x} \qquad (12.52)$$

From eqs. (12.51) and (12.52) it is evident that the radiative properties of the gas depend on the thickness of the gas layer.

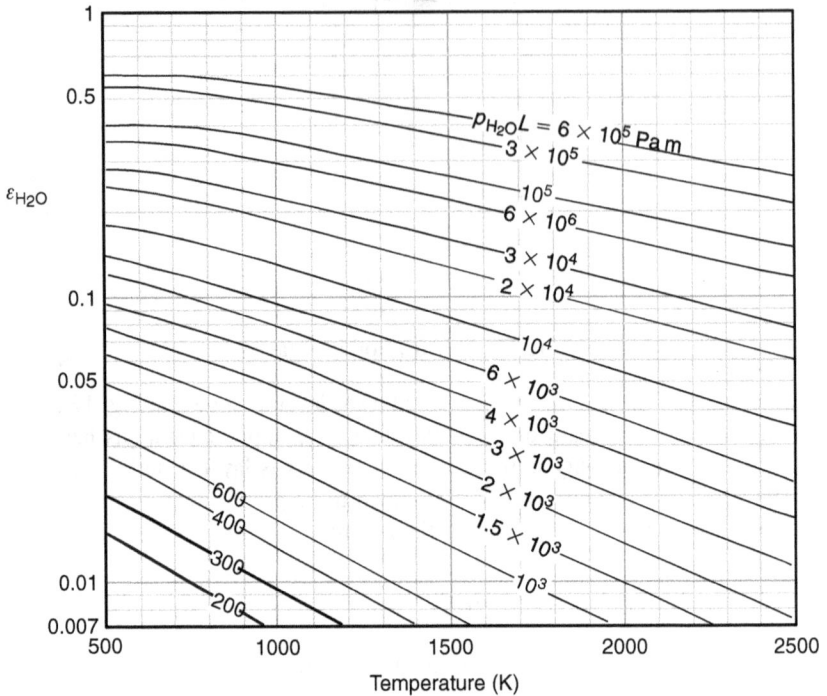

Figure 12.21: Emissivity for water vapor at a total pressure of 1 atm.

12.12 Mean beam length, equivalent beam length

Consider now a case where the gas is occupying the space between the walls of a room. The walls are assumed to emit diffusely. The beams or rays, which are transmitted through the gas, will travel different distances which means that the x-value in eqs. (12.51) and (12.52) is not the same for all the beams. Thus, it becomes difficult to work with these formulas.

Hottel [10] has developed an approximative method for description of the absorption and emission characteristics for gray gases. This method is quite simple and sufficiently accurate for simple engineering calculations or estimations. In Figs. 12.21 and 12.22, the emissivity for water vapor and CO_2, respectively, are given graphically.

In these figures, length L is included, and this is a characteristic length for the case considered. It is called the *mean beam length* or the *equivalent beam length*. In Table 12.2, L is provided for some cases (see Hottel [10] and Eckert and Drake [5]).

For geometries, where no information about the mean beam length is available, a decent approximation can be obtained from

$$L = 3.6V/A \qquad (12.53)$$

where V is the gas volume and A the surface area.

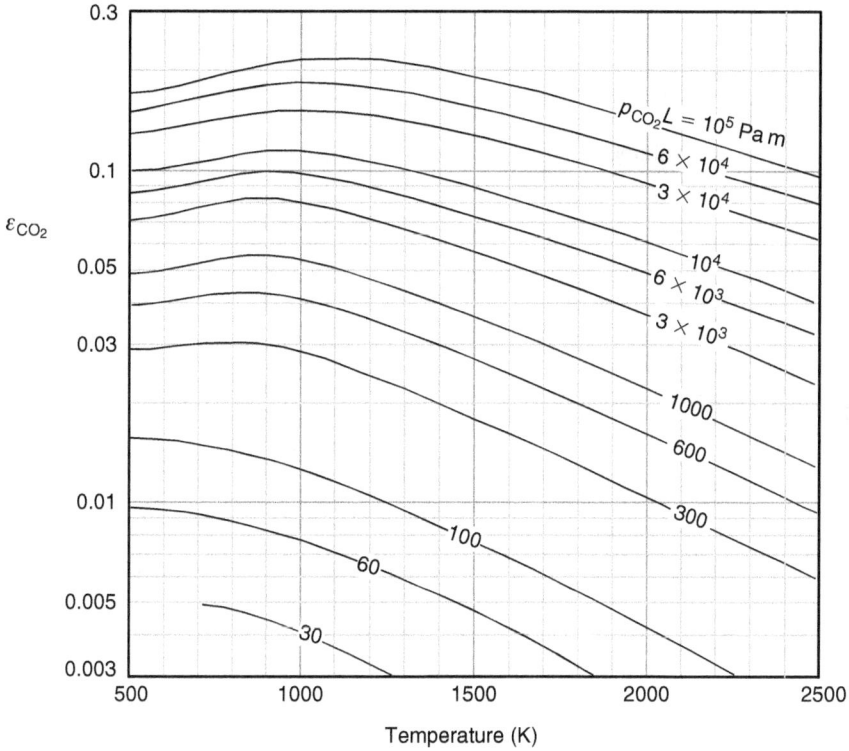

Figure 12.22: Emissivity for carbon dioxide at a total pressure of 1 atm.

In Figs. 12.21 and 12.22, the total pressure of the gas mixture is 1 atm; p_{H_2O} and p_{CO_2} are the partial pressures of water vapor and carbon dioxide, respectively. For total pressures different from 1 atm, corrections have to be introduced.

Figures 12.23 and 12.24 provide correction factors for water vapor and CO_2, respectively, to be used if the total pressure is different from 1 atm.

Another correction factor is necessary as a gas mixture contains both water vapor and carbon dioxide. The emissivity for the gas mixture is then calculated according to

$$\varepsilon_{\text{total}} = \varepsilon_{H_2O} + \varepsilon_{CO_2} - \Delta\varepsilon \qquad (12.54)$$

where the correction factor $\Delta\varepsilon$ is found from Fig. 12.25.

12.13   Radiative heat exchange between a gas and a room with black walls

Consider a gas volume of uniform temperature T_g which is enclosed by a room with black walls having the temperature T_w. The net radiative heat exchange between

Table 12.2: Equivalent mean beam length for gas radiation. Radiation from the whole gas volume.

Gas volume	Characteristic dimension	L
Volume between two infinite planes	Distance between the planes, a	$1.8a$
Circular cylinder with height equal to diameter—radiation to center of the base	Diameter, D	$0.77D$
Circular cylinder with height equal to diameter—radiation to whole surface	Diameter, D	$0.60D$
Cube—radiation to all side walls	Side, a	$0.60a$
Sphere—radiation to whole surface	Diameter, D	$0.65D$
Infinite circular cylinder—radiation to the concave surface	Diameter, D	$0.95D$
Volume enclosing an infinite long tube bundle—radiation to all tubes	Tube diameter, D	*Quadratic arrangement:* $S = 2D$: $3.5(S - D)$
	Distance between tube centers	*Triangular arrangement:* $S = 2D$: $3.0(S - D)$ $S = 3D$: $3.8(S - D)$

the gas and the walls is given by

$$\frac{\dot{Q}}{A} = \text{emitted gas radiation} - \text{absorbed wall radiation}$$

$$\frac{\dot{Q}}{A} = \varepsilon_g(T_g)\sigma T_g^4 - \alpha_g(T_w)\sigma T_w^4 \qquad (12.55)$$

where $\varepsilon_g(T_g)$ is the emissivity of the gas mixture at temperature T_g. $\varepsilon_g(T_g)$ is determined from Figs. 12.21–12.25. $\alpha_g(T_w)$ is the gas absorptivity for the wall blackbody radiation of temperature T_w. $\alpha_g(T_w)$ depends on both T_w and T_g. For a gas mixture of carbon dioxide and water vapor, an empirical relation for $\alpha_g(T_w)$ exists. This reads as

$$\alpha_g(T_w) = \alpha_{CO_2} + \alpha_{H_2O} - \Delta\alpha \qquad (12.56)$$

where

$$\alpha_{CO_2} = C_{CO_2}\varepsilon_{CO_2} \left(\frac{T_g}{T_w}\right)^{0.65} \qquad (12.57)$$

$$\alpha_{H_2O} = C_{H_2O}\varepsilon_{H_2O} \left(\frac{T_g}{T_w}\right)^{0.45} \qquad (12.58)$$

$$\Delta\alpha = \Delta\varepsilon \text{ at } T_w$$

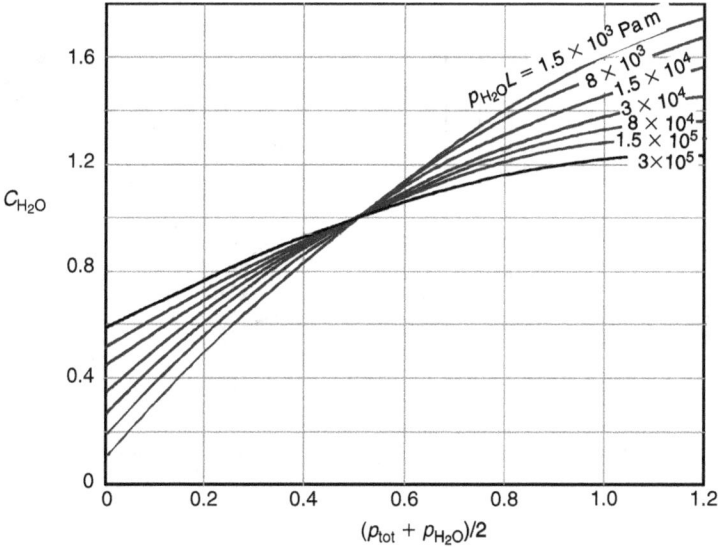

Figure 12.23: Correction factor for the water vapor emissivity as the total pressure deviates from 1 atm, $\varepsilon_{H_2O,p} = C_{H_2O,p}\varepsilon_{H_2O,p=1\,atm}$ ($\varepsilon_{H_2O,p=1\,atm}$, from Fig. 12.21).

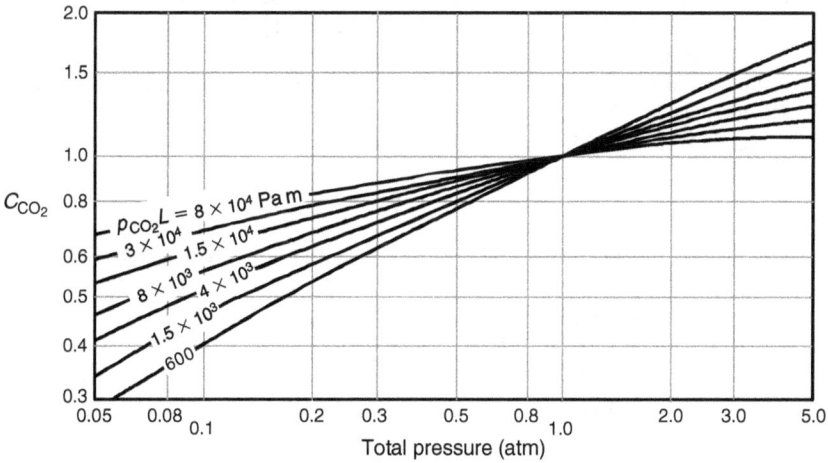

Figure 12.24: Correction factor for the carbon dioxide emissivity as the total pressure deviates from 1 atm, $\varepsilon_{CO_2,p} = C_{CO_2,p}\varepsilon_{CO_2,p=1\,atm}$ ($\varepsilon_{CO_2,p=1\,atm}$, from Fig. 12.22).

The values of $\varepsilon_{CO_2}\varepsilon_{H_2O}$ and ε_{CO_2} are taken from Figs. 12.21 and 12.22 with the abscissa equal to T_w. The parameter in these figures must be $p_{CO_2} \cdot L \cdot (T_w/T_g)$ and $p_{H_2O} \cdot L \cdot (T_w/T_g)$, respectively.

For more details, see Hottel [10].

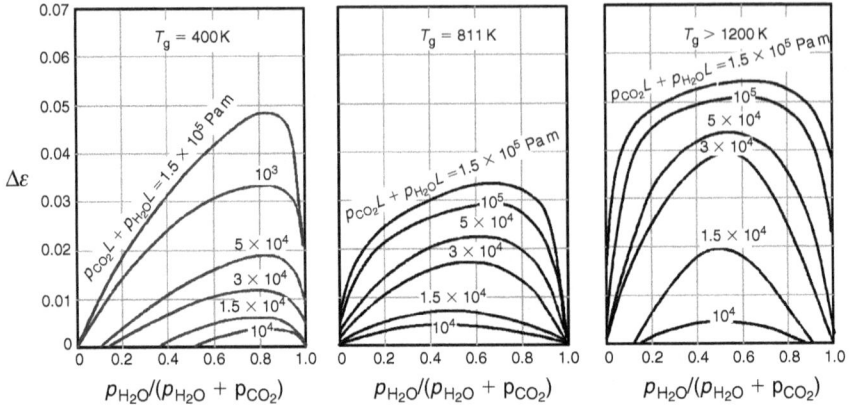

Figure 12.25: Correction factor $\Delta\varepsilon$ for gas mixtures containing both water vapor and CO_2.

References

[1] M. Planck, The Theory of Heat Radiation, Dover, New York (1959).

[2] J. Stefan, Über die Beziehung zwischen der Wärmestrahlung und der Temperatur, Sitzber. Akad. Wiss. Wien, 79(2), 391–428 (1879).

[3] L. Boltzmann, Ableitung des Stefan'schen Gesetzesdie betreffend die Abhängigkeit der Wärmestrahlung von der Temperatur aus der elektromagnetischen Lichttheorie, Ann. Physik, 22(2), 291–294 (1884).

[4] R. Siegel and J.R. Howell, Thermal Radiation Heat Transfer, 4th ed., Taylor & Francis, Philadelphia, PA (2001).

[5] E.R.G. Eckert and R.M. Drake Jr., Analysis of Heat and Mass Transfer, McGraw-Hill, New York (1972).

[6] J.R. Howell, A Catalog of Radiation Configuration Factors, McGraw-Hill, New York (1982).

[7] H.Y. Wong, Handbook of Essential Formulae and Data on for Engineers, Longman, London (1976).

[8] M.F. Modest, Radiative Heat Transfer, 2nd ed., Academic Press, New York (2003).

[9] J.P. Holman, Heat Transfer, 10th ed., McGraw-Hill, New York (2009).

[10] H.C. Hottel, Radiant heat transmission, in Heat Transmission (Ed. W.H. McAdams), 3rd ed., McGraw-Hill, New York (1954), Chapter 4.

13   Condensation

13.1   Introduction

In many engineering applications, the heat transfer process involves phase change. For example, in heat and power plants, water (liquid) is converted to steam (gas) and the boiling process must be known in details to enable efficient and reliable design of furnaces. As the steam has delivered work in the steam turbine, it is condensed to water in a condenser and then water (liquid) is pumped to the furnace to repeat and complete the thermodynamic process. Thus the condensation of steam to water is also important. Besides heat and power generation, phase change processes occur in refrigeration units, heat pumps, air-conditioning units, etc. This chapter treats condensation while boiling/evaporation is left to another chapter.

13.2   General statements

Condensation may occur in two different ways. At dropwise condensation, the vapor is condensed to small liquid droplets of different sizes. At film condensation, the vapor is condensed continuously in a film or layer which covers the surface of the cold body. In engineering applications, it is most common that film condensation occurs. To maintain dropwise condensation, the cold surface needs to be treated in such a way that the condensate layer will be nonwetted. However, for dropwise condensation, the heat transfer coefficient is 5 to 10 times higher than for film condensation. The reason is that the heat flux across the curved surfaces of the droplets is greater than that across a film with identical mass. The heat transfer between the vapor and the noncovered part of the surface is bigger (no film present) and the condensate is removed as droplets are merged and leave the surface fast.

The pioneering work on film condensation was carried out by Nusselt [1], while for dropwise condensation the pioneering work was by Jakob [2].

In the following sections, mainly film condensation will be considered, but at the end a brief discussion of dropwise condensation is provided.

13.3   Film condensation along a vertical surface

Consider a vertical wall along which condensation occurs as shown in Fig. 13.1. The vapor, which is assumed to be pure, has the saturation temperature t_s. The wall

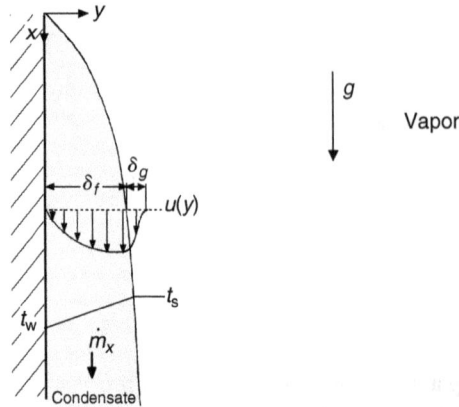

Figure 13.1: Sketch of film condensation along a vertical surface.

surface has the temperature t_w which is lower than the saturation temperature and thus condensation takes place. Due to the action of gravity forces, the liquid in the layer closest to the wall will flow down. The vapor closest to the liquid layer due to friction forces will be forced to move down, but the main part of the vapor will remain at rest. In principle, the temperature and flow fields will look as sketched in Fig. 13.1, where δ_f is the thickness of the condensate layer and δ_g is the thickness of the vapor layer being affected by the motion of the condensate. Index f refers to the liquid, while index g refers to the vapor (gas). In most applications, the thicknesses δ_f and δ_g are small compared to the dimensions of the surface and commonly the boundary layer concept (see Chapter 7) is applied in the analysis.

For both the liquid and vapor layer, the following equations are then valid:

Mass conservation equation

$$\frac{\partial u}{\partial x} + \frac{\partial v}{\partial y} = 0 \tag{13.1}$$

Momentum equation

$$\rho\left(u\frac{\partial u}{\partial x} + v\frac{\partial u}{\partial y}\right) = -\frac{\partial p}{\partial x} + \rho g + \mu\frac{\partial^2 u}{\partial y^2} \tag{13.2}$$

Temperature field equation

$$\rho c_p\left(u\frac{\partial t}{\partial x} + v\frac{\partial t}{\partial y}\right) = \lambda\frac{\partial^2 t}{\partial y^2} \tag{13.3}$$

If it is assumed that the pressure is constant across the liquid and vapor layers and that this pressure is equal to the pressure in the resting vapor, one has

$$\frac{\partial p}{\partial x} = \rho_e g \tag{13.4}$$

where ρ_e is the vapor pressure at large distances from the surface.

Equations (13.1)–(13.4) need boundary conditions at the wall, at the interface between the liquid and vapor, as well as in the vapor phase at large distances from the wall. These conditions are not formulated here, but instead the simplifying assumptions by Nusselt [1] are introduced.

Nusselt's simplifying assumptions

- The convective transport is neglected in both the liquid and vapor layers
- $\rho_f \gg \rho_g, \rho_f \gg \rho_e, \rho_g = \rho_e$
- The flow field in the vapor layer is neglected
- The temperature in the vapor layer is constant and equal to t_s

With these assumptions, eqs. (13.2) and (13.3) for the liquid layer can be written as

$$0 = \rho_f g + \mu_f \frac{\partial^2 u_f}{\partial y^2} \tag{13.5}$$

$$0 = \lambda_f \frac{\partial^2 t_f}{\partial y^2} \tag{13.6}$$

The boundary conditions to eq. (13.5) are given by

$$y = 0: \quad u_f = 0$$

and that no shear force is acting between the liquid and vapor at their interface

$$y = \delta_f: \quad \mu_f \frac{\partial u_f}{\partial y} = 0$$

The solution to eq. (13.6) shows that the temperature varies linearly across the liquid layer. A heat balance for an increment dx in the vertical downward direction reads

$$\lambda_f \frac{t_s - t_w}{\delta_f} dx = h_{fg} \frac{d\dot{m}}{dx} dx \tag{13.7}$$

where h_{fg} (J/kg) is the latent heat and $\dot{m}$, the rate of mass flow for the condensate per unit width. $\dot{m}$ is found from

$$\dot{m} = \int_0^{\delta_f} \rho_f u_f \, dy \tag{13.8}$$

The solution to eq. (13.5) with the boundary conditions as above reads

$$u_f = \frac{\rho_f g}{\mu_f}\left(y\delta_f - \frac{y^2}{2}\right) \tag{13.9}$$

With eq. (13.9), the rate of condensate mass flow per unit width can be found from eq. (13.8). One finds

$$\dot{m} = \frac{\rho_f^2 g \delta_f^3}{3\mu_f} \tag{13.10}$$

If eq. (13.10) is differentiated and the result is inserted in eq. (13.7), a differential equation for the liquid layer thickness δ_f is achieved. The solution to this equation gives

$$\delta_f = \left(\frac{4\mu_f \lambda (t_s - t_w)x}{h_{fg} g \rho_f^2}\right)^{1/4} \tag{13.11}$$

The heat transfer coefficient α is found from the expressions below:

$$q_w = \alpha(t_s - t_w) = \lambda_f \frac{t_s - t_w}{\delta_f} = \frac{d\dot{m}}{dx} h_{fg} \tag{13.12}$$

The second and third expressions in eq. (13.12) give

$$\alpha = \frac{\lambda_f}{\delta_f} \tag{13.13}$$

With eq. (13.11), eq. (13.13) can be written as

$$\alpha = \left(\frac{h_{fg} g \rho_f^2 \lambda_f^3}{4\mu_f(t_s - t_w)x}\right)^{1/4} \tag{13.14}$$

The average value of α is calculated in the conventional way, i.e.,

$$\bar{\alpha} = \frac{1}{L}\int_0^L \alpha\, dx \tag{13.15}$$

where L is the height of the vertical surface (plate). After some calculations, eqs. (13.15) and (13.14) give

$$\bar{\alpha} = 0.943\left(\frac{h_{fg} g \rho_f^2 \lambda_f^3}{\mu_f(t_s - t_w)L}\right)^{1/4} \tag{13.16}$$

The Nusselt number, which represents the heat transfer coefficent in a dimensionless form, is here defined as

$$\overline{Nu} = \frac{\bar{\alpha}L}{\lambda_f} \tag{13.17}$$

and can easily be found using eq. (13.16).

13.4 The Reynolds number

The analysis presented in the previous section assumed that the flow in the condensate layer is laminar. If the Reynolds number based on the condensate layer thickness exceeds a certain value, turbulent flow will occur. The Reynolds number is defined as

$$Re_{\delta_f} = \frac{u_m \delta_f}{v_f} = \frac{\rho_f u_m \delta_f}{\mu_f} = \frac{\dot{m}}{\mu_f} \tag{13.18}$$

(Note that $\dot{m}$ (kg/s m) is the rate of mass flow per unit width.)

Substituting eqs. (13.10) and (13.11) in eq. (13.18), one finds

$$Re_{\delta_f} = \frac{1}{3} \frac{\rho_f^2 g}{\mu_f^2} \left(\frac{4\mu_f \lambda_f (t_s - t_w)x}{h_{fg} g \rho_f^2} \right)^{3/4} = 0.943 \left(\frac{\rho_f^{2/3} g^{1/3} \lambda_f (t_s - t_w)x}{h_{fg} \mu_f^{5/3}} \right)^{3/4} \tag{13.19}$$

Experiments, see Ref. [3], have shown that if $Re_{\delta_f} > 6$, waves are created at the interface between the liquid and the vapor. The flow is still laminar, but the heat transfer coefficient is higher than that in eq. (13.16). Kutateladze [4] has suggested the following equation for the average heat transfer coefficient.

$$\bar{\alpha}_L = \frac{\lambda_f}{(v_f^2/g)^{1/3}} \left(1.47 Re_{\delta_f}^{0.22} - \frac{1.3}{Re_{\delta_f}} \right)^{-1} \tag{13.20}$$

Equation (13.20) is valid in the range of $6 < Re_{\delta_f} \leq 450$.

If $Re_{\delta_f} > 450$ transition to turbulent flow occurs in the condensate layer and the heat transfer coefficient will increase further. Experiments by Labuntsov [5] suggest the following correlation:

$$\bar{\alpha}_L = \frac{\lambda_f}{(v_f^2/g)^{1/3}} \frac{Re_{\delta_f}}{(2188 + 41(Re_{\delta_f}^{0.75} - 89.5)/Pr_f^{0.5})} \tag{13.21}$$

In the literature, another definition of the Reynolds number also exists. This latter Reynolds number is denoted by Re_{δ_f}' and is defined as

$$Re_{\delta_f}' = \frac{4\dot{m}_{tot}}{\mu_f P} \tag{13.22}$$

where $\dot{m}_{tot}$ (kg/s) is the total rate of mass flow of the condensate and P is the perimeter given as

$$P = \begin{cases} b & \text{width of a vertical plate (surface)} \\ \pi D & \text{perimeter for a vertical pipe (tube)} \end{cases}$$

For the case of a vertical plate, one has

$$Re_{\delta_f}' = 4 Re_{\delta_f} \tag{13.23}$$

An alternate correlation for turbulent cases has been suggested by Kirkbride [6]. It reads

$$\overline{\alpha} = 0.0077 \mathrm{Re}_{\delta_f}^{\prime 0.4} \left(\frac{\lambda_f^3 \rho_f^2 g}{\mu_f^2} \right)^{1/3} \tag{13.24}$$

13.5   Improvements of the Nusselt film theory for condensation

It has been found that eq. (13.16), for laminar film condensation, underestimates the heat transfer coefficient by about 20%, see Ref. [7]. The numerical constant 0.943 is commonly replaced by the value 1.13. Subcooling of the liquid film, i.e., $t_f < t_s$, can be considered if the latent heat is replaced the equation below, see Ref. [8].

$$h_{fg}' = h_{fg}(1 + 0.68 \mathrm{Ja}) \tag{13.25}$$

where Ja is the so-called Jakob number defined as

$$\mathrm{Ja} = \frac{c_{p_f}(t_s - t_w)}{h_{fg}} \tag{13.26}$$

The modified version of eq. (13.16) then reads

$$\overline{\alpha} = 1.13 \left(\frac{h_{fg}' g \rho_f^2 \lambda_f^3}{\mu_f(t_s - t_w)L} \right)^{1/4} \tag{13.27}$$

Equation (13.27) is also reasonably well applicable for film condensation on the upper side of a plate being inclined an angle ϕ with the horizontal plane if g in eq. (13.27) is replaced by $g \sin \phi$.

In the formulas presented so far, the thermophysical properties of the liquid phase should be evaluated at the temperature $t_{ref} = (t_s + t_w)/2$, while the latent heat should be evaluated at the saturation temperature t_s.

13.6   Film condensation on the outer surface of a horizontal tube and horizontal tube bundles

The Nusselt analysis of film condensation for a vertical surface (plate) can be extended to film condensation on the outer surface of a horizontal tube (pipe), see Fig. 13.2a, and the result for the average heat transfer coefficient is given by eq. (13.28), see Ref. [9]

$$\overline{\alpha} = 0.728 \left(\frac{h_{fg}' g \rho_f^2 \lambda_f^3}{\mu_f(t_s - t_w)D} \right)^{1/4} \tag{13.28}$$

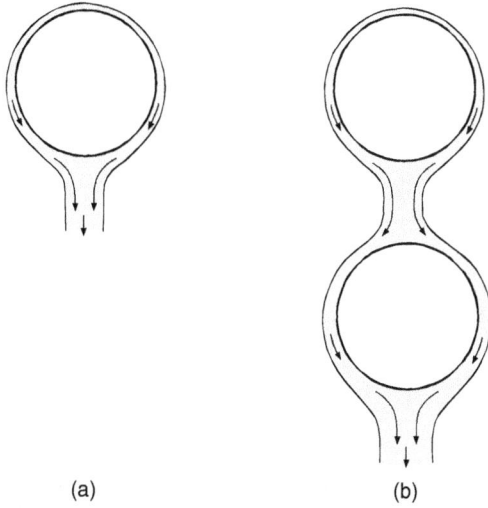

Figure 13.2: Film condensation: (a) single tube, (b) tube bundle in-line arrangement.

For horizontal tubes placed in N rows straight below each other, i.e., in-line arrangement, as shown in Fig. 13.2b, the corresponding heat transfer coefficient is found from

$$\bar{\alpha} = 0.728 \left(\frac{h'_{fg} g \rho_f^2 \lambda_f^3}{N \mu_f (t_s - t_w) D} \right)^{1/4} \tag{13.29}$$

Arrangement of the type shown in Fig. 13.2b occurs frequently in condensers. The reason why the average heat transfer coefficient is decreased by increasing the number of tube rows N is due to the fact that the thickness of the condensate layer is increased from one tube row to another as the condensate falls down. The thermal resistance will thus increase.

13.7 Condensation inside tubes

The relations presented so far assume that the vapor phase is stationary or at rest or that the vapor velocity is negligible. In some applications, e.g., condensers for refrigeration processes and air-conditioning units, the vapor is condensing inside tubes and vapor has a nonnegligible flow velocity. In a vertical pipe or tube where the vapor is flowing upward, the vapor retards the condensate downward flow. This leads to a thicker condensate layer and lower heat transfer coefficient. If the vapor is flowing downward, the condensate layer thickness decreases and a higher heat transfer coefficient is achieved. For this latter case, the following formula is

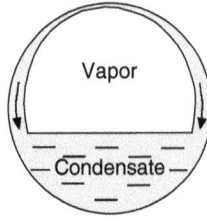

Figure 13.3: Film condensation in a horizontal tube at low vapor velocities.

sometimes applied, see Refs. [10–14].

$$\overline{\mathrm{Nu}}_D = \frac{\overline{\alpha}D}{\lambda_f} = 0.065 \left(\frac{c_{p_f}\rho_f D^2 f}{2\lambda_f \mu_f \rho_g} \frac{G_1^2 + G_1 G_2 + G_2^2}{3} \right)^{1/2} \tag{13.30a}$$

where G_1 is the vapor mass velocity ($\dot{m}_g/A$), at the inlet, and G_2 the corresponding value at the outlet. f is an apparent interface friction factor for single-phase pipe flow with an average mass velocity for the vapor G_g, which is calculated as

$$G_g = \frac{(G_1^2 + G_1 G_2 + G_2^2)^{1/2}}{3} \tag{13.30b}$$

The thermophysical properties for the liquid in eq. (13.30a) should be determined at the reference temperature t_{ref}:

$$t_{\mathrm{ref}} = \frac{1}{4}t_s + \frac{3}{4}t_w \tag{13.31}$$

In Fig. 13.3 is shown how the condensation process occurs in a horizontal tube for small vapor velocities.

The condensate (liquid) flows from the upper part of the tube toward the bottom and then together with the vapor flows in the axial direction of the tube. If the vapor velocity is so low that the condition

$$\mathrm{Re}_{g,i} = \left(\frac{\rho_g u_{m_g} D}{\mu_g} \right)_i < 35 \times 10^3$$

is satisfied (index i means tube inlet), Ref. [13] gives the following expression for the average heat transfer coefficient:

$$\overline{\alpha}_D = 0.555 \left(\frac{h'_{fg} g \rho_f(\rho_f - \rho_g)\lambda_f^3}{\mu_f(t_s - t_w)D} \right)^{1/4} \tag{13.32}$$

where h'_{fg} is calculated according to eq. (13.25).

At higher vapor velocities, the two-phase flow will be annular as depicted in Fig. 13.4.

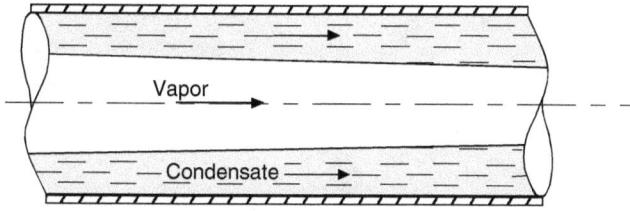

Figure 13.4: Film condensation in a horizontal tube at high vapor velocities.

The vapor occupies the central part, but the diameter of the vapor flow is decreasing in the flow direction because the condensate layer will grow in thickness. In Ref. [14], the following expression for the Nusselt number has been suggested

$$\overline{\mathrm{Nu}} = \frac{\overline{\alpha}D}{\lambda_f} = 0.026 \mathrm{Pr}_f^{1/3} \left(\frac{D}{\mu_f}(G_g \left(\frac{\rho_f}{\rho_g} \right)^{1/2} + G_f) \right)^{0.8} \tag{13.33}$$

where the mass velocities are given by $G_g = \dot{m}_g/A$ and $G_f = \dot{m}_f/A$.

The accuracy in eq. (13.33) is however limited. Other flow regimes than those shown in Figs. 13.3 and 13.4 are also possible, see Refs. [9, 15].

13.8   Influence of noncondensable gases

In the given formulas for the heat transfer coefficients, it is assumed that the vapor is pure, i.e., it does not contain any noncondensable gases. If such gases, e.g., air, are present in the vapor (even at small amounts), the heat transfer coefficient is reduced considerably. The reason is that as the vapor condenses, the noncondensable gases are present at the surface and the vapor has to diffuse across the vapor–gas mixture. The noncondensable gases act as a thermal resistance which reduces the heat transfer coefficient. The resistance to the diffusion process of the vapor creates a reduction in the vapor partial pressure which in turn lowers the condensation temperature (saturation temperature). In applications, it is already at the design stage of condensers attempted for to make sure that noncondensable gases are vented away.

13.9   Dropwise condensation

If the cold surface does not wet the condensate (liquid), dropwise condensation takes place. The heat transfer coefficient will then be higher than for film condensation. A principle sketch for dropwise condensation is shown in Fig. 13.5.

Small drops are created at some positions on the surface, i.e., nucleation sites. The drops are growing, hit each other and merge to bigger drops. The big drops fall down along the surface and leave the corresponding surface dry. New small drops

Figure 13.5: Comparison of dropwise condensation (left) and film condensation (right).

are created on the dry part and the process is repeated. Nonwetting surfaces can be established in different ways, see below.

1. By adding proper chemicals to the vapor
2. Treat the surface with proper chemicals, e.g., oils
3. Cover the surface with a Teflon layer
4. Cover the surface with a thin layer of gold or any other noble metal.

If drops can be created, the heat transfer rate might be 5–10 times higher than for film condensation. The difficulty in practical applications is that the surface is ageing and loses its nonwetting property and then film condensation will occur.

Additional information about dropwise condensation can be found in Refs. [16–18].

References

[1] W. Nusselt, Die Oberflachenkondensation des Wasserdampfes, VDI Z., 60, 541–569 (1916).

[2] M. Jakob, Heat transfer in evaporation and condensation—II, Mech. Engn., 58, 729–739 (1936).

[3] S.J. Friedman and C.O. Miller, Liquid films in the viscous flow region, Ind. Eng. Chem., 33, 885–891 (1941).

[4] S.S. Kutateladze, Fundamentals of Heat Transfer, Academic Press, New York (1963).

[5] D.A. Labuntsov, Heat transfer in film condensation of pure steam on vertical surfaces and horizontal tubes, Teploenergetika, 4, 72–80 (1957).

[6] C.C. Kirkbride, Heat transfer by condensing vapors on vertical tubes, Trans. AIChE, 30, 170–186 (1934).

[7] W.H. McAdams, Heat Transmission, 3rd ed., McGraw-Hill, New York (1954).

[8] W.M. Rohsenow, Heat transfer and temperature distribution in laminar film condensation, Trans. ASME, 78, 1645–1648 (1956).

[9] W.M. Rohsenow, Film condensation, in Handbook of Heat Transfer (Eds. W.M. Rohsenow and J.P. Hartnett), McGraw-Hill, New York (1973).

[10] E.F. Carpenter and A.P. Colburn, The effect of vapour velocity on condensation inside tubes, Proc. Gen. Dis. Heat Transfer, IME/ASME, 20–26 (1951).

[11] H.Y. Wong, Handbook of essential formulae on heat transfer for engineers, Longman, London (1977).

[12] J.G. Collier and J.R. Thome, Convective Boiling and Condensation, 3rd ed., Oxford University Press, Oxford (1994).

[13] J.C. Chato, Laminar condensation inside horizontal and inclined tubes, J. Am. Soc. Heating Refrig. Engn., 4, 52–60 (1962).

[14] W.W. Akers, H.A. Deans and O.K. Crosser, Condensing heat transfer within horizontal tubes, Chem. Engn. Prog. Symp. Ser., 55(29), 171–176 (1958).

[15] J.G. Collier, Convective boiling and condensation, 2nd ed., McGraw-Hill, New York (1981).

[16] P. Griffith, Drop condensation, in Handbook of Heat Transfer (Eds. W.M. Rohsenow and J.P. Hartnett), McGraw-Hill, New York (1973).

[17] I.T. Tanasawa, Dropwise condensation: The way to practical application, Proc. 5th IHTC, 6, 393–405, 1978.

[18] D.W. Woodruff and J.W. Westwater, Steam condensation on various gold surfaces, ASME J. Heat Transfer, 103, 685–692 (1981).

Further reading

M.A. Kedzierski, J.C. Chato and T.J. Rabas, Condensation, in Heat Transfer Handbook (Eds. A. Bejan and A.D. Kraus), Wiley, New York (2003).

14   Boiling and evaporation

14.1   Introduction

Boiling and evaporation of a liquid may occur in such a manner that vapor is created at the interface beween the liquid and the vapor. This occurs if heat (greater than or equal to the latent heat) is supplied to the liquid surface by, e.g., radiation. In engineering applications, the heat is usually supplied to the liquid through a surface enclosing the liquid. The vapor is produced in the form of bubbles that are created at the surface and grow at the hot solid surface. The bubbles leave the surface as they have reached a certain size and then rise in the liquid. Due to the fact that the process of boiling and evaporation is complicated, most of the existing knowledge is based on experiments.

14.2   General

Consider the container in Fig. 14.1, where heat is supplied through the bottom surface.

Assume that the liquid initially has a temperature below the saturation temperature. After some time vapor bubbles appear at the hot bottom surface. These bubbles separate from the surface as they have reached a certain size and are then moving upward in the liquid. After a certain distance they lose their identity and are condensed in the cold liquid which has not yet reached the saturation temperature.

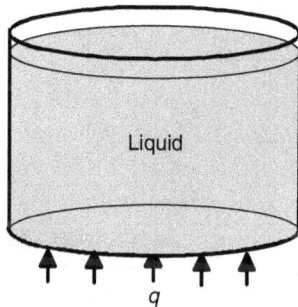

Figure 14.1: Container with a liquid.

Not until the liquid has been heated sufficiently, the vapor bubbles are able to reach the free liquid surface and a flow of vapor leaves the free surface. The process in which the vapor condenses in the liquid relatively close to the hot surface is called local boiling or subcooled boiling. The process in which the created vapor is rising through the liquid up to the free surface is boiling with net evaporation.

If the liquid has a macroscopic motion, the boiling process will be affected. In Fig. 14.1, there is no forced motion of the liquid. The process is then called pool boiling. If the liquid instead is flowing in a tube and heat is supplied through the tube wall the process is called forced convective boiling. Many investigations have concerned pool boiling but also many studies of forced convective boiling have been performed.

14.3 Nukiyama's experiment and the so-called boiling curve

In 1934 Nukiyama [1] performed the experiment presented in Fig. 14.2.

Saturated water at atmospheric pressure was boiled by using an electrically heated wire which acted as both a heater and a resistance thermometer. By calibrating the resistance for a Nicrome wire as function of the temperature before the experiment it was possible to register the heat flux and the temperature from the measured current (ampere) and the voltage (volts). As is evident from Fig. 14.2, Nukiyama found that as the heat flux was increased, the wire temperature increased moderately. Suddenly at a certain value of the heat flux, the Nicrome wire melted abruptly (the wire temperature increased severely). After this event, Nukiyama repeated the experiment but now with a platinum wire. The procedure was repeated but at the heat flux the Nicrome wire melted, the platinum wire became white but did not melt. As the heat flux was reduced (the electrical power was reduced) the wire temperature was reduced in a continuous way, as shown in Fig. 14.2, until the heat flux became below a certain value. Then all of a sudden, the wire temperature was reduced considerably and the original curve, $q = f(t)$, was followed, see the arrows in Fig. 14.2.

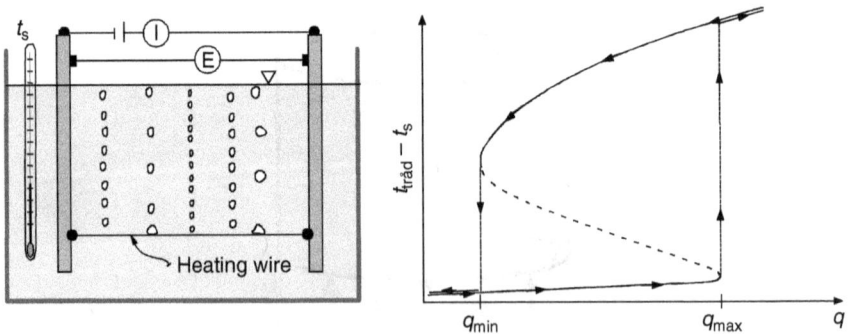

Figure 14.2: Nukiyama's experiment.

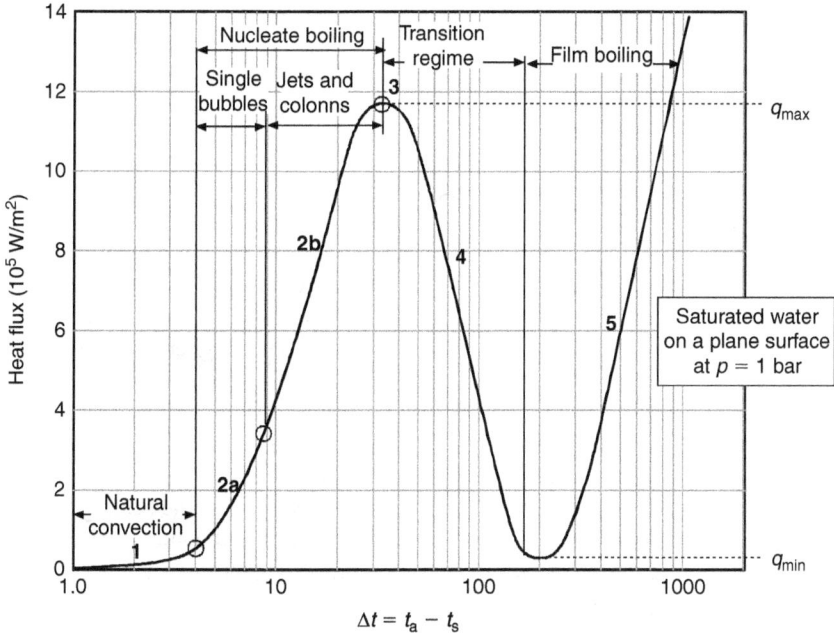

Figure 14.3:   Boiling curve for saturated water at atmospheric pressure.

Nukiyama suspected that the hysterisis which is recognized as the heat flux, q, is the controlling parameter (as in Fig. 14.2) will not occur if the temperature difference $\Delta t = t_{wire} - t_s$ (t_s = saturation temperature) is the controlling parameter.

In 1937, Drew and Mueller [2] succeeded to perform an experiment where Δt is the independent parameter. Organic liquids boiled on the outer surface of a tube while water vapor at elevated pressure was condensing inside the tube. The experiment verified Nukiyama's theory.

Figure 14.3 shows a complete boiling curve for saturated water at atmospheric pressure.

14.4   Description of the boiling curve

Figure 14.3 shows the boiling curve for water of atmospheric pressure (heat flux q (W/m^2) as function of the temperature difference ($t_w - t_s$)). Commonly the boiling curve is divided into regimes where the physical mechanisms are different.

1. *Natural convection*
   In this regime the liquid is not boiling despite the surface temperature is a few degrees above the saturation temperature. Instead heating by natural convection occurs (movement due to density variation).
2. *Nucleate boiling*
   This regime is divided into two subregimes.

a. Single bubbles, i.e., vapor bubbles are created here and there but as long as t_w is only moderately higher than t_s the bubbles appear individually.
b. Jets and colonns. Vapor bubbles are created in a large number at several places on the heating surface. The bubbles coalesce and create streaks of vapor.

3. *Maximum heat flux (burnout, critical heat flux, boiling crisis)*
From a practical point of view it is desirable to be close to q_{max} because the heat transfer coefficient is high. However, for cases where q (the heat flux) is the controlling parameter it is dangerous to be close to q_{max} due to the unstable situation (see Nukiyama's experiment). If $q > q_{max}$, t_w will increase considerably which may lead to material deterioration if t_w becomes higher than the melting temperature of the material.

4. *Transition regime*
If $\Delta t = (t_w - t_s)$ is the controlling parameter, a continuous process is achieved and q is decreasing with increasing t_w after q_{max} has been passed. The reason why q decreases is that a vapor film more or less will cover the heating surface. As the surface is completely covered by the vapor film, the minimum heat flux q_{min} is reached. The vapor film has a low thermal conductivity and accordingly the heat flux becomes low.

5. *Film boiling*
As a stable vapor film has been established and q increases with increasing Δt again. The mechanism for the heat transfer at film boiling and the regular removal of vapor bubbles have similarities with the film condensation process outlined in Chapter 13. However, the heat transfer coefficient is lower at film boiling compared to film condensation because the heat has to be transferred across a vapor layer with low thermal conductivity in contrast to film condensation where the heat transfer is across a liquid layer with higher thermal conductivity.

Figure 14.4 shows sketches of the nucleate and film boiling processes. In Fig. 14.4a the regime with single small bubbles is depicted while in Fig. 14.4b

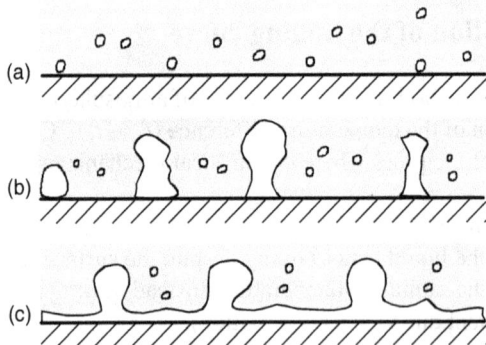

Figure 14.4: Principal sketch of nucleate boiling ((a) and (b)) and film boiling (c).

the regime with vapor jets or vapor colonns is shown. Here, the bubbles hit each other, merge, and leave the hot surface in the form of jets or slugs. In Fig. 14.4c, film boiling occurs and the hot surface is covered by a vapor layer. The liquid is thus not in direct contact with the surface. The vapor film surface is unstable and bubbles leave it and flow up in the liquid.

14.5   Temperature distribution in the liquid phase for saturated pool boiling

Consider again the container in Fig. 14.1. Closest to the hot bottom the temperature of the liquid is changing significantly while the major part of the liquid has a temperature only slightly above the saturation temperature, see Fig. 14.5.

At net evaporation the bubbles, created at the interface between the liquid and the hot bottom, are rising and leave at the liquid–vapor interface.

Superheating of the liquid phase is necessary for the boiling process. This will briefly be discussed.

First, consider a vapor bubble in thermal equilibrium with a liquid of uniform temperature. For this condition, a bubble with a certain radius can survive, i.e., not disappear due to condensation and not grow by further evaporation of the liquid.

Now, consider the mechanical equilibrium for the bubble according to Fig. 14.6.

Inside the bubble the pressure is p_{bubble} and in the liquid it is p_{liquid}. The difference between these pressures is balanced by the surface tension force. One finds

$$\pi r^2 (p_{bubble} - p_{liquid}) = 2\pi r \sigma \qquad (14.1)$$

Note that the units of the surface tension σ is N/m.

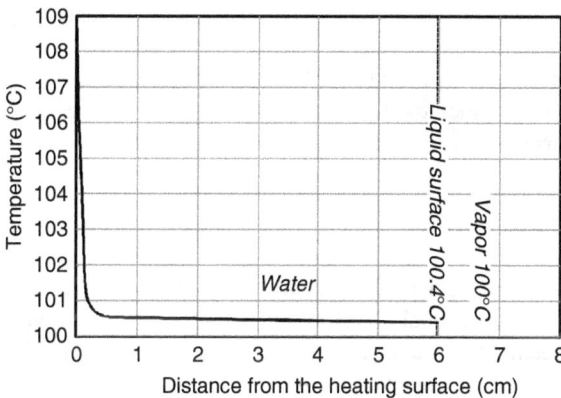

Figure 14.5: Temperature distribution in the liquid phase for pool boiling.

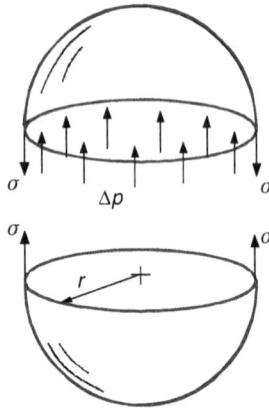

Figure 14.6: Mechanical equilibrium for a vapor bubble.

From eq. (14.1), the bubble radius is found as

$$r = \frac{2\sigma}{(p_{\text{bubble}} - p_{\text{liquid}})} \tag{14.2}$$

Because the thermal equilibrium requires that the liquid and vapor temperatures are equal and that the vapor temperature inside the bubble must be equal to the saturation temperature at the pressure p_{bubble}, it is found that the liquid must be superheated, i.e., $(t_{\text{liquid}} > t_s$ at $p_{\text{liquid}})$.

It is easily recognized that the equilibrium bubble with the radius from eq. (14.2) is unstable. If the radius is less than the value of eq. (14.2), the surface tension force is greater than the force due to the pressure difference. The vapor in the bubble condenses at the higher pressure and the bubble collapses. If the radius of the vapor bubble is greater than the value of eq. (14.2), the liquid close to the bubble surface will evaporate and the bubble will grow.

From eq. (14.2) (and steam tables) one realizes that greater superheating will give a smaller equilibrium radius.

The precise manner in which bubbles initially are created on the hot surface is far from being fully understood. Surface roughness, material properties, wetting or non-wetting surfaces are very important, see Refs. [3, 4].

14.6   Nucleate boiling

Experimental data for nucleate boiling are shown in Fig. 14.7.

Often in the international literature, an equation (14.9) derived by Rohsenow [5] is used for estimating the heat flux for nucleate boiling. The fundamental idea is based on single-phase correlations of the structure

$$\text{Nu} = \text{function}(\text{Re}, \text{Pr})$$

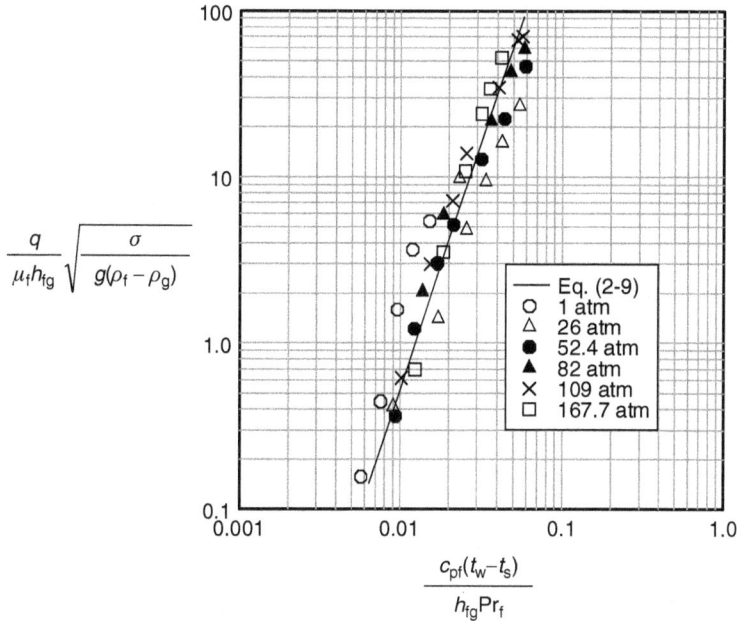

Figure 14.7: Nucleate boiling data for water.

What is unique here is the determination of the characteristic velocity and length scale. The characteristic length scale is determined from

$$L_k = \left(\frac{\sigma}{g(\rho_f - \rho_g)}\right)^{1/2} \tag{14.3}$$

which is related to the wavelength for the most unstable wave at the interface between the vapor and the liquid, see Ref. [6].

The characteristic velocity is determined from

$$u_f = \frac{q_w}{\rho_f h_{fg}} \tag{14.4}$$

This is the liquid velocity toward the hot surface as the surface is supplied with liquid as a substitute for the produced vapor.

By applying the definition of the Reynolds and Nusselt numbers, i.e.,

$$\text{Re} = \frac{\rho_f u_f L_k}{\mu_f} \tag{14.5}$$

$$\text{Nu} = \frac{\alpha L_k}{\lambda_f} \tag{14.6}$$

Table 14.1: Constants s and C_{sl}.

Surface – liquid	C_{sl}	s
Nickel – water	0.006	1.0
Platinum – water	0.013	1.0
Copper – water	0.013	1.0
Brass – water	0.006	1.0
Chrome – benzene	0.010	1.7
Chrome – ethanol	0.0027	1.7

and the assumption

$$\mathrm{Nu} = \frac{1}{C_{sf}} \mathrm{Re}^{1-n} \mathrm{Pr}_f^m \tag{14.7}$$

the following result is obtained

$$\frac{c_{p_f}(t_w - t_s)}{h_{fg}} = C_{sf} \left[\frac{q_w}{\mu_f h_{fg}} \left(\frac{\sigma}{g(\rho_f - \rho_g)} \right)^{1/2} \right]^n \cdot \left(\frac{\mu_f c_{p,f}}{\lambda_f} \right)^{1+m} \tag{14.8}$$

With $n = 1/3$ and $1 + m = s$ the so-called *Rohsenow equation* (14.9) is obtained.

$$\frac{c_{p_f}(t_w - t_s)}{h_{fg} \mathrm{Pr}_f^s} = C_{sl} \left[\frac{q_w}{\mu_f h_{fg}} \left(\frac{\sigma}{g(\rho_f - \rho_g)} \right)^{1/2} \right]^{0.33} \tag{14.9}$$

C_{sl} is a constant which depends of the combination of liquid and surface material. In Table 14.1, some values of C_{sl} and the exponents are given for a few common combinations of liquid and surface material.

As is found in Fig. 14.7, the precision in eq. (14.9) is not always high.

The surface tension σ depends on temperature. Commonly the temperature dependence can be regarded as linear, i.e.,

$$\sigma = C_1 - C_2 t \tag{14.10}$$

It is important to note that in eq. (14.10), t must be in °C.

In Table 14.2 the constants C_1 and C_2 are given for some common substances and the valid temperature range. More information can be found in Ref. [7].

Equation (14.9) and similar correlations have some disadvantages. They require accurate values of the thermal physical properties and are relatively complex and cumbersome to use. In addition, the uncertainty in C_{sl}, which describes the surface–liquid interaction, is high. As alternatives to eq. (14.9) simpler expressions have appeared in the literature. These involve the so-called reduced pressure (actual

Table 14.2: Constants C_1 and C_2 in eq. (14.10).

Substance	Temperature interval (°C)	C_1 (N/m)	C_2 (N/m °C)
Water	10–100	75.83×10^{-3}	0.1477×10^{-3}
Pentane	10–30	18.25×10^{-3}	0.1102×10^{-3}
Naftalene	100–200	42.84×10^{-3}	0.1107×10^{-3}
Ethanol	10–100	24.05×10^{-3}	0.0832×10^{-3}
CCl$_4$	15–105	24.49×10^{-3}	0.1224×10^{-3}

pressure divided by the critical pressure) and the molecular weight of the substance as well as the surface roughness. Correlations of this type are reasonable as density, surface tension, and latent heat can be expressed as functions of the reduced pressure. Cooper [8, 9] has suggested the following relation:

$$\alpha = A \cdot P_R^{(0.12-0.4343 \ln R_P)} \cdot (-0.4343 \ln P_R)^{-0.55} \cdot M^{-0.5} \cdot q_w^{0.67} \qquad (14.11)$$

where M is the molecular weight and $P_R = p/p_{critical}$ is the reduced pressure. The recommended value of the constant A is 55. R_P in eq. (14.11) is the surface roughness in μm. For a surface without a specified surface roughness, R_P is set as $R_P = 1 \, \mu$m. For horizontal copper cylinders, Cooper recommended that the α-value calculated by eq. (14.11) should be multiplied by 1.7. Equation (14.11) is recommended for water, refrigerants, and organic liquids for which the thermal physical properties are less well known.

Gorenflo [10, 11] suggested an alternative method for calculation of the heat transfer coefficient in the nucleate boiling regime. It is based on a reference heat transfer coefficient α_0 valid a certain reference state, namely $P_{R0} = 0.1$, $R_{P0} = 0.4 \, \mu$m, and $q_{w0} = 2 \times 10^4$ W/m^2. Reference values for α_0 are given in Table 14.3.

For calculation of the heat transfer coeffcient at other states, the following expression is used.

$$\alpha = \alpha_0 \cdot F_{PF} \cdot \left(\frac{q_w}{q_{w0}}\right)^b \cdot \left(\frac{R_p}{R_{p0}}\right)^{0.133} \qquad (14.12)$$

F_{PF} is a pressure correction factor given in eq. (14.13)

$$F_{PF} = 1.2P_R^{0.27} + 2.5P_R + \frac{P_R}{1 - P_R} \qquad (14.13)$$

The exponent b is determined according to

$$b = 0.9 - 0.3P_R^{0.3} \qquad (14.14)$$

Table 14.3: Reference values α_0, Gorenflo [10, 11].

Liquid	P_{crit} (bar)	α_0 (W/(m² K))	Liquid	P_{crit} (bar)	α_0 (W/(m² K))
Methane	46.0	7000	R114	32.6	3800
Ethane	48.8	4500	R115	31.3	4200
Propane	42.4	4000	R123	36.7	2600
Butane	38.0	3600	R134a	40.6	4500
n-Pentane	33.7	3400	R152a	45.2	4000
i-Pentane	33.3	2500	R226	30.6	3700
Hexane	29.7	3300	R227	29.3	3800
Heptane	27.3	3200	RC318	28.0	4200
Benzene	48.9	2750	R502	40.8	3300
Toluene	41.1	2650	Chlormethane	66.8	4400
Diphenyl	38.5	2100	Tetraflouromethane	37.4	4750
Ethanol	63.8	4400	Water	220.64	5600
n-Propanol	51.7	3800	Ammonia	113.0	7000
i-Propanol	47.6	3000	Carbon dioxide	73.8	5100
n-Butanol	49.6	2600	Sulfur hexaflouride	37.6	3700
i-Butanol	43.0	4500	Oxygen (on Cu)	50.5	9500
Acetone	47.0	3950	Oxygen (on Pt)	50.5	7200
R11	44.0	2800	Nitrogen (on Cu)	34.0	10000
R12	41.6	4000	Nitrogen (on Pt)	34.0	7000
R13	38.6	3900	Argon (on Cu)	49.0	8200
R13B1	39.8	3500	Argon (on Pt)	49.0	6700
R22	49.9	3900	Neon (on Cu)	26.5	20000
R23	48.7	4400	Hydrogen (on Cu)	13.0	24000
R113	34.1	2650	Helium	2.28	2000

The surface roughness R_p must be in μm and is set to $0.4\,\mu$m for an unknown surface roughness. Equations (14.13) and (14.14) are recommended for all liquids except water and helium. For water, eqs. (14.15) and (14.16) should be used.

$$F_{PF} = 1.73 P_R^{0.27} + \left(6.1 + \frac{0.68}{1 - P_R} \right) P_R^2 \qquad (14.15)$$

$$b = 0.9 - 0.3 P_R^{0.15} \qquad (14.16)$$

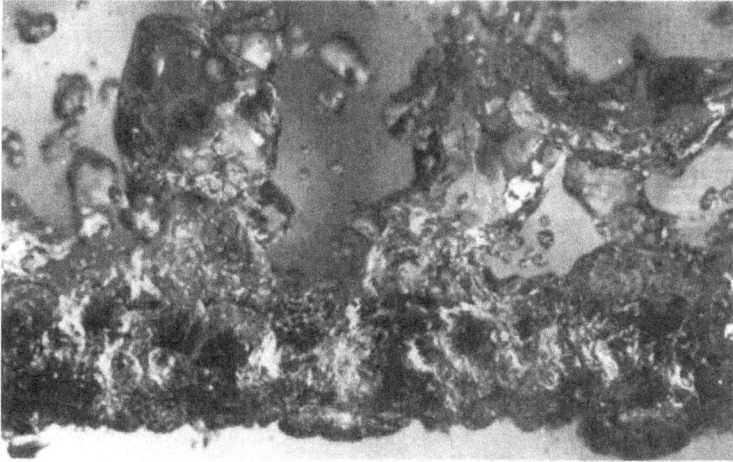

Figure 14.8: Boiling of methanol in the transition regime.

14.6.1   Maximum heat flux (critical heat flux) q_{max}

14.6.1.1   Transition regime for the boiling curve and so-called Taylor instability

In Fig. 14.3, the so-called boiling curve was shown for saturated water at atmospheric pressure. As boiling occurs in the transition regime, a huge amount of vapor is present on the heating surface. The vapor will rise but has no clear way to follow. The jets, which transport the vapor in the regime of slugs and colonns (lower Δt) are unstable and can not have a corresponding role in the transition regime. Thus, the vapor is rising in big clumps and liquid is forced to the surface, hit the surface, start to evaporate, and new vapor clumps are formed. The process is partly shown in Fig. 14.8 for methanol.

During the boiling process in the transition regime, the heating surface is almost covered by vapor and for film boiling the surface is completely covered by a vapor film. In both cases, the unstable situation of having a heavier liquid above a lighter vapor prevails.

In Fig. 14.9, two common examples where a heavy liquid is falling down into a lighter one are shown. In Fig. 14.9a, it is shown what happens as a honey can is turned upside down while Fig. 14.9b shows how water vapor in air is condensing on a cold tube surface.

The heavier fluid falls down at a node of the wave while the lighter fluid rises at the other node. The process is called *Taylor instability* after G.I. Taylor who proposed such a process. The so-called Taylor wavelength λ_T is the length of the fastest growing wave and which is dominating the process. For a plane surface λ_T can be determined with dimensional analysis. For λ_T the following relation should in principle be valid

$$\lambda_T = \text{function}(\sigma, g(\rho_f - \rho_g)) \tag{14.17}$$

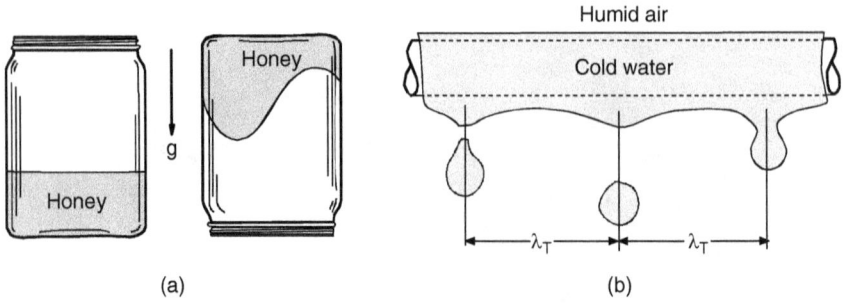

Figure 14.9: Taylor instability for two common cases.

because the wave is created as an effect of the force balance between surface tension, inertia, and gravity forces. Based on the fundamentals of dimensional analysis one may write

$$\lambda_T = \text{const} \cdot \sigma^a \cdot g^b (\rho_f - \rho_g)^c \qquad (14.18)$$

With the valid units one has

$$(\text{m}) = (\text{N} \cdot \text{m}^{-1})^a \cdot (\text{m} \cdot \text{s}^{-2})^b \cdot (\text{kg} \cdot \text{m}^{-3})^c$$

However, $1\,\text{N} = 1\,\text{kg} \cdot \text{m} \cdot \text{s}^{-2}$ and then one finds

$$a = \frac{1}{2}, \, b = c = \frac{-1}{2} \qquad (14.19)$$

$$\lambda_T = \text{constant} \sqrt{\frac{\sigma}{g(\rho_f - \rho_g)}} \qquad (14.20)$$

A more complete analysis shows, see Ref. [12], that the constant in eq. (14.20) becomes

$$\text{constant} = \begin{cases} 2\pi\sqrt{3} \text{ for one-dimensional waves} \\ 2\pi\sqrt{6} \text{ for two-dimensional waves} \end{cases} \qquad (14.21)$$

In the so-called transition regime, the vapor is rising up in the liquid at the nodes for the Taylor waves and at q_{max} (see Fig. 14.3) the vapor is rising in form of jets. These jets are most commonly arranged in a quadratic arrangement as shown in Fig. 14.10.

The wavelength for the most fundamental mode is λ_{T_1}, see Fig. 14.10.

14.6.2   Helmholtz instability for the vapor jets

In Fig. 14.11, an example of Helmholtz instability is shown. A flag is exposed to a wind and the flag becomes in a state of continuous collapse because the pressure is high where the velocity is low and vice versa.

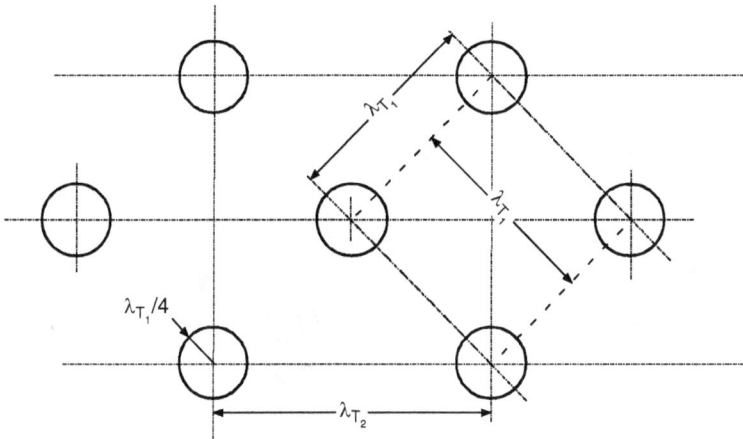

Figure 14.10: Arrangement of vapor jets at q_{max}.

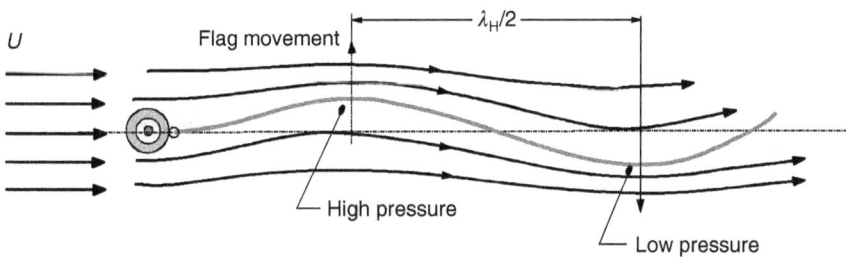

Figure 14.11: A flag in a wind exemplifying Helmholtz instability.

A similar instability phenomenon for a vapor jet is shown in Fig. 14.12. However, in this case there is also the surface tension between the vapor and the liquid. The surface tension is counteracting the pressure forces which cause the collapse. However, the vapor velocity has to reach a certain value before the vapor jet becomes instable.

Lamb [13] has given the relation between the vapor velocity u_g and the wavelength λ_H as

$$u_g = \sqrt{\frac{2\pi\sigma}{\rho_g \lambda_H}} \tag{14.22}$$

A real liquid–vapor interface is usually not as regular as shown in Fig. 14.12 but consists of several wavelengths superimposed on each other. If one of these is more pronounced than others can not be easily stated.

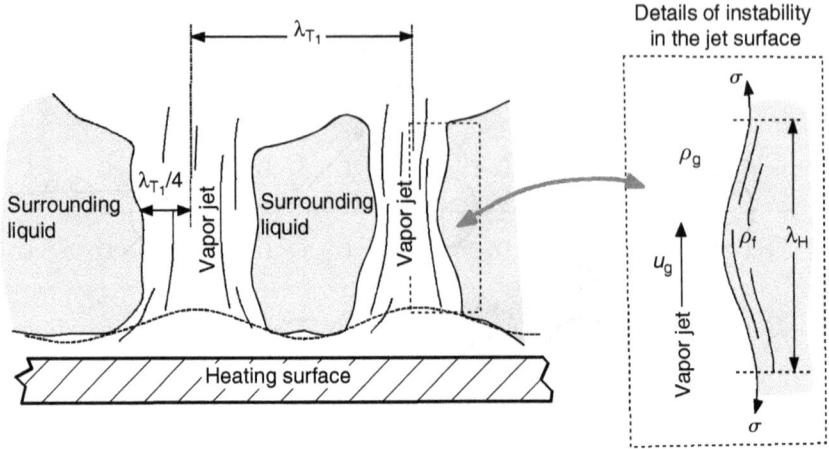

Figure 14.12: Helmholtz instability for vapor jets.

14.6.3   Estimations of q_{max}

14.6.3.1   General expression
The heat flux must be equal to the latent heat as the liquid is saturated from the beginning. One then has

$$q_{max} = \rho_g h_{fg} u_g \frac{A_j}{A_h} \tag{14.23}$$

where A_j is the cross-sectional area of a vapor jet and A_h that part of the heating surface which is surrounding the vapor jet and secure the feed of liquid. For every case, the wavelength λ_H (which is triggering the Helmholtz instability) and the area ratio A_j/A_h must be determined.

14.6.3.2 q_{max} for a horizontal plane surface
The original and first analysis of q_{max} was performed by Zuber [14]. He assumed that the vapor jet radius was $\lambda_{T_1}/4$ (see Fig. 14.10). With Fig. 14.10, the area ratio A_j/A_h can be determined as

$$\frac{A_j}{A_h} = \frac{\pi(\lambda_{T_1}/4)^2}{\lambda_{T_1}^2} = \frac{\pi}{16} \tag{14.24}$$

Lienhard and Dhir [15] assumed that $\lambda_H = \lambda_{T_1}$ and one then finds

$$q_{max} = 0.149\rho_g^{1/2}h_{fg}\left(g(\rho_f - \rho_g)\sigma\right)^{1/4} \tag{14.25}$$

The determination of q_{max} is commonly referred to Kutateladze as given in Ref. [12] and Zuber [14]. Kutateladze determined q_{max} by studying instability phenomena or flooding in desiltation colonns. In the analyses by Kutateladze and

Table 14.4: Y_1 as a function of Y_2 for some different cases, from Ref. [12].

Heating surface	Characteristic length	Valid range	$Y_1 = f(Y_2)$
Large sphere	Radius R	$Y_2 \geq 4.26$	$Y_1 = 0.84$
Small sphere	Radius R	$0.15 \leq Y_2 \leq 4.26$	$Y_1 = 1.734/\sqrt{Y_2}$
Horizontal cylinder	Radius R	$Y_2 \geq 0.15$	$Y_1 = 0.89 + 2.27e^{-3.44\sqrt{Y_2}}$

Zuber, respectively, the constant was found to be 0.131 and the corresponding relation for q_{max} is usually given a subscript z after Zuber. One then has

$$q_{max_z} = 0.131\rho_g^{1/2}h_{fg}\left(g(\rho_f - \rho_g)\sigma\right)^{1/4} \qquad (14.26)$$

14.6.4 q_{max} for pool boiling at different geometries

For other geometries than a plane horizontal heating surface, a characteristic length, L, is included. q_{max} then has a general form given by

$$q_{max} = \text{function}(\rho_g, \rho_f - \rho_g, g, \sigma, L, h_{fg})$$

Two dimensionless variables are commonly introduced. The first one is

$$Y_1 = \frac{q_{max}}{q_{max_z}} \qquad (14.27)$$

where q_{max_z} is taken from eq. (14.26). The second variable is

$$Y_2 = \frac{L}{\sqrt{\sigma/(g(\rho_f - \rho_g))}} \qquad (14.28)$$

The dimensionless variable Y_2 is equal to the square root of a dimensionless number called the Bond number, Bo. The Bond number can physically be interpreted as the ratio between the body force due to density differences and the capillary force. In Table 14.4, relations $Y_1 = f(Y_2)$ are given for a few cases.

14.7 Film boiling

For film boiling, i.e., a vapor film is present around the heating surface, some formulas are available for the heat transfer coefficient (Nusselt number) for cylindrical and spherical surfaces. Such formulas have been established in analogy with film condensation. So, for instance, one has

$$
\mathrm{Nu}_D = \frac{\alpha D}{\lambda_g} = c_1 \cdot \left(\frac{\rho_g(\rho_f - \rho_g)gh'_{fg}D^3}{\mu_g\lambda_g(t_w - t_s)} \right)^{1/4}
\tag{14.29}
$$

where $c_1 = 0.62$ for cylinders and 0.67 for spheres. For h'_{fg} one has (compare with eq. (13.25))

$$
h'_{fg} = h_{fg}(1 + 0.4\mathrm{Ja})
\tag{14.30}
$$

In eq. (14.30), Ja is the Jakob number defined according to

$$
\mathrm{Ja} = c_{p_g}\frac{(t_w - t_s)}{h_{fg}}
\tag{14.31}
$$

In some cases, the temperature t_w of the heating surface will be so high that also thermal radiation has to be considered.

14.8   Minimum heat flux, q_{min}

In Fig. 14.3, it is shown that the heat flux has a minimum when the surface is completely covered by a vapor film. The expression for q_{min} is also available in the literature but is not given here. It can be found in Ref. [12].

14.9   Influence of various parameters on the boiling curve

14.9.1   Subcooling

A liquid enclosed in a heated container will not stay at a temperature below the saturation temperature very long. Before the liquid reaches the saturation temperature or if the warm liquid is continuously replaced by cold liquid (e.g., by forced flow), the subcooling will affect the boiling curve. It has been found that the nucleate boiling regime is not very much affected but the values of q_{max} and q_{min} increase linearly with the subcooling, $\Delta t_{sub} = t_s - t_{bulk}$. The influence on the transition regime is less known.

14.9.2   Gravity

The influence of the gravity or other body forces is of interest as the boiling process also appears in rotating or accelerated systems. Reduction of the gravity is important for boiling processes in space applications. Because the gravity acceleration g is included in most expressions, its role is evident.

14.9.3 Surface roughness

A heating surface might be treated in various ways to find out the importance of the surface roughness. Some studies, cited in Ref. [12], indicate that q_{max} is almost independent of the surface roughness and whether the surface is clean or oxidized. The film boiling regime is not affected significantly by surface properties which is understandable as the liquid phase is not in direct contact with the solid surface. The nucleate boiling regime is however affected by the surface roughness.

14.10 Forced convective boiling for immersed bodies

For the maximum heat flux q_{max}, it is by dimensional analysis possible to show that the influence of the flow field has the form

$$\frac{q_{max}}{\rho_g h_{fg} U_\infty} = \text{function}\left(\frac{\text{We}_L, \rho_f}{\rho_g}\right) \tag{14.32}$$

where We is the so-called Weber number defined according to

$$\text{We}_L = \frac{\rho_g U_\infty^2 L}{\sigma} \tag{14.33}$$

The Weber number may be interpreted as the ratio between inertia forces and surface tension forces. For a circular cylinder in cross flow it has been found, see Ref. [16],

Low velocities

$$\frac{q_{max}}{\rho_g h_{fg} U_\infty} = \frac{1}{\pi}\left[1 + \left(\frac{4}{\text{We}_D}\right)^{1/3}\right] \tag{14.34}$$

High velocities

$$\frac{q_{max}}{\rho_g h_{fg} U_\infty} = \frac{(\rho_f/\rho_g)^{3/4}}{169\pi} + \frac{(\rho_f/\rho_g)^{1/2}}{19.2\pi \, \text{We}_D^{1/3}} \tag{14.35}$$

If the low- or high-velocity formula is valid, it is determined by the conditions

$$\frac{q_{max}}{\rho_g h_{fg} U_\infty} \begin{cases} < \dfrac{0.275}{\pi}\left(\dfrac{\rho_f}{\rho_g}\right)^{1/2} + 1 & \text{high velocity} \\[3mm] > \dfrac{0.275}{\pi}\left(\dfrac{\rho_f}{\rho_g}\right)^{1/2} + 1 & \text{low velocity} \end{cases} \tag{14.36}$$

14.11   Forced convective boiling in tubes

14.11.1   Two-phase flow (liquid–gas)

In many engineering applications two or more substances are flowing as a mixture. The substances might be at different states, i.e., the mixture can include a combination of a solid phase, a liquid phase, and/or a gaseous phase. In such cases one speaks about multiphase flow. Most common is however that only two phases are present and one then has a two-phase flow. In this section only two-phase flows with gas and liquid are considered.

Two-phase flow of a liquid and a gas (or a vapor) occurs in pipelines where oil and gas are flowing and in evaporators or condensers where a liquid and its vapor are flowing. The oil–gas flow is a two-component two-phase flow because the oil and gas have different molecular structure as well as phase. The flow of a liquid and its vapor in an evaporator tube or in a condenser is a one-component two-phase flow. The characteristic features of these two types of two-phase flow might be very different.

The ratio between the gas flow rate (or the vapor flow rate) and the liquid flow rate is an essential parameter for description of the two-phase flow. The flow regime is affected significantly by this parameter.

The flow regimes which may appear in horizontal two-phase flow of a gas–liquid mixture are shown in Fig. 14.13. The pictures are ordered by increasing ratio between the gas and the liquid flow rates. At small gas (vapor) flow rates, the gas phase appears mainly as small bubbles that are located at the upper part of the tube. As the gas flow rate is increasing, the bubbles are coalescing and gas plugs are created. These also occupy the upper part of the tube. For further increase in the gas flow rate, the plugs are joining and a continuous gas flow will be present above the liquid. The regime is called stratified flow. At higher gas flow rates, waves may be established at the gas–liquid interface. These waves may at some situations be quite big and slugs of liquid occupying the whole cross section may be formed. For annular flow the gas flow rate is so high that the liquid phase is forced to flow in a thin layer along the tube wall while the gas phase occupies the central part of the tube. It may happen that liquid droplets penetrate into the gas phase. At very high gas flow rates, the liquid may appear as a spray or a dispersed phase in the gas.

In Fig. 14.14, the flow regimes for a vertical upward two-phase flow of a gas and liquid are shown. In this case, the gravity force is in the direction of the tube and the flow types become symmetric which means that the wavy and stratified flow types, present in a horizontal tube, do not occur.

Many methods have been suggested for determination of the flow regime occurring at certain conditions. Figure 14.15 shows a so-called modified Baker-plot from Ref. [17]. This figure provides an approximative determination of the flow type as function of the liquid and gas flow rates. G_g on the vertical axis in Fig. 14.15 is the mass velocity for the gas phase, i.e., $\dot{m}_g/A_{\text{cross section}}$. G_f on the horizontal axis is the mass velocity for the liquid phase, i.e., $\dot{m}_f/A_{\text{cross section}}$. The parameter λ for the

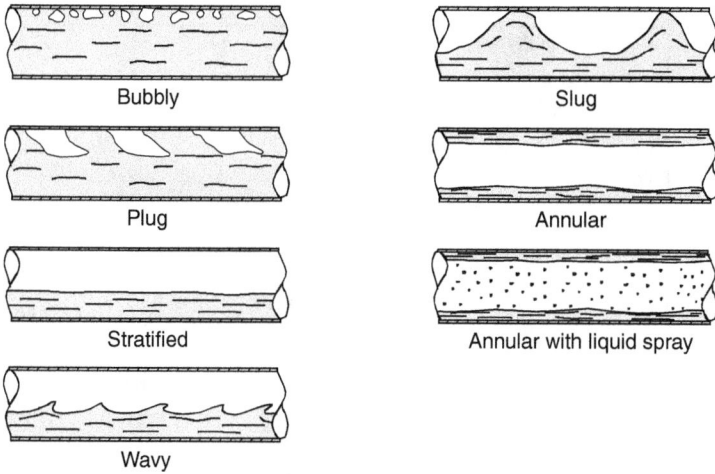

Figure 14.13: Flow regimes for horizontal two-phase flow of gas and liquid in tubes.

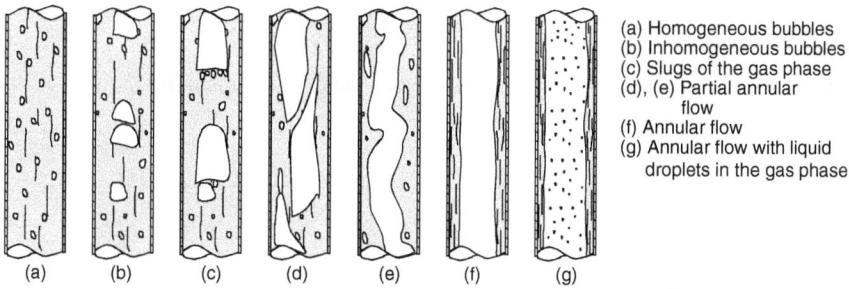

(a) Homogeneous bubbles
(b) Inhomogeneous bubbles
(c) Slugs of the gas phase
(d), (e) Partial annular flow
(f) Annular flow
(g) Annular flow with liquid droplets in the gas phase

Figure 14.14: Flow regimes for vertical gas–liquid flow.

vertical axis is determined by

$$\lambda = \left(\frac{\rho_g}{\rho_{air}} \cdot \frac{\rho_f}{\rho_{H_2O}} \right)^{1/2} \qquad (14.37)$$

while the parameter ψ for the horizontal axis is determined by

$$\psi = \frac{\sigma_{H_2O}}{\sigma} \left(\frac{\mu_f}{\mu_{H_2O}} \left(\frac{\rho_{H_2O}}{\rho_f} \right)^2 \right)^{1/3} \qquad (14.38)$$

For vertical upward flow, a diagram presented by Hewitt and Roberts [18] is commonly used. This is provided in Fig. 14.16.

At adiabatic flows or if no liquid is evaporated, the flow type is relatively similar in the main flow direction. Large pressure drops may however cause sufficient changes in the gas phase density and then the flow regime may change.

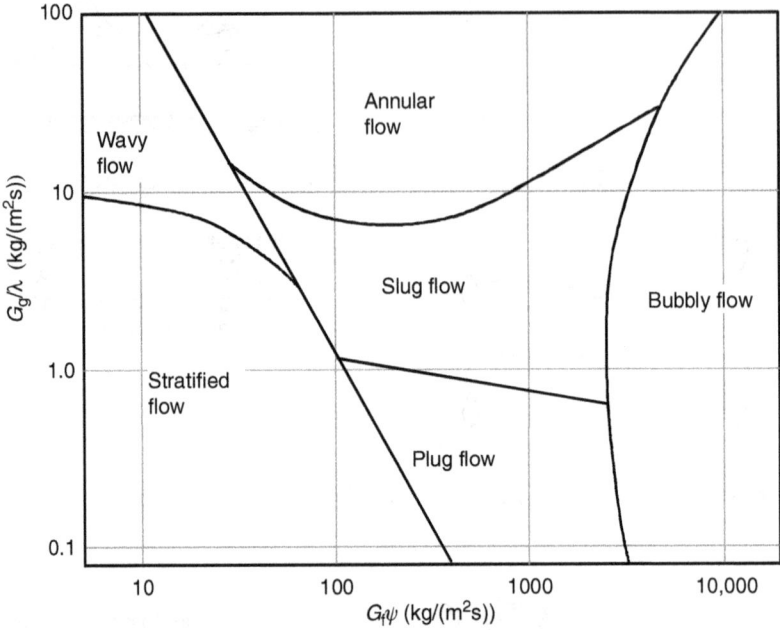

Figure 14.15: Baker-plot for determination of the flow regime in horizontal tubes.

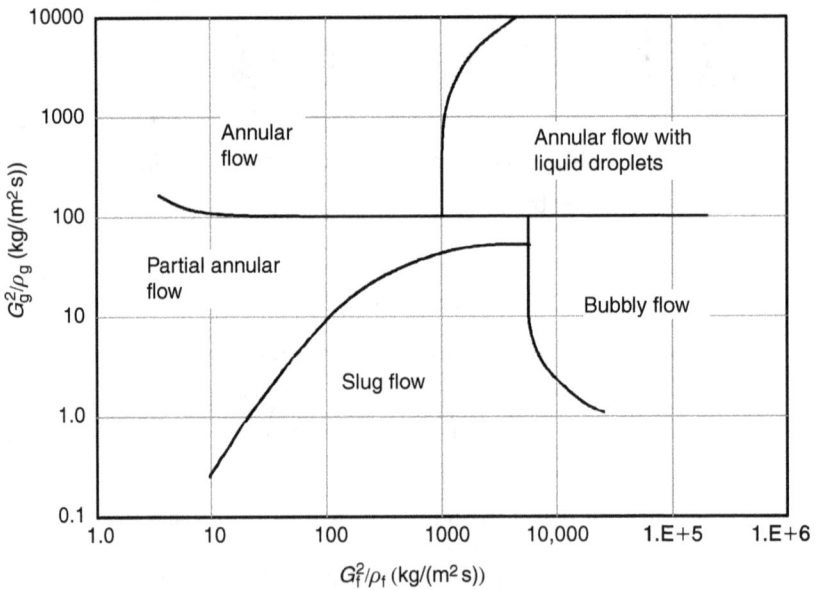

Figure 14.16: Flow regimes for vertical upward two-phase flow.

14.12   Some definitions and relations for two-phase flows

In two-phase gas–liquid flow, the flow velocity is not the same for the two phases. The difference is called slip. The property void fraction, denoted by ε, is commonly used to determine the ratio between the gas volume and the mixture volume, i.e.,

$$\varepsilon = \frac{V_g}{V_g + V_f} = \frac{V_g}{V} \tag{14.39}$$

where V_g is the gas volume and V_f the liquid volume.

Two mass fraction numbers are used. These are defined as

$$X_F = \frac{\dot{m}_g}{\dot{m}_g + \dot{m}_f} = \frac{\dot{m}_g}{\dot{m}} \tag{14.40}$$

$$X_S = \frac{m_g}{m_g + m_f} = \frac{m_g}{m} \tag{14.41}$$

where $\dot{m}_g$ and $\dot{m}_f$ are the mass flow rates for the gas and liquid phases, respectively. m_g and m_f represent mass in a control volume for the gas and liquid phases, respectively (X_F is the flowing mass quality and X_S the static mass quality).

If no velocity difference between the phases exists, one has $X_F = X_S$. For a mixture of a liquid and its vapor, the mass fraction corresponds to the vapor fraction.

The mass average velocity G is defined according to

$$G = \frac{\dot{m}}{A}$$

The mass flow rates of the gas and the liquid are given by

$$\dot{m}_g = GAX_F$$

$$\dot{m}_f = GA(1 - X_F)$$

The phase velocity ratio u_R is defined according to

$$u_R = \frac{u_g}{u_f} \tag{14.42}$$

where u_g is the gas phase average velocity and u_f is the liquid phase average velocity.

In a two-phase flow, either of the phases may be assumed to flow in the tube and occupy the whole cross-sectional area but at the real mass flow rate, pressure, and temperature. The corresponding flow velocities are called superficial velocities and are given by

$$u_{fS} = \frac{\dot{m}_f}{\rho_f A} = \frac{G(1 - X_F)}{\rho_f} \tag{14.43}$$

$$u_{gS} = \frac{\dot{m}_g}{\rho_g A} = \frac{GX_F}{\rho_g} \tag{14.44}$$

In eqs. (14.43) and (14.44), index S means superficial velocity.
With the definitions above, the following relations can be derived:

$$\frac{u_g}{u_f} = \frac{\rho_f}{\rho_g}\frac{(1-\varepsilon)X_F}{\varepsilon(1-X_F)} \tag{14.45}$$

$$\frac{\rho_g}{\rho_f} = \frac{(1-\varepsilon)X_S}{\varepsilon(1-X_S)} \tag{14.46}$$

$$u_{fS} = u_{gS}\frac{(1-X_F)}{X_F}\frac{\rho_g}{\rho_f} \tag{14.47}$$

14.13   Pressure drop for two-phase flow

The pressure drop for two-phase flow is commonly higher than the pressure drop
for either of the phases at the same mass flow rate. The pressure drop is affected by
friction, changes in elevation (gravity), and acceleration of the fluid.

A method to calculate the pressure drop for two-phase flows was suggested by
Lockhardt and Martinelli [19]. It is commonly applied. The method is based on
experimental data for isothermal flow of air and various liquids at incompressible
conditions. Only the frictional part is considered. The flow field is characterized in
four categories depending on the Reynolds number, which is based on the superficial
velocities for the different phases. Depending on the Reynolds number, laminar or
turbulent flow may prevail in each phase. Turbulent flow may occur if the Reynolds
number is greater than 2000.

The Reynolds numbers are calculated as

$$Re_f = \frac{u_{fS}D}{\nu_f} = \frac{G(1-X_F)D}{\mu_f} \tag{14.48}$$

$$Re_g = \frac{u_{gS}D}{\nu_g} = \frac{GX_FD}{\mu_g} \tag{14.49}$$

The pressure drop for the two-phase flow is related to the pressure drop of the
corresponding single-phase flow. In Fig. 14.17, four curves are shown and their
validity regimes are indicated.

The pressure drops Δp_f and Δp_g are calculated as

$$\Delta p_f = f_f\frac{L}{D}\frac{\rho_f u_{fS}^2}{2} \tag{14.50}$$

$$\Delta p_g = f_g\frac{L}{D}\frac{\rho_g u_{gS}^2}{2} \tag{14.51}$$

f_f and f_g are determined from standard formulas for friction factors in tube or pipe
flow, see Chapters 8 and 9 as well as Ref. [20].

Δp_{TF} which appears in the variable or quantity on the y-axis in Fig. 14.17 is the
pressure drop for the two-phase flow.

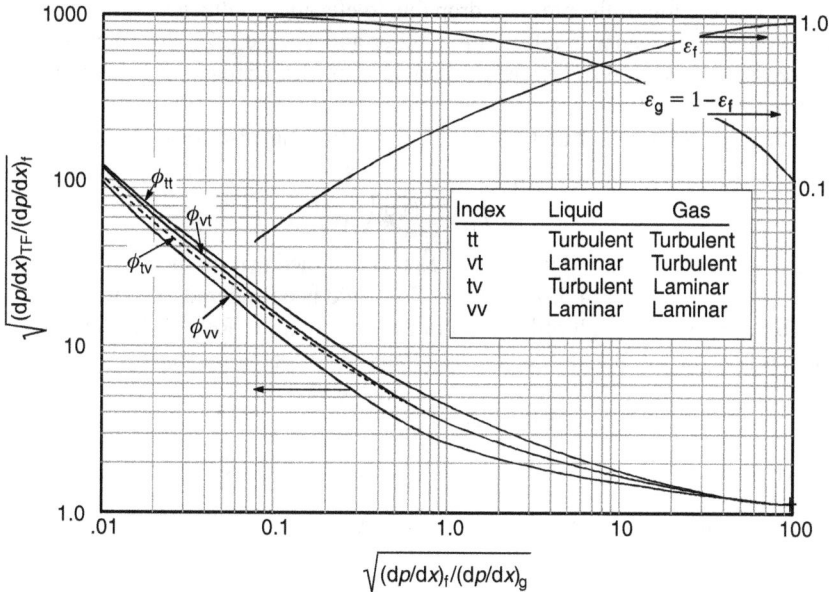

Figure 14.17: Diagram for calculation of pressure drop at two-phase flow.

The theoretical background for the Lockhardt–Martinelli method has been presented in Refs. [21, 22].

The variable on the horizontal axis (or the square of it) is called the Martinelli parameter, i.e.,

$$X^2 = \frac{(dp/dx)_f}{(dp/dx)_g} \qquad (14.52)$$

The variable on the vertical axis (or rather the square of it) is called the two-phase multiplier,

$$\phi_f^2 = \frac{(dp/dx)_{TF}}{(dp/dx)_f} \qquad (14.53)$$

Empirically, the relation between ϕ and X has been found to follow

$$\phi_f^2 = 1 + \frac{C}{X} + \frac{1}{X^2} \qquad (14.54)$$

where the constant C has the following values

$C = 20$ if turbulent flow prevails in the liquid as well as in the gas (tt)
$C = 12$ if the liquid flow is viscous (laminar) and the gas flow is turbulent (vt)
$C = 10$ if the liquid flow is turbulent and the gas flow is laminar (tv)
$C = 5$ if laminar flow prevails in the liquid as well as in the gas (vv).

For calculation of the pressure drop due to elevation (influence of gravity) and acceleration, the reader is referred to Collier and Thome [22]. For calculation of the pressure drop in nozzles, bends, and sudden area changes, the reader is also referred to Ref. [22].

The pressure drop per unit length due to friction can also be written as

$$\left(\frac{dp}{dx}\right)_{TF} = \phi_{LO}^2 \left(\frac{dp}{dx}\right)_{LO} \tag{14.55}$$

where index LO means "liquid only," i.e., it is assumed that for the quantities at the right hand side of eq. (14.55) the whole flow field is considered to be liquid.

To determine the multiplier ϕ_{LO}^2, Friedel's correlation (Ref. [23]) is commonly used. It is regarded to be reasonably accurate. The correlation reads

$$\phi_{LO}^2 = A_1 + \frac{3.24 A_2 A_3}{Fr^{0.045} We^{0.035}} \tag{14.56}$$

where

$$A_1 = (1 - X_F)^2 + X_F^2 \left(\frac{\rho_f f_{GO}}{\rho_g f_{LO}}\right), A_2 = X_F^{0.78}(1 - X_F)^{0.224},$$

$$A_3 = \left(\frac{\rho_f}{\rho_g}\right)^{0.91} \left(\frac{\mu_g}{\mu_f}\right)^{0.19} \left(1 - \frac{\mu_g}{\mu_f}\right)^{0.7}$$

$$Fr = \frac{G^2}{gD\bar{\rho}^2}, We = \frac{G^2 D}{\bar{\rho}\sigma}$$

$$\bar{\rho} = \frac{m}{V}(\text{where } V \text{ is the volume})$$

The single-phase friction factors f_{GO} and f_{LO} are calculated as if the whole flow was gas or liquid, respectively.

14.14   Heat transfer and temperature distributions

Figure 14.18 shows the thermal process as a liquid is flowing in a vertical tube which is heated by a uniform heat flux q_w (W/m²).

The vapor fraction X_F is increasing in the vertical direction until the wall dries out. Then the wall temperature is increased suddenly. If the heat flux q_w is sufficiently high the tube wall may melt before dry out has occurred.

Figure 14.19 shows how the flow and heat transfer regimes in Fig. 14.18 are distributed in terms of heat flux level and the positions along the tube. Note that at sufficiently high heat fluxes, q_w, burn out can occur anywhere along the tube.

14.14.1   Chen's method to determine the heat transfer coefficient and heat flux

For annular two-phase flow (regions E and F in Fig. 14.18), a method to calculate the heat transfer coefficient and the heat flux q_w has been developed by Chen [24].

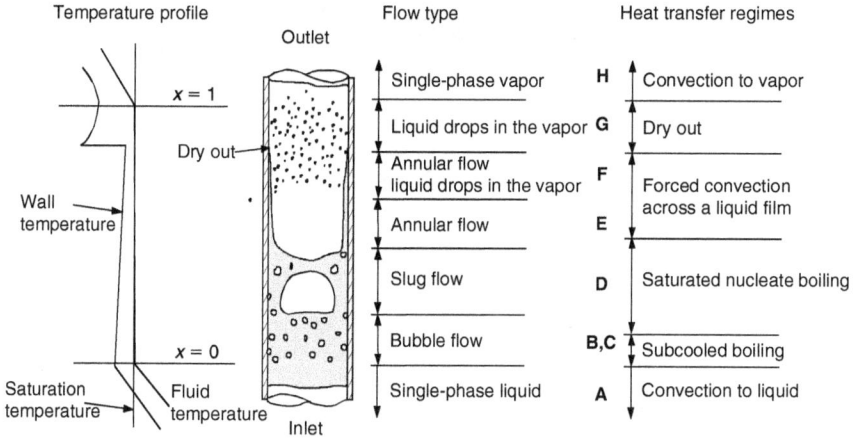

Figure 14.18: Development of two-phase flow in a vertical tube with $q_w =$ constant. From Ref. [22].

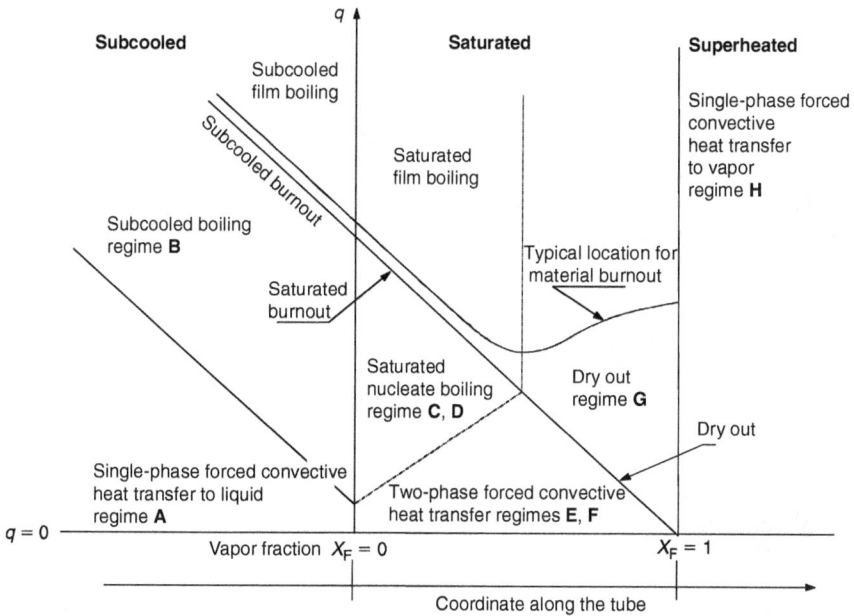

Figure 14.19: Influence of q_w on the character of the two-phase flow along the tube.

The method has been found to be reasonable, accurate, and applicable for water and certain organic substances.

The suggested method is based on the assumption that both nucleate boiling and forced two-phase convective heat transfer occur and that their contributions are

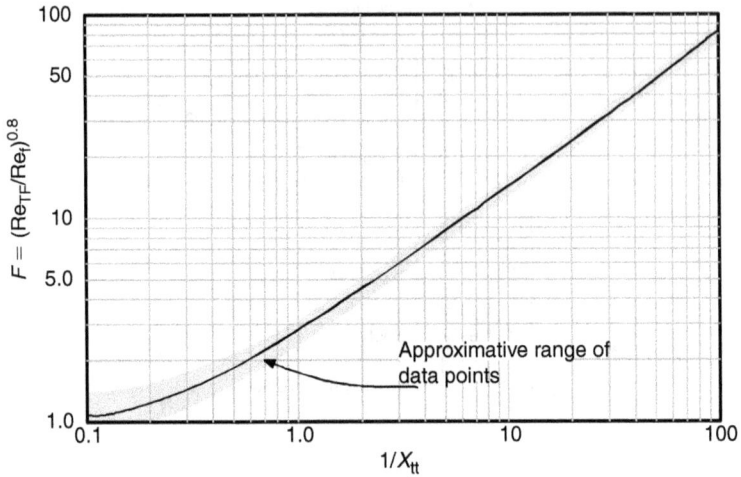

Figure 14.20: F as a function of $1/X_{tt}$.

additive. The following formula is used:

$$\alpha_{TF} = S\alpha_{KK} + F\alpha_C \qquad (14.57)$$

where F is the so-called enhancement factor for the convective heat transfer while S is a suppression factor for the nucleate boiling.

The following procedure is applied as S, F, q_w and the heat transfer coefficients in the formula above are determined.

14.14.1.1   Calculations

1. Determine the so-called Martinelli parameter X_{tt} (index tt means turbulent–turbulent)

$$X_{tt} = \left(\frac{1 - X_F}{X_F}\right)^{0.9} \left(\frac{\rho_g}{\rho_f}\right)^{0.5} \left(\frac{\mu_f}{\mu_g}\right)^{0.1} \qquad (14.58)$$

where X_F is the vapor fraction. The Martinelli parameter is defined according to

$$X_{tt} = \sqrt{\frac{(dp/dx)_f}{(dp/dx)_g}} \qquad (14.59)$$

2. Determine the empirical function F for X_{tt} from Fig. 14.20. F is defined according to

$$F = \left(\frac{Re_{TF}}{Re_f}\right)^{0.8} \qquad (14.60)$$

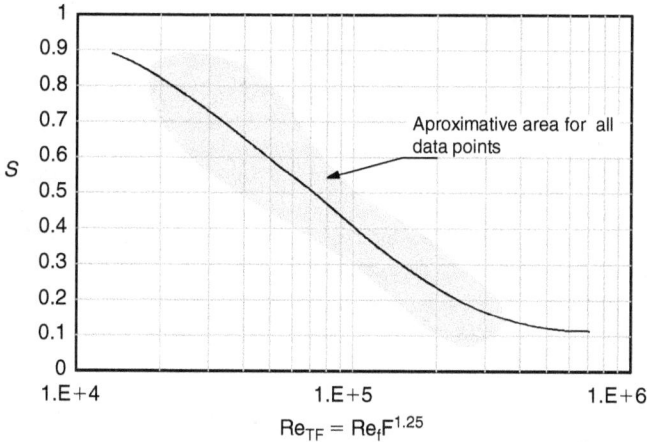

Figure 14.21: S as function of Re_{TF}.

An approximative formula for determination of F is given by

$$F = \begin{cases} 1 & \text{if } \dfrac{1}{X_{tt}} \leq 0.1 \\[3mm] 2.35 \left(\dfrac{1}{X_{tt}} + 0.213 \right)^{0.736} & \text{if} \dfrac{1}{X_{tt}} > 0.1 \end{cases} \qquad (14.62)$$

3. Calculate the heat transfer coefficient α_C by the Dittus–Boelter equation for single-phase flow, see, e.g., Chapter 9 or Ref. [21]. The physical properties should be for saturated liquid and the following formula is valid

$$\alpha_C = 0.023 \mathrm{Re}_f^{0.8} \mathrm{Pr}_f^{0.4} \frac{\lambda_f}{D} \qquad (14.63)$$

where $\mathrm{Re}_f = G(1 - X_F)D/\mu_f$

4. Determine the empirical factor S from Fig. 14.21.

An approximative formula for determination of S is given by

$$S = \frac{1}{1 + 2.53 \cdot 10^{-6} \mathrm{Re}_{TF}^{1.17}} \qquad (14.64)$$

5. Determine a nucleate boiling coefficient α_{KK}, see Forster and Zuber [25], according to

$$\alpha_{KK} = 0.00122 \frac{\lambda_f^{0.79} c_{pf}^{0.45} \rho_f^{0.49}}{\sigma^{0.5} \mu_f^{0.29} h_{fg}^{0.24} \rho_g^{0.24}} \Delta t_s^{0.24} \Delta p_s^{0.75} \qquad (14.65)$$

where $\Delta t_s = t_w - t_s$ and $\Delta p_s = p_s(t_w) - p_s(t_s)$.

6. Calculate the heat transfer coefficient α_{TF} for the two-phase flow case according to

$$\alpha_{TF} = S\alpha_{KK} + F\alpha_C \qquad (14.66)$$

for a number of Δt_m.

7. Draw a figure showing $q = \alpha_{TF}\Delta t_m$ as a function of Δt_m. At $q = q_w$, Δt_m can be found.

   With this calculation procedure the wall temperature can be determined for an arbitrary position along the tube as q_w is known.

14.15   Additional correlations

Recently additional correlations have appeared. They have a similar structure to that suggested by Chen [24]. Gungor and Winterton [26] studied boiling of water and a number of refrigerants as well as ethylene glycol for vertical tubes (opposite to and in the direction of the gravity) and horizontal tubes. For saturated boiling their correlation reads

$$\alpha_{TF} = S\alpha_{KK} + E\alpha_C \qquad (14.67)$$

where α_C is determined by the Dittus–Boelter Equation (14.63) for the liquid flow. The enhancement factor E is determined from

$$E = 1 + 2.4 \cdot 10^4 \mathrm{Bo}^{1.16} + 1.37\left(\frac{1}{X_{tt}}\right)^{0.86} \qquad (14.68)$$

In eq. (14.68), Bo is the so-called boiling number defined as

$$\mathrm{Bo} = \frac{q_w}{(G \cdot h_{fg})} \qquad (14.69)$$

The damping (suppression) factor S is determined according to

$$S = \frac{1}{1 + 1.15 \cdot 10^{-6}E^2\mathrm{Re}_f^{1.17}} \qquad (14.70)$$

The nucleate boiling coefficient in eq. (14.67) is determined according to Cooper's equation (14.11).

If the Froude number, $Fr_f = G^2/(\rho_f^2 gD)$, is less than 0.05 so-called stratified flow in horizontal tubes appears. E is calculated according to eq. (14.68) and should be multiplied by

$$Fr_f^{0.1-2Fr_f} \qquad (14.71)$$

while S according to eq. (14.71) should be multiplied by

$$Fr_f^{0.5} \qquad (14.72)$$

In 1992, Steiner and Taborek [27] suggested a new general correlation for boiling in vertical tubes. The general equation reads

$$\alpha_{TF} = \left(\alpha_{KK}^n + \alpha_C^n\right)^{1/n} \qquad (14.73)$$

In the work by Steiner and Taborek, n is chosen as $n = 3$. The nucleate boiling coefficient is determined by the method for saturated pool boiling, e.g., Gorenflo's method, eq. (14.2) which is multiplied by a correction factor. For the heat transfer coefficient α_C the so-called Gnielinski equation, see Ref. [24], is recommended together with a two-phase multiplier or enhancement factor. More details can be found in Refs. [22, 27].

14.16   Maximum heat flux

For convective boiling in tubes it has been found, see Figs. 14.18 and 14.19, that two critical heat fluxes exist, at dry out and at burn out (compare q_{max} for pool boiling). q_{max} at burn out is more severe because it occurs at higher heat fluxes and gives a higher temperature increase. The maximum heat flux at burn out is discussed in Refs. [22, 28, 29].

More information about two- and multiphase flows can be found in Refs. [21, 30, 31].

References

[1] S. Nukiyama, The maximum and minimum values of the heat transmitted from metal to boiling water under atmospheric pressure, J. Jpn. Soc. Mech. Eng., 37, 367–374 (1934) (translated to English in Int. J. Heat Mass Transfer, 9, 1419–1433 (1966)).

[2] T.B. Drew, C. Mueller, Boiling, Trans. AICHE, 33, 449–473 (1937).

[3] Y.Y. Hsu, On the size range of active nucleation cavities on a heating surface, Trans. ASME J. Heat Transfer, 84, 207–216 (1962).

[4] W.M. Rohsenow and J.P. Hartnett, Handbook of Heat Transfer, McGraw-Hill, New York (1973).

[5] W.M. Rohsenow, A method of correlating heat transfer data for surface boiling of liquids, Trans. ASME, 74, 969–975 (1952).

[6] P.B. Whalley, Boiling, Condensation and Gas–Liquid Flow, Oxford University Press, New York (1982).

[7] J.J. Jasper, The surface tension of pure liquid compounds, J. Phys. Chem. Ref. Data, 1(4), 841–1010 (1972).

[8] M.G. Cooper, Saturated nucleate pool boiling—a simple correlation, Proc. First UK National Heat Transfer Conference, Chem. Eng. Prog. Symp. Series No. 86, 2, 785–793 (1984).

[9] M.G. Cooper, Heat flow rates in saturated nucleate pool boiling—a wide ranging examination using reduced properties, Adv. Heat Transfer, 16, 157–239, Academic Press (1984).

[10] D. Gorenflo, P. Sokol and S. Caplanis, Pool boiling heat transfer from single plain tubes to various hydrocarbons, Int. J. Refrig., 13, 286–292 (1990).

[11] D. Gorenflo, Pool Boiling, VDI-Heat Atlas, VDI-Verlag, Düsseldorf (1993).

[12] J.H. Lienhard, A Heat Transfer Textbook, Prentice Hall, Englewood Cliffs, New Jersey (1981).

[13] H. Lamb, Hydrodynamics, 6th ed., Dover Publications, New York (1945).

[14] N. Zuber, Hydrodynamic aspects of boiling heat transfer, AEC Rep. AECU-4439, Physics and Mathematics (1959).

[15] J.H. Lienhard and V.K. Dhir, Extended hydrodynamic theory of the peak and minimum pool boiling heat fluxes, NASA CR-2270 (1973).

[16] J.H. Lienhard and R. Eichhorn, Peak boiling heat flux on cylinders in a cross flow, Int. J. Heat Mass Transfer, 19, 1135–1142 (1976).

[17] K.J. Bell, J. Taborek and F. Fenoglio, Interpretation of horizontal in-tube condensation heat transfer correlations with a two-phase flow regime map, Chem. Eng. Prog. Symp. Series No. 102, 66, 150–163 (1970).

[18] G.F. Hewitt and D.N. Roberts, Studies of two-phase flow patterns by simultaneous flash and X-ray photography, AERE-M2159 (1969).

[19] R.W. Lockhardt and R.D. Martinelli, Proposed correlation of data of isothermal, two-phase, two-component flow in pipes, Chem. Eng. Prog., 45, 39–48 (1949).

[20] F.M. White, Fluid Mechanics, 6th ed., McGraw-Hill, New York (2008).

[21] D. Chisholm, A theoretical basis for the Lockhardt–Martinelli correlation for two-phase flow, Int. J. Heat Mass Transfer, 10, 1767–1778 (1967).

[22] J.G. Collier and R. Thome, Convective boiling and condensation, 3rd ed., Oxford University Press, Oxford (1994).

[23] L. Friedel, Improved friction pressure drop correlations for horizontal and vertical two-phase pipe flow, European Two-Phase Flow Group Meeting, Paper E2, Ispra, Italy (1979).

[24] J.C. Chen, A correlation for boiling heat transfer to saturated fluids in convective flow, ASME-Paper Preprint 63-HT-34, 6th ASME-AICHE Heat Transfer Conference, Boston (1963).

[25] H.K. Forster and N. Zuber, Bubble dynamics and boiling heat transfer, AIChE J., 1, 532–535 (1955).

[26] K.E. Gungor and R.H.S. Winterton, A general correlation for flow boiling in tubes and annuli, Int. J. Heat Mass Transfer, 29, 351–358 (1986).

[27] D. Steiner and J. Taborek, Flow boiling heat transfer in vertical tubes correlated by an asymptotic model, Heat Transfer Eng., 13, 43–69 (1992).

[28] Y.Y. Hsu and R.W. Graham, Transport Processes in Boiling and Two-Phase Systems, Hemisphere Publ. Corp., Washington, DC (1976).

[29] Y. Katto, A generalized correlation of critical heat flux for forced convection boiling in vertical uniformly heated round tubes, Int. J. Heat Mass Transfer, 21, 1527–1542 (1978).

[30] G.F. Hewitt. Heat Exchanger Design Handbook, Hemisphere Publ. Corp., Washington (1983).

[31] J.R. Thome, Enhanced Boiling Heat Transfer, Hemisphere Publ. Corp. New York (1990).

Further reading

J.R. Thome, Boiling, Chapter 9, in Heat Transfer Handbook (Eds. A. Bejan and A.D. Kraus), Wiley (2003).

15 Heat exchangers

15.1 Introduction

Heat exchangers are equipment being used for transfer of heat between two or more fluids at different temperatures. Several types of heat exchangers have been developed and are being used in power plants, such as refrigerators and automotive heat exchangers, heat pumps, air-conditioning, and chemical process industries. In the so-called shell-and-tube heat exchangers and vehicle radiators, heat is primarily transferred by convection and conduction from a hot fluid to a cold one which are separated by a metallic wall. In evaporators and condensers, the heat transfer due to evaporation and condensation is the primary mechanism. In some heat exchangers, e.g., cooling towers the hot fluid (water) is cooled by direct mixing with the cold fluid (air). Design and sizing of heat exchangers are complicated engineering work. Convective heat transfer, pressure drop, estimation of the thermal performance, and economical issues are important at the final design. For big units in power plants or chemical process industries, the cost (investment and operation) might be an important issue, while in applications for space and aircraft, the weight and size (compactness) might be most important. In this chapter, the functioning and classification of heat exchangers are presented as well as methods for analysis, sizing, and rating are provided.

15.2 Classification of heat exchangers

Heat exchangers may be classified in various ways. Here the method introduced in Ref. [1] is followed. Heat exchangers are distinguished by:

a) How the heat transfer process occurs
b) Compactness (heat transfer area per unit volume)
c) Design principle
d) Flow process
e) Mechanism for the heat exchange.

15.2.1 Heat transfer process

Heat exchangers may operate by direct contact with the fluids or by indirect contact. If direct contact prevails, the heat transfer occurs between two immiscible fluids,

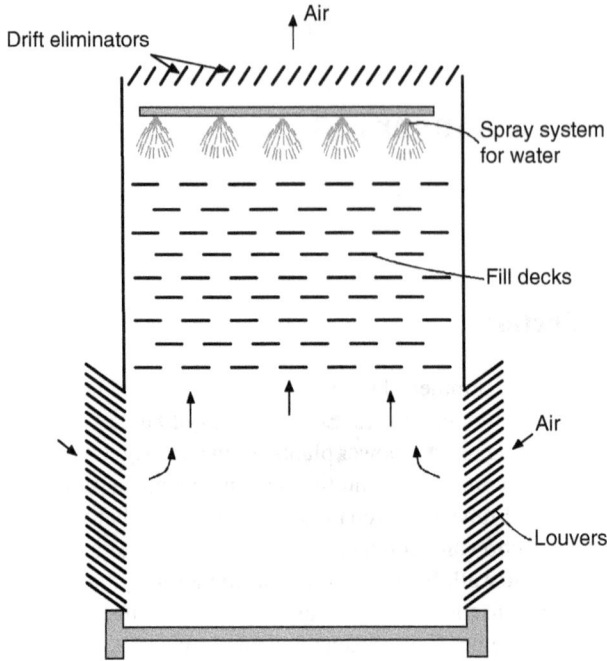

Figure 15.1: Cooling tower where the air movement is by natural convection.

e.g., a gas and a liquid which are forced into contact. Cooling towers are such examples.

Cooling towers are often applied to cool waste heat from industrial processes. Commonly, both natural convection and forced convection towers are used. Figure 15.1 shows a cooling tower in which natural convection prevails. Water is sprayed directly into the airstream which is moving upward by natural convection.

The falling water droplets are cooled by convection and also by evaporation of the liquid water. In the tower, there are several decks which among other things slow down the droplet downward motion and thereby the exposure time to the cold airstream is increased. Cooling towers may have heights of more than 100 m. For cases with forced convection, the airstream is forced through the tower by fans. The fans might be located at the top of the tower and then the air is sucked through the tower. In other designs, the fans are at the bottom of the tower and the air is pressed through the tower. The increased air circulation will increase the cooling capacity.

In heat exchangers having indirect contact, the hot and cold fluids are separated by an impermeable surface. The fluids are not mixed. This is the most common heat exchanger type and sometimes these are referred to as surface heat exchangers.

Figure 15.2: Compact heat exchanger. (Radiator for a private car.)

15.2.2 Compactness

The ratio between the heat transfer area (on one side of the heat exchanger) and the volume is used as a measure of the compactness of the heat exchanger. If this ratio, A/V, is greater than $700\,\mathrm{m^2/m^3}$, the heat exchanger is said to be compact. Radiators in private cars have typically $1100\,\mathrm{m^2/m^3}$ and in some glass–ceramic heat exchangers, A/V might be $6500\,\mathrm{m^2/m^3}$. The human lungs have $A/V \approx 20{,}000\,\mathrm{m^2/m^3}$ and are the most compact heat mass exchangers. Shell-and-tube heat exchangers, which are very common in the process industries, have $A/V \approx 70\text{–}500\,\mathrm{m^2/m^3}$ and are usually not considered as compact.

A reason to apply compact heat exchangers is that a high A/V values diminishes the heat exchanger volume. When heat exchangers are used in cars, trucks and buses, marine vehicles, aircraft and space ships, cryogenic systems as well as in air-conditioning, the weight and size (volume) and thus compactness are the key issues. In gas–liquid heat exchangers, the heat transfer coefficient on the gas side is much less than that on the liquid side. To enable the transfer of a certain heat power, the surface of the gas side must be increased. Fins or extended surfaces of various geometries are then being used. Figure 15.2 shows a typical compact heat exchanger.

Sometimes, the amount of heat transferred per unit volume $(\mathrm{W/m^3})$ is used as a measure of the compactness.

15.2.3 Types of design

Heat exchangers may also be classified as the design. So, for instance, one speaks about shell-and-tube heat exchangers, plate heat exchangers (PHEs), finned PHEs, finned tubular heat exchangers, regenerative heat exchangers, etc.

Shell-and-tube heat exchangers are the most common heat exchangers and are manufactured in a wide range of sizes, flow arrangement, etc. The manufacturing

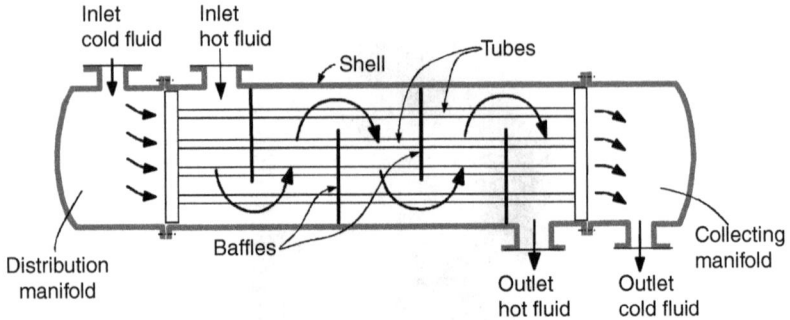

Figure 15.3: Principle sketch of a shell-and-tube heat exchanger. One shell pass and one tube pass.

method is relatively simple and if common carbon steels can be used, the heat exchanger might be very cheap. The most common layout consists of a large number of tubes placed in a cylindrical shell (thus the name shell-and-tube heat exchanger). Figure 15.3 shows a principle sketch of a simple shell-and-tube heat exchanger where one of the media is flowing in the tubes (tube side) and the other one is flowing on the outside of the tubes (shell side fluid). Typical components in such heat exchangers are the tube bundle, the shell, distribution and collection headers, and the baffles.

The baffles are used to support the tubes, guide the flow so that the outside surfaces of the tubes are mainly approached in cross flow and to indirectly increase the turbulence of the flow field. Different types of baffles exist and the design of the baffles depend on the flow velocity, permitted pressure drop on the shell side, required support for the tubes and the risk for flow-induced vibrations. The heat exchange may occur from one liquid to another, from a liquid to a gas or from a gas to another gas. Also two-phase flows on either side are possible. Liquid–liquid is the most common case, but also liquid–gas occurs frequently. In the latter case, usually fins or extended surfaces are used on the gas side, as the convective heat transfer coefficient is low there.

PHEs are built up by a number of thin plates assembled together in a package. The plates can be smooth or more commonly corrugated in some way. PHEs may normally not operate at as high pressure and high temperatures as shell-and-tube heat exchangers. This is so because often a gasket is used to seal between adjacent plates. The compactness is about 120–250 m^2/m^3. Figure 15.4 shows a typical PHE.

Finned PHEs are shown in Fig. 15.5. The compactness can be as high as 6000 m^2/m^3. Most commonly, this heat exchanger type is for gas to gas exchange. Louvered, perforated, or the so-called offset-strip fins are used to separate the plates and create flow channels. Cross flow, counterflow, or parallel flow arrangements are common.

Finned tubular heat exchangers are used as a high operating pressure prevails for one of the media or when finned surfaces are needed, e.g., for heat exchange

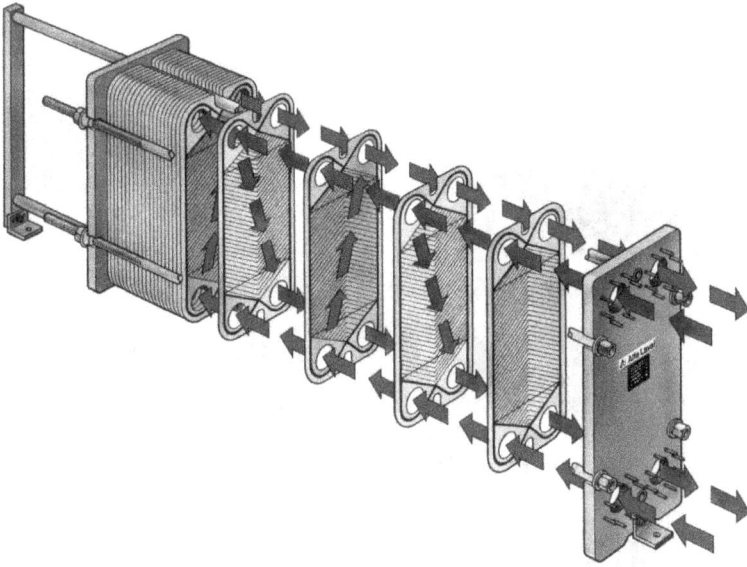

Figure 15.4: Plate heat exchanger – exploded view.

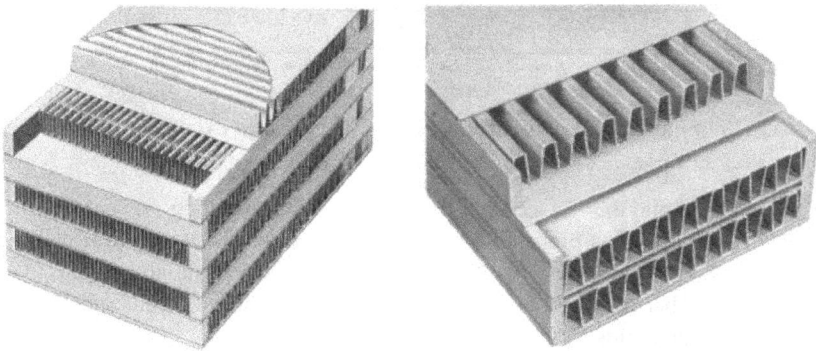

Figure 15.5: Finned PHEs.

between a liquid and a gas. Figure 15.6 shows two common configurations, one with circular tubes and another one with flat tubes. The compactness is usually less than $350 \, \text{m}^2/\text{m}^3$.

Regenerative heat exchangers might be static or dynamic. The static type consists of a porous material through which the hot and cold fluids are flowing in alternate fashion. A switch equipment controls the periodic flow of the two fluids. The hot fluid heats up the porous material which in the next sequence heats up the cold fluid and the process is repeated periodically. In the dynamic type, the heat exchanger core is rotating in such a way that part of it is periodically exposed

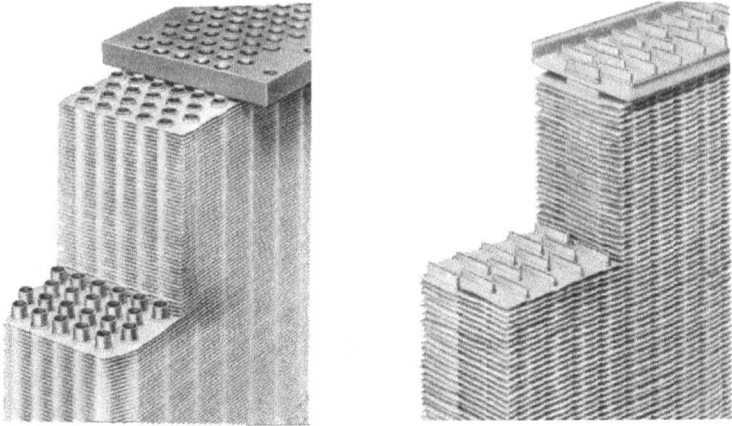

Figure 15.6: Finned tubular heat exchangers.

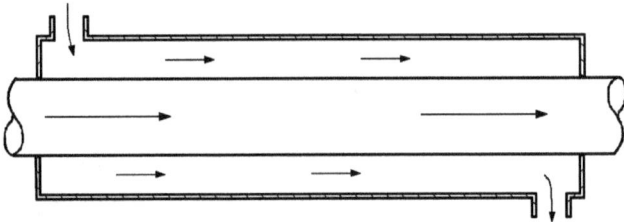

Figure 15.7: Co-current or parallel flow heat exchanger.

to the hot fluid and then successively to the cold fluid. Rotating regenerative heat exchangers are used as preheaters in heat and power plants and as heat recovery units in air-conditioning systems. The heat exchanger is most suitable for gas to gas heat exchange because only for gases the heat capacity of the core is much bigger than those of the fluids.

15.2.4   Classification based on flow process

In this section, the most common flow processes considered for classification of heat exchangers are presented.

Co-current flow or parallel flow means that the fluids are entering the heat exchanger at the same place and are flowing in the same direction along the exchanger and finally leaves the heat exchanger at the same place. Figure 15.7 shows a principle sketch of a parallel flow heat exchanger.

Counterflow in heat exchangers means that the hot and cold fluids are entering the exchanger at different places and flow in opposite directions, which is depicted in Fig. 15.8.

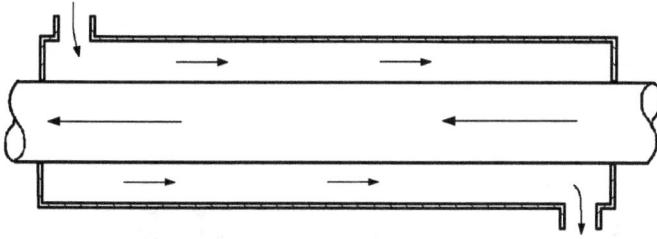

Figure 15.8: Counterflow arrangement.

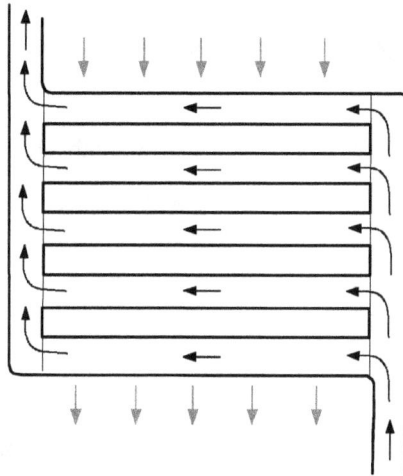

Figure 15.9: Simple cross flow arrangement.

In the so-called cross flow heat exchangers, the fluids are flowing perpendicular to each other, see Fig. 15.9.

The flow field for each media is said to be mixed or unmixed. Figure 15.10 shows a case where both the hot and cold fluids are flowing through individual channels, i.e., the fluid streams cannot pass in the transversal direction. Both fluids are unmixed in this case.

In Fig. 15.11, one medium is flowing in the tubes and cannot move in the transversal direction and this medium is unmixed. The other fluid is flowing across the tubes and is free to move in the transversal direction. This medium is therefore regarded as mixed.

Multipass flow fields are common in heat exchangers, in particular for the so-called shell-and-tube heat exchangers. Figure 15.12 shows some typical arrangements.

Figure 15.12a shows a case with one shell pass and two tube passes, while Fig. 15.12b shows an arrangement with two shell passes and four tube passes.

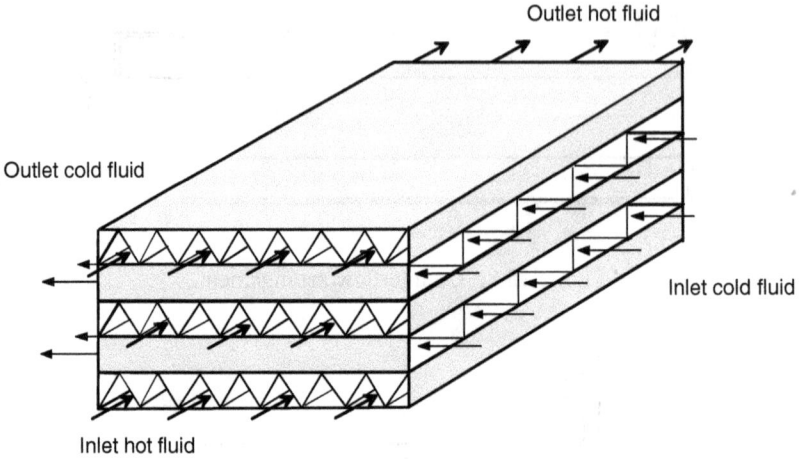

Figure 15.10: Cross flow heat exchanger. Both media are unmixed.

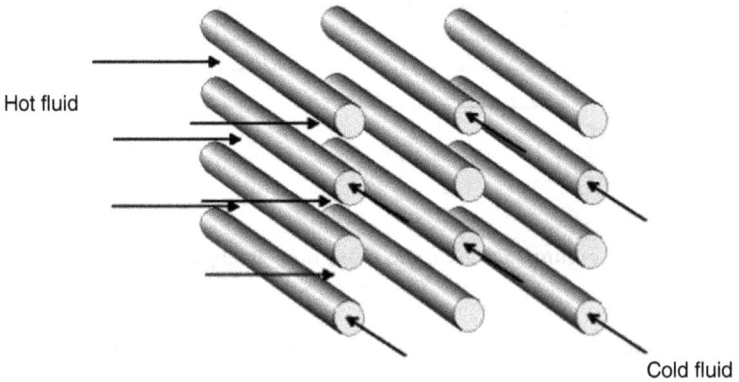

Figure 15.11: One fluid unmixed, one mixed.

15.2.5 Classification according to the mechanism for the heat transfer

The mechanism for heat transfer includes a combination of the mechanisms listed below.

a) Single-phase forced or free (natural) convection
b) Boiling or condensation
c) Radiation or combined radiation and convection.

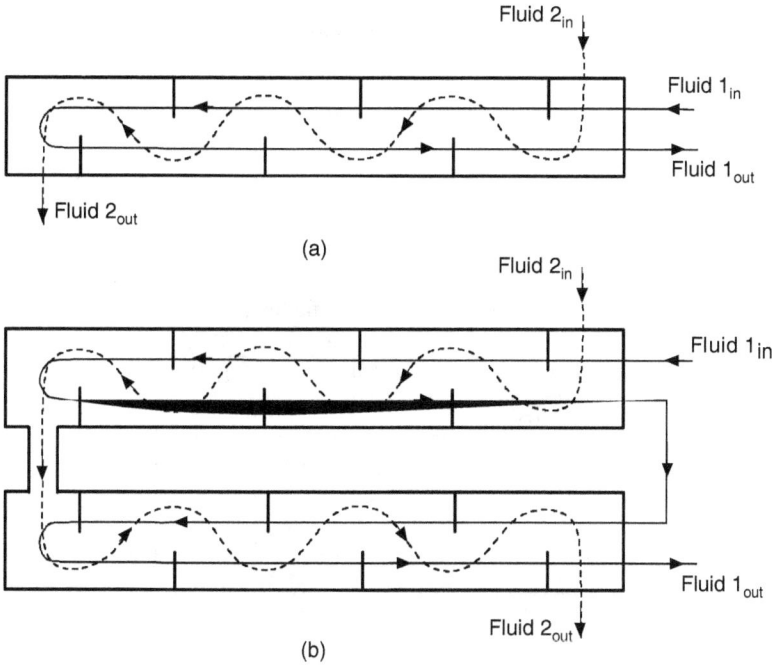

Figure 15.12: Common multipass arrangements. (a) One shell pass, two tube passes. (b) Two shell passes, four tube passes.

15.3   The overall heat transfer coefficient

Figure 15.13 shows the principle of the heat transfer from the hot fluid to the cold fluid.

The amount of heat transferred is written as

$$\dot{Q} = UA \cdot \Delta t_{\mathrm{m}} = \frac{1}{TR} \cdot \Delta t_{\mathrm{m}} \qquad (15.1\mathrm{a})$$

where U is the overall heat transfer coefficient, A the area the heat is passing and Δt_{m} a mean temperature difference between the hot and cold fluids. $TR = 1/UA$ is the resistance to heat transfer and is called the total thermal resistance. Between the hot and cold fluids, several resistances appear, see Ref. [2], and these are in series and are combined to TR as

$$TR = \frac{1}{\alpha_{\mathrm{i}} A_{\mathrm{i}}} + \frac{1}{\alpha_{F_{\mathrm{i}}} A_{\mathrm{i}}} + \frac{b_{\mathrm{w}}}{\lambda_{\mathrm{w}} A_{\mathrm{vl}}} + \frac{1}{\alpha_{F_{\mathrm{o}}} A_{\mathrm{o}}} + \frac{1}{\alpha_{\acute{o}} A_{\mathrm{o}}} \qquad (15.1\mathrm{b})$$

where α_{i} is the heat transfer coefficient on the inside, A_{i} the convective heat transfer area on the inside, $\alpha_{F_{\mathrm{i}}}$ the fouling factor on the inside, b_{w} the thickness of the intermediate solid wall, λ_{w} the thermal conductivity of the wall material, A_{vl} the

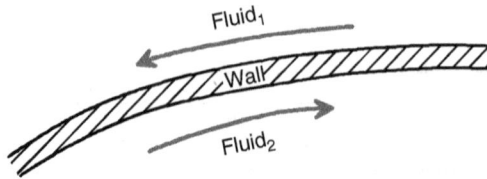

Figure 15.13: Principle sketch of the heat transfer process.

Table 15.1: Fouling factors.

Fluid	$1/\alpha_F$ (m^2 K/W)
Destilled water	1×10^{-4}
Sea water ($T < 325$ K)	1×10^{-4}
Sea water ($T > 325$ K)	2×10^{-4}
Feed water to furnaces	2×10^{-4}
Oil	9×10^{-4}
Dirty air	3.5×10^{-4}

heat conducting area, α_{F_o} the fouling factor on the outer surface, α_o is the heat transfer coefficient on the outer surface, and A_o the convective heat transfer area on the outer side. The resistance due to heat conduction in eq. (15.1) is only valid for a plane wall or if the material thickness is very small.

The heat transfer coefficients α_i and α_o are determined with methods presented in, e.g., Ref. [2] and in previous chapters of this book.

In practical applications, the heat transferring surfaces are fouled by the fluids due to various mechanisms. The fluid type itself is also important. This results in thermal resistances are evident in eq. (15.1b). Usually fouling factors $1/\alpha_{F_i}$ and $1/\alpha_{F_o}$, respectively, are introduced to account for this. Table 15.1 (data from Ref. [3]) provides some values for the fouling factor.

15.4   The LMTD method for analysis of heat exchangers

In the thermal analysis of heat exchangers, the total heat flow $\dot{Q}$ (W) is of primary interest. To get started, the overall heat transfer coefficient U is assumed to be constant in the whole heat exchanger (average value). The heat flow is then written as

$$\dot{Q} = UA \cdot \Delta t_m \qquad (15.2)$$

where A is the heat transferring area and Δt_m a proper average of the temperature difference between the hot and cold fluids. Now one has to find an expression for Δt_m so that eq. (15.2) is valid. To enable this, counterflow and parallel flow will be considered.

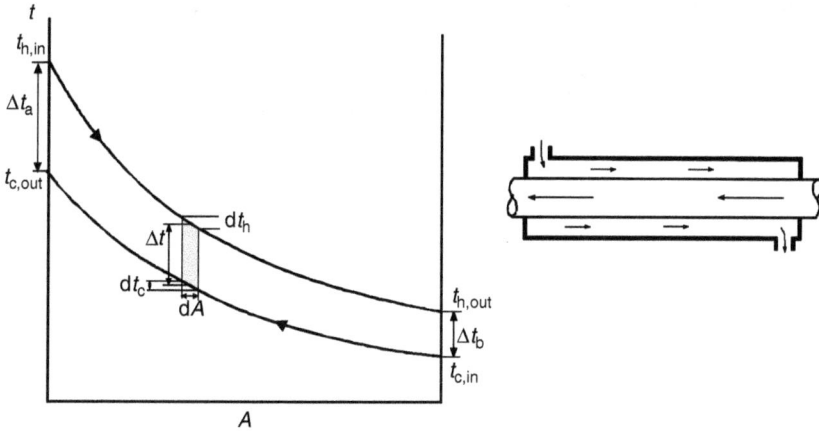

Figure 15.14: Principle sketch of temperature distributions in a counterflow heat exchanger.

15.4.1 Counterflow heat exchangers

Figure 15.14 shows in principle the temperature distributions of the hot and cold fluids in a single-pass counterflow heat exchanger.

For the element dA, one has

$$d\dot{Q} = U\, dA \cdot \Delta t = -(\dot{m}c_p)_h\, dt_h = -(\dot{m}c_p)_c\, dt_c \qquad (15.3)$$

where index h means hot fluid and index c means cold fluid.

For the heat capacity flow rates $\dot{m}c_p$, the following notations are introduced:

$$C_h = (\dot{m}c_p)_h, \quad C_c = (\dot{m}c_p)_c \qquad (15.4)$$

The total heat flow $\dot{Q}$ can be written as

$$\dot{Q} = C_h(t_{h_{in}} - t_{h_{out}}) \qquad (15.5)$$

$$\dot{Q} = C_c(t_{c_{out}} - t_{c_{in}}) \qquad (15.6)$$

Δt in eq. (15.3) can be written as

$$\Delta t = t_h - t_c \qquad (15.7)$$

The change in Δt is $d(\Delta t) = dt_h - dt_c$. With eqs. (15.3) and (15.4), one has

$$d(\Delta t) = d\dot{Q} \cdot \left(\frac{1}{C_c} - \frac{1}{C_h}\right) \qquad (15.8)$$

By using the first part of eq. (15.3), one has

$$d(\Delta t) = U \, dA \, \Delta t \left(\frac{1}{C_c} - \frac{1}{C_h} \right)$$

$$\frac{d(\Delta t)}{\Delta t} = U \, dA \left(\frac{1}{C_c} - \frac{1}{C_h} \right)$$

Integration over the whole heat exchanger gives

$$\int_{\Delta t_a}^{\Delta t_b} \frac{d(\Delta t)}{\Delta t} = \int_0^A U \, dA \left(\frac{1}{C_c} - \frac{1}{C_h} \right)$$

$$\ln \frac{\Delta t_b}{\Delta t_a} = UA \left(\frac{1}{C_c} - \frac{1}{C_h} \right)$$

With eqs. (15.2), (15.5), and (15.6), it is obtained

$$\ln \frac{\Delta t_b}{\Delta t_a} = \frac{\dot{Q}}{\Delta t_m} \left(\frac{(t_{c_{out}} - t_{c_{in}})}{\dot{Q}} - \frac{(t_{h_{in}} - t_{h_{out}})}{\dot{Q}} \right)$$

For Δt_m, one finds

$$\Delta t_m = \text{LMTD} = \frac{\Delta t_b - \Delta t_a}{\ln(\Delta t_b / \Delta t_a)}$$

or

$$\Delta t_m = \text{LMTD} = \frac{(t_{h_{out}} - t_{c_{in}}) - (t_{h_{in}} - t_{c_{out}})}{\ln((t_{h_{out}} - t_{c_{in}})/(t_{h_{in}} - t_{c_{out}}))} \qquad (15.9)$$

Equation (15.9) gives Δt_m for a counterflow heat exchanger. This temperature difference, which also will be applied for other heat exchangers, is called LMTD, the logarithmic mean temperature difference.

Observe that if the heat capacity flow rates of the fluids are equal, the temperature difference will be constant across the heat exchanger, i.e., $\Delta t = \Delta t_m = (t_{h_{out}} - t_{c_{in}}) = (t_{h_{in}} - t_{c_{out}})$.

15.4.2   Parallel flow heat exchangers

Figure 15.15 shows the principle temperature distributions in a single-pass parallel flow heat exchanger.

With the notations in Fig. 15.15, it is possible to derive, in a similar manner, an expression for Δt_m of a parallel flow heat exchanger:

$$\Delta t_m = \frac{\Delta t_b - \Delta t_a}{\ln(\Delta t_b / \Delta t_a)}$$

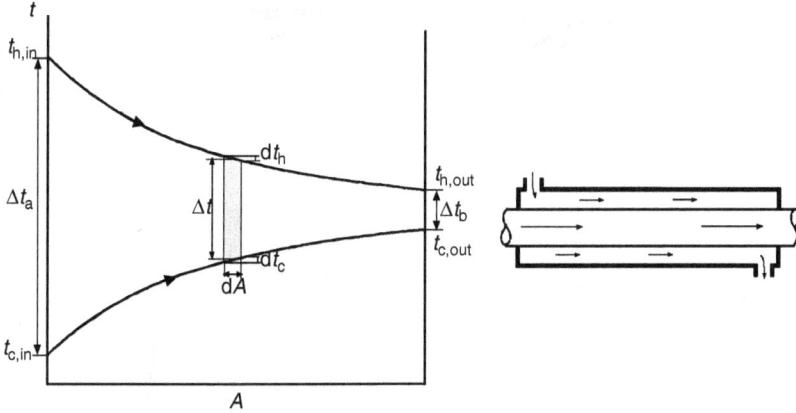

Figure 15.15: Principle sketch of a parallel flow heat exchanger.

or

$$\Delta t_m = \frac{(t_{h_{in}} - t_{c_{in}}) - (t_{h_{out}} - t_{c_{out}})}{\ln((t_{h_{in}} - t_{c_{in}})/(t_{h_{out}} - t_{c_{out}}))} \qquad (15.10)$$

15.4.3 Correction factors for LMTD for noncounterflow heat exchangers

In engineering analysis and design, commonly the LMTD according to eq. (15.9) is used independent of the heat exchanger type and the flow arrangement. The heat flow is written as

$$\dot{Q} = UA \cdot F \cdot \text{LMTD} \qquad (15.11)$$

where F $(0 < F \leq 1)$ is a correction factor which accounts for the deviation from the corresponding counterflow arrangement.

Commonly, two parameters P and R are introduced. These represent an efficiency or goodness number and the ratio between the heat capacity flow rates, respectively.

One has

$$P = \frac{t_{c_{out}} - t_{c_{in}}}{t_{h_{in}} - t_{c_{in}}} \qquad (15.12)$$

and

$$R = \frac{(\dot{m}c_p)_c}{(\dot{m}c_p)_h} \qquad (15.13a)$$

With eqs. (15.5) and (15.6), the ratio R can be written as

$$R = \frac{t_{h_{in}} - t_{h_{out}}}{t_{c_{out}} - t_{c_{in}}} \qquad (15.13b)$$

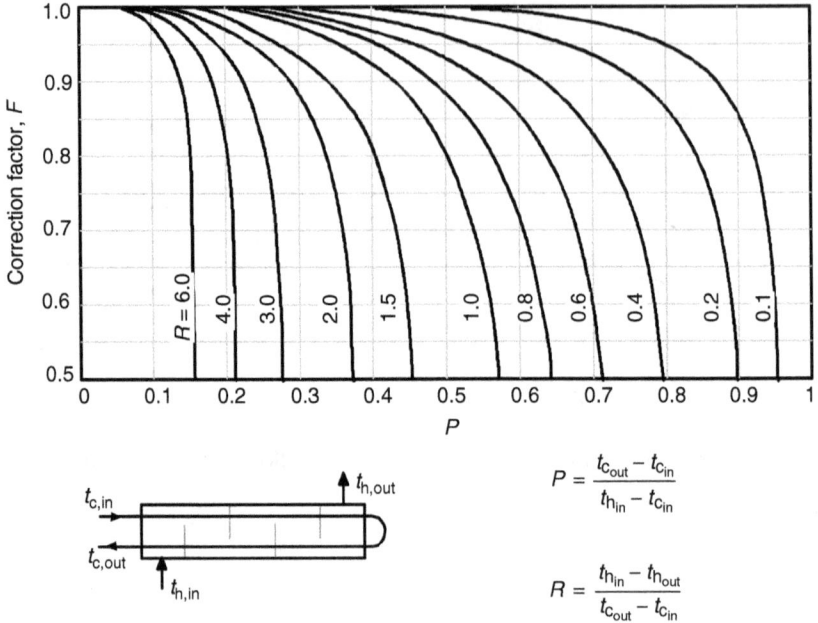

Figure 15.16: Correction factor F for a shell-and-tube heat exchanger with one shell pass and two tube passes.

The correction factor F depends on P and R as well as on the heat exchanger type. Analytical expressions are available in the international literature for a number of cases. The mathematics and the algebra to derive the expressions are quite extensive and here only some results are presented in a few figures and from these F can be determined.

Figure 15.16 gives F for a shell-and-tube heat exchanger with one shell pass and multiples of two tube passes (2, 4, 6, 8, ..., $2n$ tube passes). In Fig. 15.17, F is provided as a function of P and R for a shell-and-tube heat exchanger with two shell passes and four tube passes (or multiples of four tube passes), while Fig. 15.18 presents F for a cross flow heat exchanger where both fluids are unmixed. Figure 15.19 gives F for a cross flow heat exchanger where one fluid is unmixed while the other is mixed.

For technical applications, it is improper to use a heat exchanger with $F \leq 0.75$. If $F > 0.75$ cannot be achieved for a certain design, one should select another heat exchanger type. For $F < 0.75$, it is clear from the graphs that the curves become almost vertical, which means that small variations in the temperatures or flow rates will result in great differences in the performance of the heat exchanger.

More detailed information concerning the correction factor F can be found in Refs. [4–7].

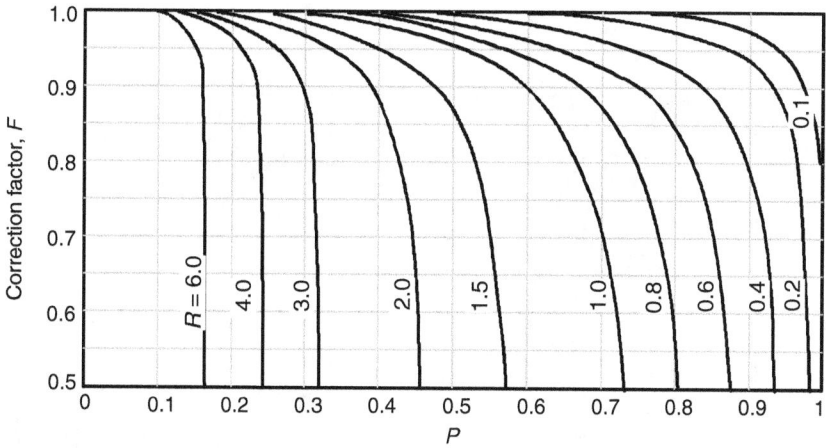

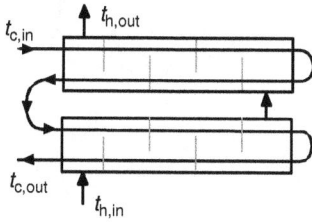

$$P = \frac{t_{c_{out}} - t_{c_{in}}}{t_{h_{in}} - t_{c_{in}}}$$

$$R = \frac{t_{h_{in}} - t_{h_{out}}}{t_{c_{out}} - t_{c_{in}}}$$

Figure 15.17: Correction factor F for a shell-and-tube heat exchanger with two shell passes and four tube passes.

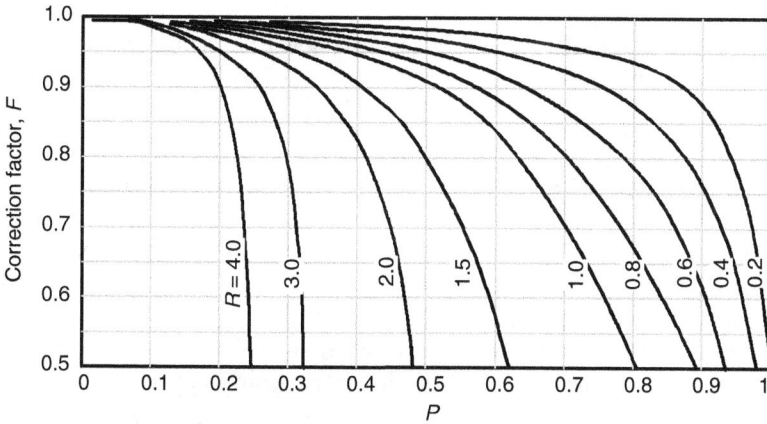

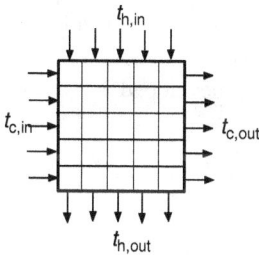

$$P = \frac{t_{c_{out}} - t_{c_{in}}}{t_{h_{in}} - t_{c_{in}}}$$

$$R = \frac{t_{h_{in}} - t_{h_{out}}}{t_{c_{out}} - t_{c_{in}}}$$

Figure 15.18: Correction factor F for a cross flow heat exchanger with both fluids unmixed.

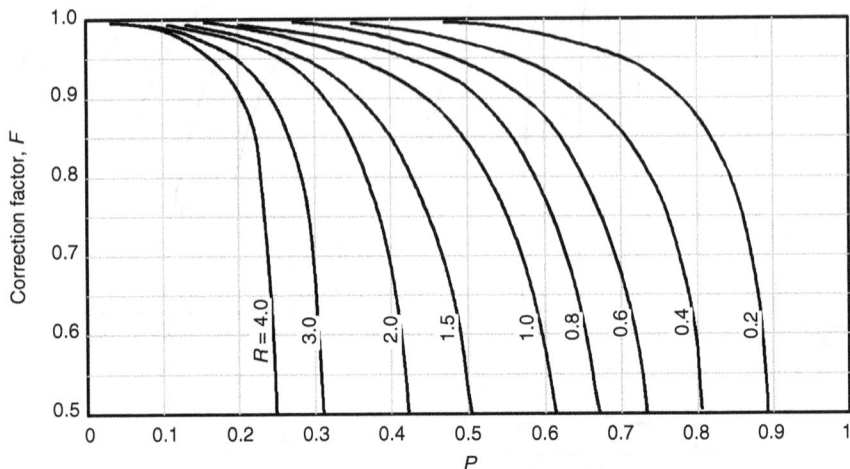

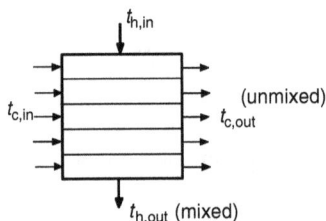

Figure 15.19: Correction factor F for a cross flow heat exchanger with one fluid unmixed.

15.5   The ε–NTU method for analysis of heat exchangers

In the analysis of the performance of a certain heat exchanger, the amount of heat being transferred, outlet temperatures of the fluids, and the pressure drops are of most interest. At the design and sizing stage of a heat exchanger, the heat transferring area and other dimensions are determined in such a way that a pre-scribed heat flow can be transferred and that the pressure drops are within permitted limits.

If the inlet and outlet temperatures of the hot and cold fluids are given, the LMTD method is quite suitable. In other cases, the so-called ε–NTU method is more appropriate. This method was originally developed by Kays and London (1955), see Ref. [8].

The efficiency or effectiveness ε is defined as

$$\varepsilon = \frac{\text{real amount of heat}}{\text{maximum possible transferrable amount of heat}} = \frac{\dot{Q}}{\dot{Q}_{max}} \qquad (15.14)$$

The theoretical maximum transferrable heat is that heat amount the fluid with the lowest heat capacity flow rate receives/gives up if its outlet temperature becomes equal to the inlet temperature of the other fluid. This means $\dot{Q}_{max} = C_{min}(t_{h_{in}} - t_{c_{in}})$. With the notations already introduced, one has:

$$\varepsilon = \frac{C_h(t_{h_{in}} - t_{h_{out}})}{C_{min}(t_{h_{in}} - t_{c_{in}})} = \frac{C_c(t_{c_{out}} - t_{c_{in}})}{C_{min}(t_{h_{in}} - t_{c_{in}})} \qquad (15.15)$$

where C_{min} is the smallest value of C_h and C_c.

From eqs. (15.14) and (15.15), it follows that

$$\dot{Q} = \varepsilon C_{min}(t_{h_{in}} - t_{c_{in}}) \qquad (15.16)$$

NTU, the number of transfer units is defined as

$$NTU = \frac{UA}{C_{min}} \qquad (15.17)$$

This dimensionless number (originally introduced by W. Nusselt) expresses the ratio between the heat capacity of the heat exchanger (W/K) and the smallest heat capacity flow rate $C_{min} = (\dot{m}c_p)_{min}$.

Consider now a counterflow heat exchanger. Equations (15.17), (15.2), and (15.9) give

$$NTU = \frac{\dot{Q}/\text{LMTD}}{C_{min}} \qquad (15.18)$$

The temperature differences in LMTD, eq. (15.9), can be rewritten as

$$(t_{h_{out}} - t_{c_{in}}) - (t_{h_{in}} - t_{c_{out}}) = (t_{h_{out}} - t_{h_{in}}) - (t_{c_{in}} - t_{c_{out}})$$

$$= -\frac{\dot{Q}}{C_h} + \frac{\dot{Q}}{C_c} = \dot{Q}\left(\frac{1}{C_c} - \frac{1}{C_h}\right) \qquad (15.19)$$

$$\frac{(t_{h_{out}} - t_{c_{in}})}{(t_{h_{in}} - t_{c_{out}})} = \frac{-(t_{h_{in}} - t_{h_{out}}) + (t_{h_{in}} - t_{c_{in}})}{(t_{h_{in}} - t_{c_{in}}) + (t_{c_{in}} - t_{c_{out}})}$$

$$= \frac{-\dot{Q}/C_h + \dot{Q}/\varepsilon C_{min}}{\dot{Q}/\varepsilon C_{min} - \dot{Q}/C_c} = \frac{C_c(C_h - \varepsilon C_{min})}{C_h(C_c - \varepsilon C_{min})} \qquad (15.20)$$

With eqs. (15.9), (15.18), (15.19), and (15.20), one obtains

$$NTU = \frac{1}{C_{min}} \frac{\ln((C_c/C_h) \cdot (C_h - \varepsilon C_{min})/(C_c - \varepsilon C_{min}))}{(1/C_c) - (1/C_h)} \qquad (15.21)$$

Table 15.2: ε–NTU relations for some common heat exchanger types (HEX).

HEX type	ε	
Parallel flow	$\varepsilon = \dfrac{1 - \exp[-\mathrm{NTU}(1 + C)]}{1 + C}$	Figure 15.20b
Counterflow	$\varepsilon = \dfrac{1 - \exp[-\mathrm{NTU}(1 - C)]}{1 - C\exp[-\mathrm{NTU}(1 - C)]} \quad C < 1$	Figure 15.20a
	$\varepsilon = \dfrac{\mathrm{NTU}}{1 + \mathrm{NTU}} \quad C = 1$	
(Shell-and-tube HEX)		
1 Shell pass $2, 4, 6, \ldots$ tube passes	$\varepsilon_1 = 2\left\{1 + C + (1 + C^2)^{1/2} \times \dfrac{1 + \exp[-\mathrm{NTU}(1 + C^2)^{1/2}]}{1 - \exp[-\mathrm{NTU}(1 + C^2)^{1/2}]}\right\}^{-1}$	Figure 15.21a
n Shell passes $2n, 4n, \ldots$ tube passes	$\varepsilon_n = \left[\left(\dfrac{1 - \varepsilon_1 C}{1 - \varepsilon_1}\right)^n - 1\right]\left[\left(\dfrac{1 - \varepsilon_1 C}{1 - \varepsilon_1}\right)^n - C\right]^{-1}$	Figure 15.21b
Cross flow (single pass)		
Both fluids unmixed	$\varepsilon \approx 1 - \exp\left[C^{-1}(\mathrm{NTU})^{0.22}\left\{\exp\left[-C(\mathrm{NTU})^{0.78}\right] - 1\right\}\right]$	Figure 15.21c
Both fluids mixed	$\varepsilon = \mathrm{NTU}\left[\dfrac{\mathrm{NTU}}{1 - \exp(-\mathrm{NTU})} + \dfrac{C(\mathrm{NTU})}{1 - \exp[-C(\mathrm{NTU})]} - 1\right]^{-1}$	Figure 15.21d
$C_{\min}$ unmixed $C_{\max}$ mixed	$\varepsilon = C^{-1}\left(1 - \exp[-C\{1 - \exp(-\mathrm{NTU})\}]\right)$	Figure 15.21f
$C_{\min}$ mixed $C_{\max}$ unmixed	$\varepsilon = 1 - \exp\left(-C^{-1}\{1 - \exp[-C(\mathrm{NTU})]\}\right)$	Figure 15.21e
All heat exchangers $C = 0$	$\varepsilon = 1 - \exp(-\mathrm{NTU})$	

$C = C_{\min}/C_{\max}$.

Assume first that $C_{\min} = C_c$ which means that $C_{\max} = C_h$. After some algebraic manipulations, one receives

$$\varepsilon = \frac{1 - \exp[-(1 - C_{\min}/C_{\max})\mathrm{NTU}]}{1 - C_{\min}/C_{\max}\exp[-(1 - C_{\min}/C_{\max})\mathrm{NTU}]} \tag{15.22}$$

If $C_{\min}$ is set to C_h, one will obtain exactly the same result.

Similar calculations can be carried out for other heat exchanger configurations and results are presented in the international literature.

In Tables 15.2 and 15.3, formulas $\varepsilon = \text{function}(\mathrm{NTU}, C_{\min}/C_{\max})$ and $\mathrm{NTU} = \text{function}(\varepsilon, C_{\min}/C_{\max})$, respectively, are given for several cases. In Figs. 15.20 and 15.21, some solutions are presented as diagrams. These are suitable for simple calculations and estimations.

Table 15.3: ε–NTU relations for some common heat exchangers (HEX).

HEX type	NTU	
Parallel flow	$NTU = -\dfrac{\ln[1-\varepsilon(1+C)]}{1+C}$	Figure 15.20b
Counterflow	$NTU = \dfrac{1}{C-1}\ln\left(\dfrac{\varepsilon-1}{\varepsilon C-1}\right)\quad C<1$	Figure 15.20a
	$NTU = \dfrac{\varepsilon}{1-\varepsilon}\quad C=1$	
(Shell-and-tube HEX)		
1 Shell pass 2, 4, 6, ... tube passes	$NTU = -(1+C^2)^{-1/2}\ln\left(\dfrac{E-1}{E+1}\right)\quad E = \dfrac{2/\varepsilon_1 - (1+C)}{(1+C^2)^{1/2}}$	Figure 15.21a
n Shell passes 2n, 4n, ... tube passes	Use the expression for one shell pass but with $\varepsilon_1 = \dfrac{F-1}{F-C}$, where $F = \left(\dfrac{\varepsilon C-1}{\varepsilon-1}\right)^{1/n}$	Figure 15.21b
Cross flow (one pass)		
C_{min} unmixed C_{max} mixed	$NTU = -\ln\left[1+C^{-1}\ln(1-\varepsilon C)\right]$	Figure 15.21f
C_{min} mixed C_{max} unmixed	$NTU = -C^{-1}\ln[C\ln(1-\varepsilon)+1]$	Figure 15.21e
All heat exchangers $C=0$	$NTU = -\ln(1-\varepsilon)$	

$C = C_{min}/C_{max}$.

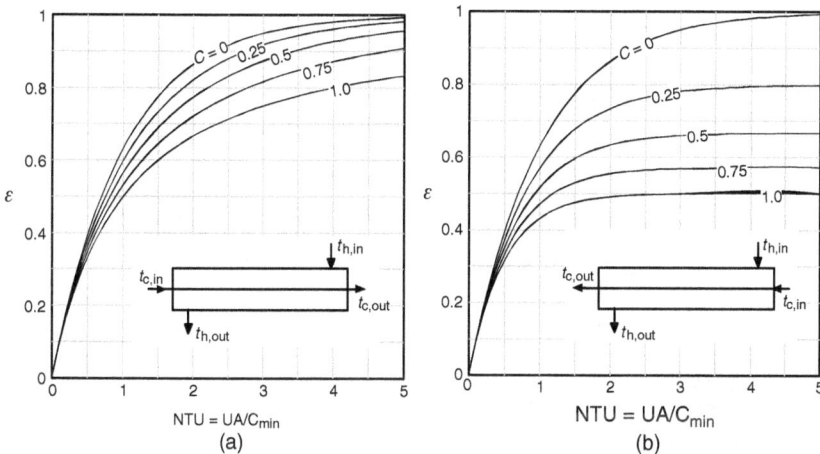

Figure 15.20: (a) ε–NTU for a counterflow heat exchanger. (b) ε–NTU for a parallel flow heat exchanger.

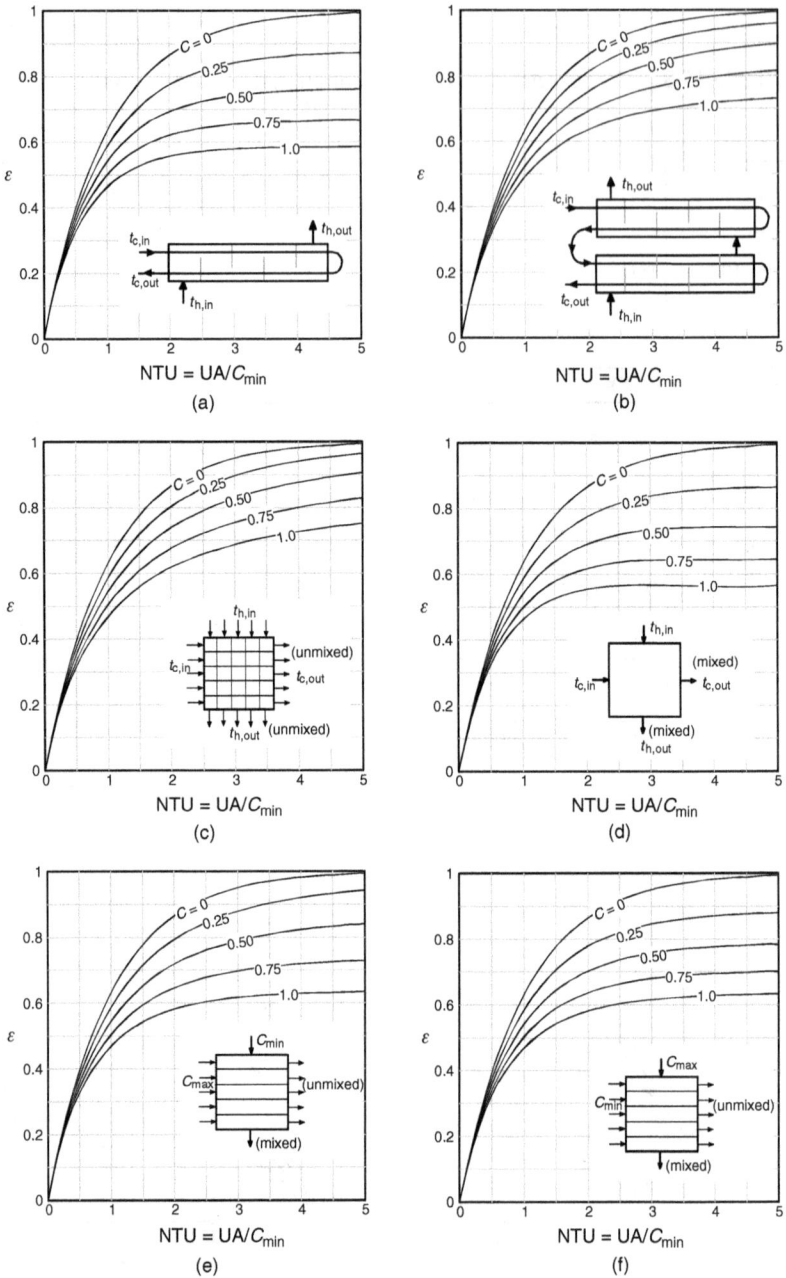

Figure 15.21: (a) ε–NTU for shell-and-tube heat exchangers with one shell pass and two tube passes. (b) ε–NTU for shell-and-tube heat exchangers with two shell passes and four tube passes. (c) ε–NTU for cross flow heat exchangers with both fluids unmixed. (d) ε–NTU for cross flow heat exchangers with both fluids mixed. (e) ε–NTU for cross flow heat exchangers where the fluid with C_{min} is mixed while the other is unmixed. (f) ε–NTU for cross flow heat exchangers where the fluid with C_{max} is mixed while the other is unmixed.

Please note that in Figs. 15.20 and 15.21 and in the Tables 15.2 and 15.3, C is the heat capacity flow rate ratio, i.e., $C = C_{min}/C_{max}$.

15.6   Condensers and evaporators (boilers)

For cases with condensers and evaporators, the fluid being condensed or evaporated has a constant temperature (saturation temperature). If the effectiveness ε should be finite, the heat capacity flow rate C (C_h and C_c) must be infinite because the temperature difference is zero. C_{max} is thus infinite and the ratio $C_{min}/C_{max} = 0$.

15.7   Compact heat exchangers

15.7.1   Heat transfer and friction factor

In Section 2, it was said that a heat exchanger is compact when the ratio between the heat transferring area and the volume (A/V) was larger than $700\,m^2/m^3$. Commonly, at least one of the fluids in such heat exchangers is a gas. Many different configurations exist and heat transfer and pressure drop data are available in, for example, Ref. [8]. In this section, some configurations will be presented and the associated heat transfer and pressure drop data are given. Figure 15.22 shows the heat transfer coefficient and friction factor on the gas side for a so-called tubular heat exchanger with plane lamellas (also called plate fin-and-tube heat exchangers), while Fig. 15.23 shows corresponding data for a heat exchanger with circular or annular fins. Observe that the given data are only valid for specified dimensions which are given in each figure. In these figures, the Stanton number St and the Reynolds number are defined as

$$St = \frac{\alpha}{Gc_p}, \quad Re = \frac{GD_h}{\mu} \qquad (15.23)$$

where G is the mass velocity which is given by

$$G = \frac{\dot{m}}{A_{min}} \qquad (15.24)$$

where $\dot{m}$ is the mass flow rate and A_{min} the minimum cross flow area. The hydraulic diameter is defined as

$$D_h = 4\frac{A_{min}L}{A} \qquad (15.25)$$

where A is the total heat transferring area and L the depth of the heat exchanger in the main flow direction.

Tubular heat exchangers in staggered arrangement and with plane fins, as shown in Fig. 15.22, are common in, for example, air-conditioning units and refrigeration equipment. Several correlations for heat transfer and friction factor are available

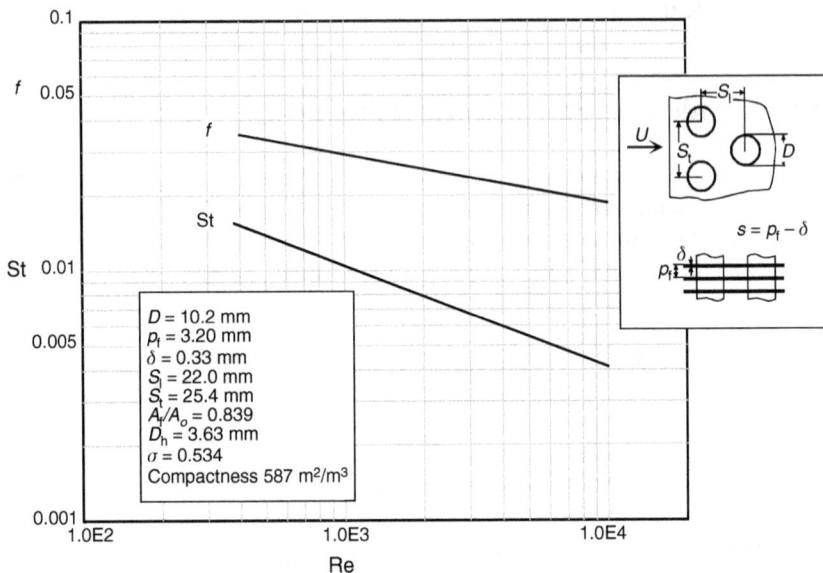

Figure 15.22: Stanton number, St, and friction factor, f, for a finned tubular heat exchanger. (Based on Ref. [8].)

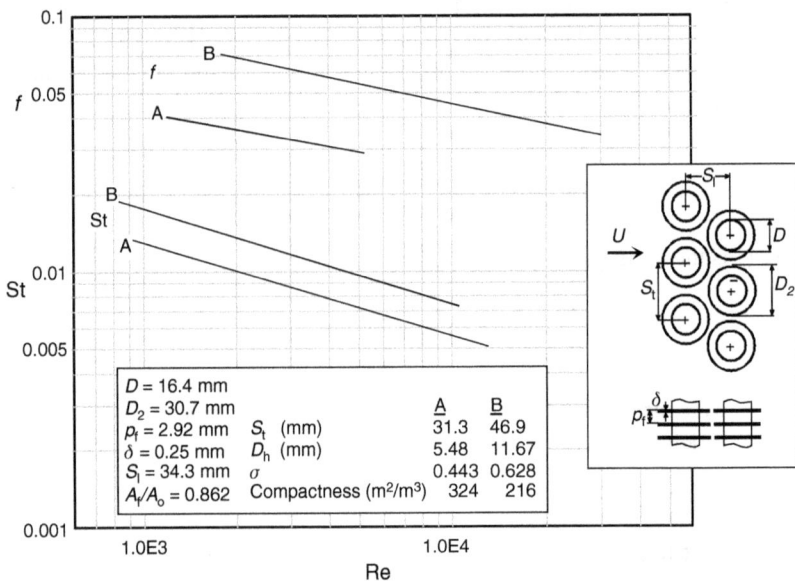

Figure 15.23: Stanton number, St, and friction factor, f, for a tubular heat exchanger with plane fins. (From Ref. [8].)

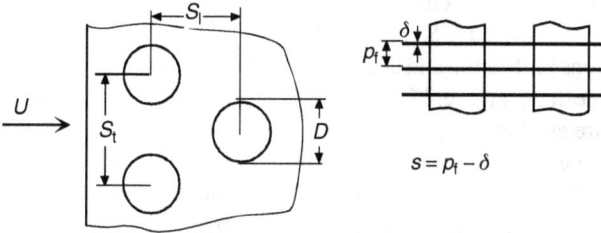

Figure 15.24: Notations for tubular heat exchangers with plane fins.

in the literature, see, for example, Eriksson and Sundén [9]. Gray and Webb [10] suggest the following correlation:

$$St = 0.14 Re_D^{-0.328} \left(\frac{S_t}{S_l}\right)^{-0.502} \left(\frac{s}{D}\right)^{0.0312} Pr^{-2/3} \tag{15.26}$$

This equation is valid for four or more tube rows in the main flow direction. For fewer tube rows, N, a correction is recommended as

$$\frac{St_N}{St} = 0.991 \left(2.24 Re_D^{-0.092} \left(\frac{N}{4}\right)^{-0.031}\right)^{0.607(4-N)} \tag{15.27}$$

The Reynolds number is calculated as

$$Re_D = \frac{GD}{\mu} \tag{15.28}$$

where D is the outer diameter of the tubes. Other notations are given in Fig. 15.24.

The friction factor is splitted up in two parts which can be related to the pressure drop over the fins, f_f, and the pressure drop over the tubes, f_t. The friction factor f_f is calculated as

$$f_f = 0.508 Re_D^{-0.521} (S_t/D)^{1.318} \tag{15.29}$$

and f_t can be calculated as the correlation for tube bundles by Zukauskas and Ulinskas as reported in Ref. [11]. A somewhat simpler correlation has been presented by Jakob as reported in Ref. [10]. This relation reads

$$f_t = \frac{4}{\pi} \left(0.25 + \frac{0.118}{(S_t/D - 1)^{1.08}} Re_D^{-0.16}\right)(S_t/D - 1) \tag{15.30}$$

The total friction factor is calculated as

$$f = f_f \frac{A_f}{A_o} + f_t \left(1 - \frac{A_f}{A_o}\right)\left(1 - \frac{\delta}{p_f}\right) \tag{15.31}$$

where A_f is the fin area and A_o the total area on the gas side, i.e., the fin area plus the tube area.

The correlation by Gray and Webb is valid for $400 \leq \text{Re} \leq 24{,}700$, $1.97 \leq S_t/D \leq 2.55$, $1.70 \leq S_l/D \leq 2.58$, and $0.08 \leq s/D \leq 0.64$.

Correlations for friction and heat transfer on the gas side in tubular heat exchangers with plane fins, Figure 15.23, are available in Ref. [11]. Data for other fin geometries are available in Ref. [12].

By using the heat transfer data (α from St) for the gas side, from, for example, Figs. 15.22, 15.23, and eq. (15.26), and the heat transfer coefficient for inside tube flow (from, for example, Ref. [2]), the thermal resistance ($1/UA$) for the heat exchanger can be calculated. The LMTD method or the ε–NTU method can then be applied for design and sizing or analysis of the heat exchanger.

As the thermal resistance on the gas side is determined, the fin efficiency has to be considered, see, Ref. [2]. The total resistance (except fouling factors) is written as

$$\frac{1}{UA} = \frac{1}{\phi_o A_o \alpha_o} + \frac{b_w}{\lambda_w A_{vl}} + \frac{1}{\alpha_i A_i} \tag{15.32}$$

(In eq. (15.32) it is assumed that only the gas side (outer surface) has fins while the inside is smooth.)

The efficiency ϕ_o on the gas side is related to the fin efficiency

$$\phi_o = 1 - \frac{A_f}{A_o}(1 - \phi) \tag{15.33}$$

where A_o is the total heat transferring area on the gas side.

15.7.2 Pressure drop in compact heat exchangers

The pressure drop on the gas side in compact heat exchangers like those in Figs. 15.22 and 15.23 is usually splitted up in three components, namely the frictional loss, acceleration of the fluid, and the inlet and outlet losses. For tubular heat exchangers with plane fins as in Fig. 15.23 and if the gas is passing the tubes in cross flow, the pressure drop is calculated as

$$\Delta p = \frac{G^2}{2\rho_{in}}\left[(1+\sigma^2)\left(\frac{\rho_{in}}{\rho_{out}}-1\right)+f\frac{A}{A_{min}}\frac{\rho_{in}}{\rho_m}\right] \tag{15.34}$$

In eq. (15.34), ρ_{in} is the density at the inlet, ρ_{out} is the density at the outlet, and G is determined by eq. (15.24). The area ratio σ is determined as

$$\sigma = \frac{A_{min}}{A_{front}} \tag{15.35}$$

The average density ρ_m is calculated as

$$\frac{1}{\rho_m} = \frac{1}{2}\left(\frac{1}{\rho_{in}} + \frac{1}{\rho_{out}}\right) \tag{15.36}$$

In eq. (15.33), the inlet and outlet losses are included in the friction factor f.

For finned PHEs, see Fig. 15.5, the pressure drop is calculated as

$$\Delta p = \frac{G^2}{2\rho_{in}}\left[\underbrace{(K_c + 1 - \sigma^2)}_{\text{inlet}} + \underbrace{2\left(\frac{\rho_{in}}{\rho_{out}} - 1\right)}_{\text{acceleration}} + \underbrace{f\frac{A}{A_{min}}\frac{\rho_{in}}{\rho_m}}_{\text{friction}} - \underbrace{(1 - K_e - \sigma^2)\frac{\rho_{in}}{\rho_{out}}}_{\text{outlet}}\right]$$

(15.37)

In eq. (15.37), K_c is a contraction coefficient at the inlet and K_e is an expansion coefficient at the outlet. Typical values of K_c and K_e are given in Ref. [8].

15.7.3   Trends in development and ongoing research

In Ref. [8], heat transfer and pressure drop data are available for a number of configurations for compact heat exchangers. However, end users of heat exchangers require increased compactness and cheaper manufacturing techniques. The heat transferring surfaces then need to be modified or further developed, and innovative new surfaces are also of interest. This also requires that new heat transfer and pressure drop data are established. In Refs. [13–17] recent R&D works are exemplified.

15.8   Shell-and-tube heat exchangers

The so-called shell-and-tube heat exchangers are the most common ones in the process industries. The heat power is commonly more than 1 MW and the heat transferring area might be up to 5000 m^2. Figure 15.25a–d shows principle sketches of some design layouts.

The advantages which are commonly associated with shell-and-tube heat exchangers are:

- great flexibility in operating conditions: phase change, condensation, evaporation;
- robust equipment;
- huge operating pressure range;
- thermal stresses can be handled by proper selection of material;
- fins can be used on the tube surfaces and then the heat transferring area is increased.

Some disadvantages associated with shell-and-tube heat exchangers are:

- risk for flow-induced vibrations,
- hard to perform an accurate design because the correlations on the shell side suffice of inaccuracies.

The tube length might be 1–20 m and the shell diameter is typically in the range of 0.25–3.1 m. The outer tube diameter is within 6–51 mm.

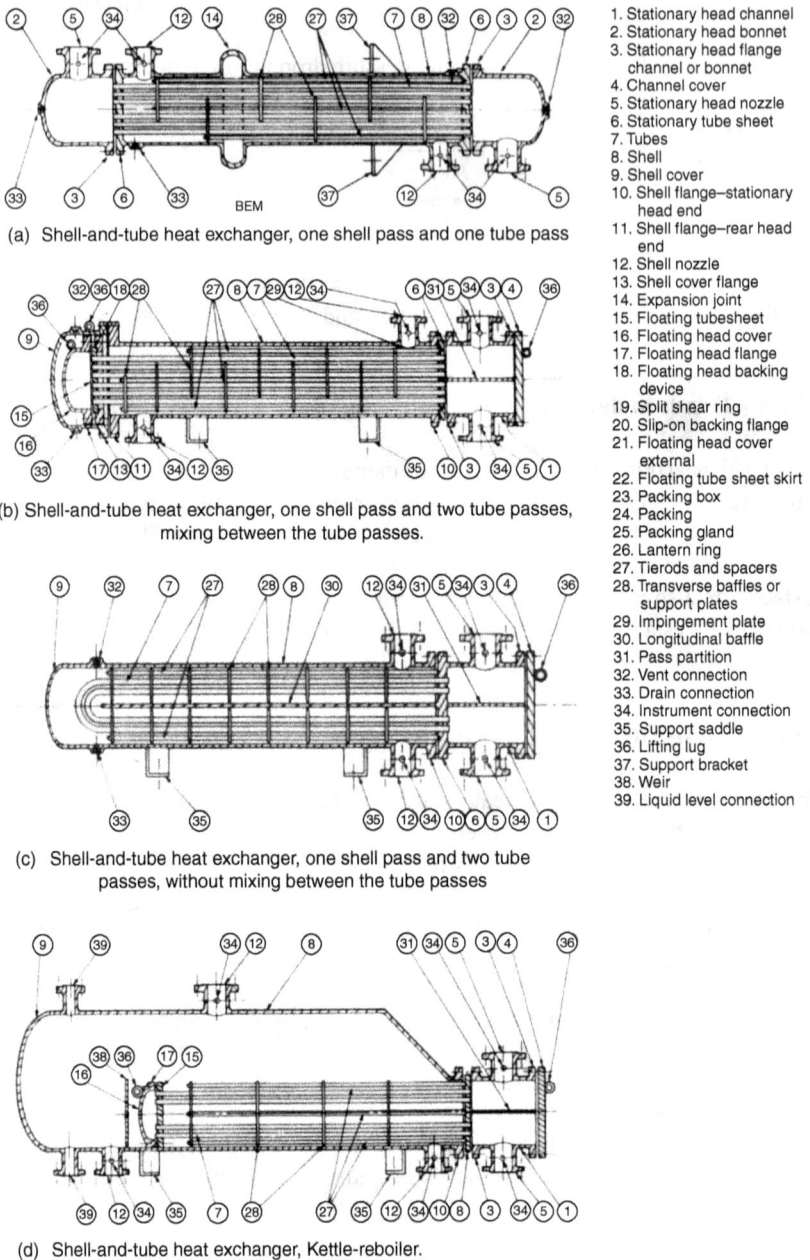

(a) Shell-and-tube heat exchanger, one shell pass and one tube pass

(b) Shell-and-tube heat exchanger, one shell pass and two tube passes, mixing between the tube passes.

(c) Shell-and-tube heat exchanger, one shell pass and two tube passes, without mixing between the tube passes

(d) Shell-and-tube heat exchanger, Kettle-reboiler.

1. Stationary head channel
2. Stationary head bonnet
3. Stationary head flange channel or bonnet
4. Channel cover
5. Stationary head nozzle
6. Stationary tube sheet
7. Tubes
8. Shell
9. Shell cover
10. Shell flange–stationary head end
11. Shell flange–rear head end
12. Shell nozzle
13. Shell cover flange
14. Expansion joint
15. Floating tubesheet
16. Floating head cover
17. Floating head flange
18. Floating head backing device
19. Split shear ring
20. Slip-on backing flange
21. Floating head cover external
22. Floating tube sheet skirt
23. Packing box
24. Packing
25. Packing gland
26. Lantern ring
27. Tierods and spacers
28. Transverse baffles or support plates
29. Impingement plate
30. Longitudinal baffle
31. Pass partition
32. Vent connection
33. Drain connection
34. Instrument connection
35. Support saddle
36. Lifting lug
37. Support bracket
38. Weir
39. Liquid level connection

Figure 15.25: Shell-and-tube heat exchangers. (a) Shell-and-tube heat exchanger, one shell pass and one tube pass. (b) Shell-and-tube heat exchanger, one shell pass and two tube passes, mixing between the tube passes. (c) Shell-and-tube heat exchanger, one shell pass and two tube passes, without mixing between the tube passes. (d). Shell-and-tube heat exchanger, Kettle reboiler.

15.8.1   Practical design aspects

Temperature differences: $t_{h_{in}} - t_{c_{out}} > 20°C$, $t_{h_{out}} - t_{c_{in}} > 5°C$.

Temperature level: If one of the fluids is at a high temperature, this fluid should be on the tube side as the number of components to be manufactured in high temperature material will be limited.

Pressure drop: The order of magnitude of Δp is typically 10–500 kPa on both the tube and shell sides but usually a little less on the shell side.

Pressure level: The fluid with the highest pressure should be on the tube side.

Viscosity: The most viscous fluid should be on the shell side.

Rate of mass flow: The fluid with the smallest rate of mass flow should be on the shell side.

Corrosion: The fluid being most corrosive should be on the tube side to minimize the damage effect.

Fouling: The fluid suspected to foul the surface should be on the tube side.

15.8.2   Heat transfer and pressure drop on the tube side

The pressure drop on the tube side consists of expansion and contraction losses at inlets and outlets, losses in U-bends or mixing chambers, and the friction losses in the straight tubes. For long tubes, the friction loss dominates. For calculation of this, standard methods, see Ref. [2], can be used. For other losses, reference is given in handbooks, e.g., Ref. [7]. The heat transfer coefficient inside the tubes can also be determined by standard methods as presented in Ref. [2] if the tube surfaces are smooth.

15.8.3   Heat transfer and pressure drop on the shell side

The flow field on the shell side is quite complicated. This is in principle conjectured in Figs. 15.3, 15.12, and 15.25. The fluid is entering the shell side through an inlet pipe and passes across a tube bundle (sometimes an impingement plate is placed just below the inlet pipe to retard the fluid and protect the tubes). The baffles direct the flows but also support the tubes. Between the baffles and the inner shell surface, gaps are present and similarly between the baffle holes and the tubes gaps exist. Leakage occurs through these gaps or openings. This means that part of the mass flow is not passing the tubes in cross flow.

Figure 15.26 shows Tinker's principle sketch of the flow field, see Ref. [18]. The nonuniformity in flow velocity and direction as well as the leakage flows makes a precise calculation of the heat transfer coefficient and the pressure drop difficult. The notations A–F in Fig. 15.26 represent:

A: leakage flow due to the gaps between baffle holes and tubes
B: main flow path, approximately cross flow.
C: bypass flow between tube bundle and inner shell surface.
E: leakage flow between baffles and inner shell surface.
F: (not marked in Fig. 15.26): bypass flow in streaks as a result of missing tubes in some regions.

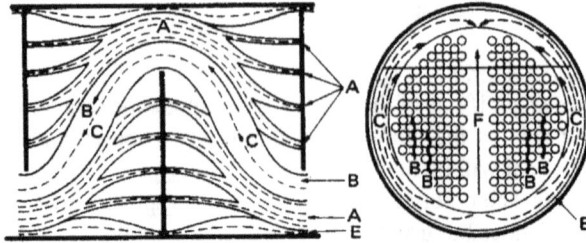

Figure 15.26: Tinker's sketch of the flow field on the shell side in a shell-and-tube heat exchanger.

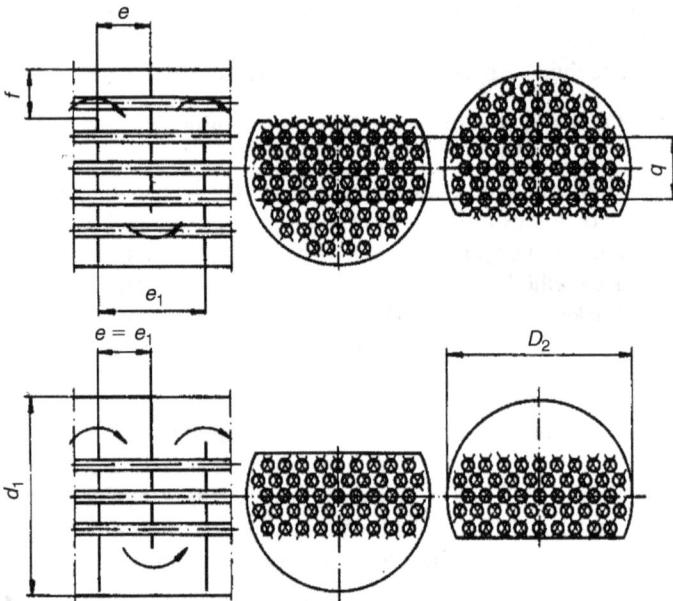

Figure 15.27: Baffle designs.

Figure 15.27 shows two examples of baffles.

The baffle distance in the axial direction is typically $0.25 \times$ shell diameter. The heat transfer coefficient α_S is calculated as

$$\alpha_s = c\alpha_{\text{tube bundle}} \tag{15.38}$$

where $\alpha_{\text{tube bundle}}$ is the heat transfer coefficient for a tube bundle in cross flow. Such coefficients can be determined as described in Refs. [2, 19, 20]. The correction factor c involves several factors considering effects of leakage, bypass flow, etc., see, for example, Ref. [7]. The order of magnitude for c is $c \approx 0.6$.

The pressure drop on the shell side is spitted up in the pressure drop contributions at the inlet and outlet, cross flow over the tube bundle, and flow in the so-called window sections. A window section is the region between the inner shell surface and the location where a baffle ends, see also Fig. 15.25. For the tube bundle cross flow, the pressure drop is calculated as

$$\Delta p_c = \Delta p_{\text{tube bundle}}(N_b - 1)R_1 \tag{15.39}$$

where $\Delta p_{\text{tube bundle}}$ is the pressure drop across the tube bundle, see Refs. [2, 11, 19, 20], N_b the number of baffles and R_1 a correction factor taking leakage and bypass flow into account, see Ref. [7].

For the inlet and outlet, the pressure drop is written as

$$\Delta p_e = \Delta p_{\text{tube bundle}} \frac{N_c + N_{cw}}{N_c} R_2 \tag{15.40}$$

where N_c is the number of tube rows in cross flow between two adjacent baffles and N_{cw} the number of tube rows in the window section, and R_2 a correction factor for the bypass flow and the specific conditions at the inlet and outlet regions. For more details, see Ref. [7].

In the window section, the pressure drop is calculated from

$$\Delta p_w = \Delta p_{\text{tkw}} N_b R_3 \tag{15.41}$$

where Δp_{tkw} is the pressure drop over the tube bundle in the window section and R_3 is a correction factor for leakage.

The total pressure drop Δp_{tot} is calculated as

$$\Delta p_{\text{tot}} = \Delta p_c + \Delta p_e + \Delta p_w \tag{15.42}$$

15.8.4 Error estimations

In the design process of shell-and-tube heat exchangers, it is important to know the accuracy in the calculations of the heat transfer coefficient α_s and the pressure drop Δp_{tot} on the shell side. α_s is estimated to be accurate within 25% while Δp_{tot} is within 40–75%.

15.8.5 Need for research

As is evident from the presentation above, the uncertainty in the estimations of the heat transfer coefficient and the pressure drop is rather high, particularly for the shell side. To improve the correlations and methods to calculate the correction factors is indeed a difficult task due to the complex geometry and the complex flow field. The risk of flow-induced vibrations at high velocities has to be considered as high velocities are also good for the thermal performance of the heat exchanger. Studies of flow-induced vibrations for idealized situations are presented in Refs. [21, 22].

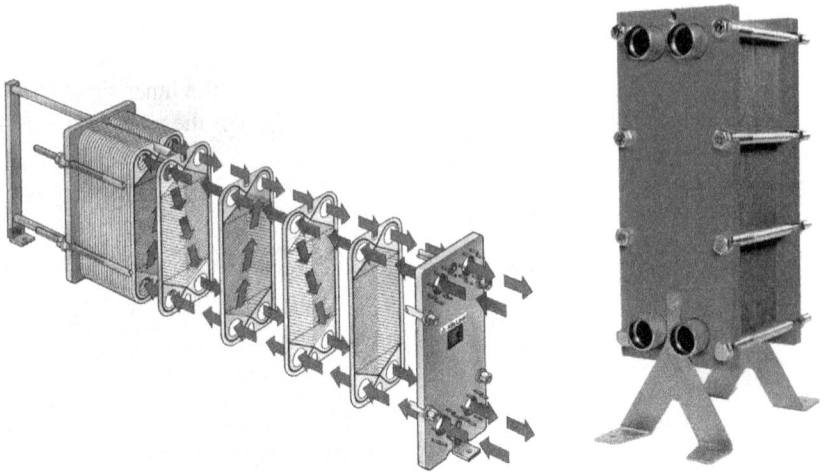

Figure 15.28: Plate heat exchanger – exploded and assembled views.

15.9 Plate heat exchangers

PHEs (plate-and-frame heat exchangers), see Fig. 15.28, is the second most common heat exchanger. PHEs have in general higher heat transfer coefficients (U-value) and higher compactness than shell-and-tube heat exchangers and in addition they are more easy to clean as the whole package can be disassembled. However, the operating pressures and temperatures are lower.

The heat transferring surface consists of a number of plates, see Figs. 15.28 and 15.29, which are assembled together in a package. Every plate is equipped with a gasket which is fixed in grooves on one side along the plate border. The gasket seals off against the adjacent plate. As is evident in Fig. 15.29, a plate has holes at the corners. With the layout of the gasket and its placement, the flow is controlled so that a stream either enters or passes the channel between two adjacent plates. As the plates are assembled to a package, see Fig. 15.28, the holes and the gaskets create a channel system. The hot and cold fluids are passing and entering every second channel. The operation is most commonly in counterflow. The distance between the plates is very small, typically a few millimeters. A big heat transfer area per unit volume is achieved.

Commonly, the plate surfaces are corrugated which creates turbulence and enhanced mixing and as a result the heat transfer coefficient is high. The corrugation also improves the stiffness of the plates which means that plate material thickness can be made small (typically 0.5–0.6 mm). The thermal resistance in the plates will then be small. The surface pattern is commonly splitted up in two main patterns, namely the so-called herringbone pattern and the washboard pattern. These are shown in Fig. 15.29. Other patterns exist and combined patterns also are available.

(a)

(b)

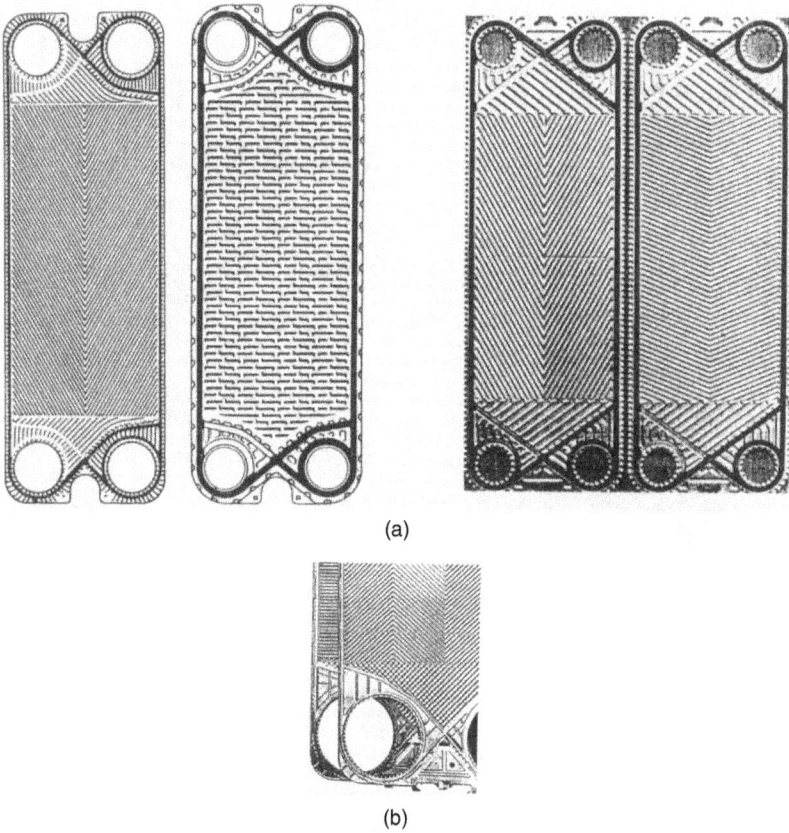

Figure 15.29: Plate pattern.

The support structure of a PHE consists of one fixed and one adjustable end plate. The thick end plates are bolted together with the plate package in between. The number of bolts depends on the operating pressure.

The plates can be coupled together in several ways. The plate package can be divided into a number of streaks for the fluids. In Fig. 15.30, a few examples of couplings are shown. Nowadays, PHEs are also manufactured with plates without gaskets. The PHEs are then brazed or welded. For brazed PHEs, vacuum brazing is applied and the contact points are then brazed together creating a strong bond. The operating pressures and temperatures can then be much higher. The disadvantage is that the brazed PHE cannot be disassembled. Figure 15.31 shows pictures of some brazed PHEs.

Sometimes a thermal length θ is introduced as heat exchangers are discussed. It is defined as

$$\theta = \frac{\Delta t}{\text{LMTD}} \tag{15.43}$$

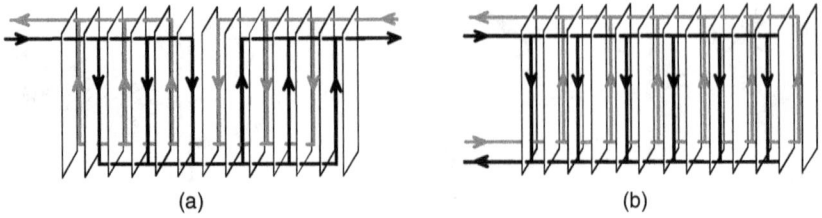

Figure 15.30: Flow streaks. (a) Two streaks $(2 \times 3/2 \times 3)$. (b) One streak $(1 \times 6/1 \times 6)$.

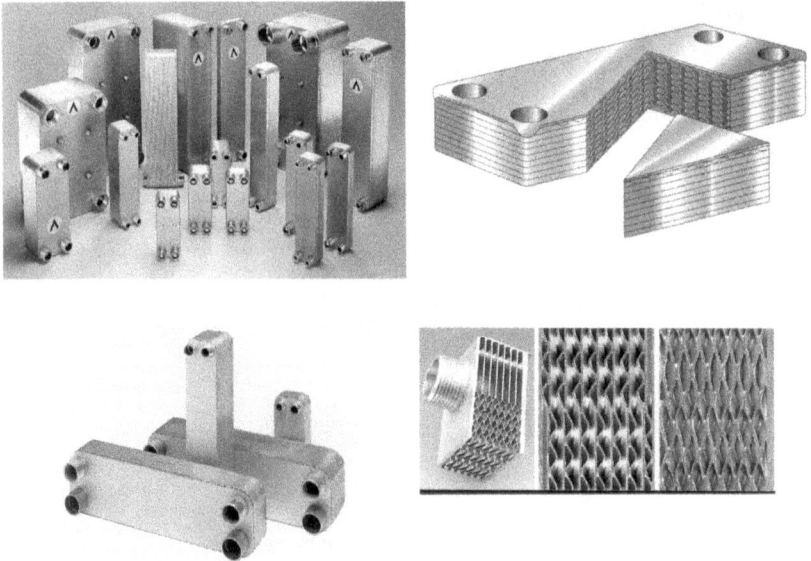

Figure 15.31: Brazed PHEs.

where Δt is the temperature difference of one of the fluids and LMTD the logarithmic mean temperature difference for an ideal counterflow heat exchanger. If Δt is equal to the temperature difference for the fluid with the smallest heat capacity flow rate, θ will be equal to NTU, i.e., the number of transfer units.

For plates with a tight pattern, the pressure drop is high and the heat transfer is very efficient. One then has a thermally long channel (high θ). If plates have a more open pattern, the pressure drop will be low and the heat transfer will be worse. One then has a thermally short channel (low θ). It is possible to assemble plates with different patterns together. The result will be something in between a long and a short channel in terms of pressure drop and heat transfer performance.

15.10 Regenerative heat exchangers

A regenerative heat exchanger has a setup of channels inside a relatively large matrix of a solid material. The hot and cold fluids pass the matrix in an alternating manner.

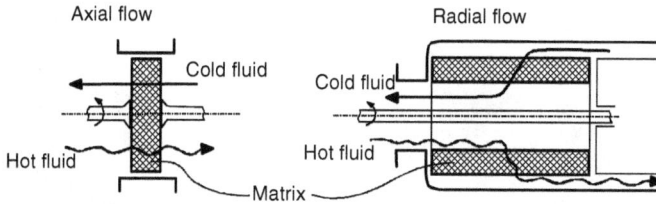

Figure 15.32: Rotating regenerative heat exchanger.

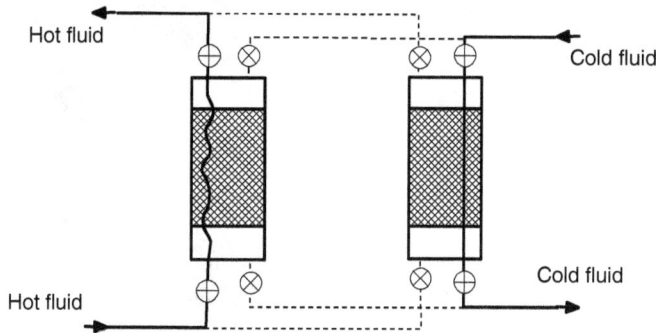

Figure 15.33: Static regenerative heat exchanger.

As the hot fluid passes the flow channels in the matrix, heat is transferred to the matrix and the temperature (internal energy) of the matrix is increased. In the next step, as the cold fluid is passing through the matrix heat is transferred to the cold fluid and the matrix is cold down.

Regenerative heat exchangers are classified as static or dynamic. The dynamic ones have moving parts. In most cases, the matrix is in the form of a planar disk or a drum, see Fig. 15.32. The matrix material is exposed sequentially to the hot and cold fluids. Most commonly, the matrix is rotating. This type is frequently occurring as preheater in heat and power plants (Ljungstrom preheater or regenerator) and for heat recovery in air-conditioning units. In a variant of this heat exchanger, the matrix is fixed and instead the inlet and outlet nozzles of the fluids are moving over the matrix cross section.

Static regenerative heat exchangers have no moving parts except for supply and delivery auxiliary equipment. Most applications have continuous flows of the hot and cold fluids and then two matrices are required and periodic switching occurs, see Fig. 15.33. At every instant of time, one matrix is in contact with the hot fluid, while the other one is exchanging heat with the cold fluid.

15.10.1 Rotating heat exchanger for heat recovery and air preheating

Figure 15.34 shows the matrix (rotor or wheel) of a typical rotating regenerative heat exchanger. The rotating wheel is built up by two foil systems arranged in a way that small channels are created. The wheel rotates relatively slow, up to 10 rpm. Usually

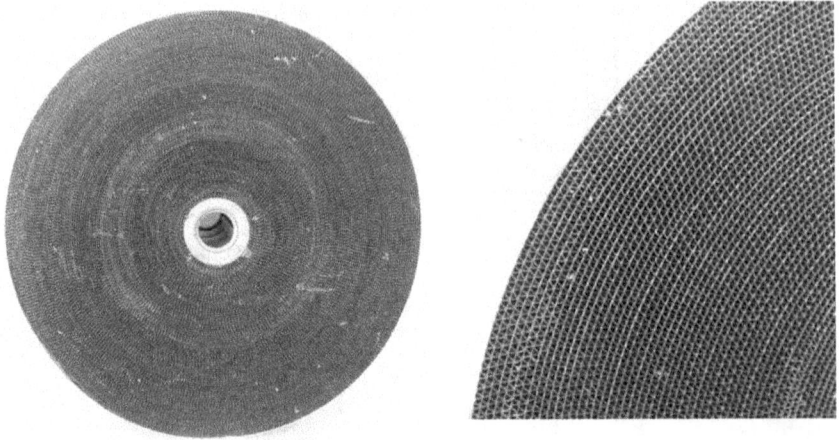

Figure 15.34: The wheel or rotor of a rotating regenerative heat exchanger.

approximately 50% of the front surface is exposed for one of the fluids, while the other half is open for the other fluid. Commonly, counterflow operation prevails. The wheel is heated up and cooled in a periodic manner. In continuous operation, the wheel approaches a certain average temperature and periodic fluctuations around this temperature occur.

The conventional theory for calculation of the heat transfer coefficient in rotating regenerative heat exchangers was established in 1930s by Hausen. Without derivation, the following equation is given for U.

$$\frac{1}{U} = (\tau_1 + \tau_2) \left(\frac{1}{\alpha_1 \tau_1} + \frac{\delta_m}{3\lambda_m} \left(\frac{1}{\tau_1} + \frac{1}{\tau_2} \right) + \frac{1}{\alpha_2 \tau_2} \right) \qquad (15.44)$$

where τ_1 and τ_2 are the time duration the matrix is in contact with fluids 1 and 2, respectively. δ_m is the material thickness and λ_m is its thermal conductivity. α_1 and α_2 are the heat transfer coefficients in the channels for fluids 1 and 2, respectively.

The analysis by Hausen was based on several assumptions and later research has suggested other methods for the analysis.

Reference [23] gives the following method to analyze a rotating regenerator in counterflow:

$$\varepsilon = \varepsilon_{\text{counter flow}} \times \left[1 - \frac{1}{9(C_r^*)^{1.93}} \right] \qquad (15.45)$$

where $\varepsilon_{\text{counter flow}}$ is the effectiveness of a nonrotating counterflow heat exchanger and

$$C_r^* = \frac{(mc_p)_{\text{hex}}\omega}{C_c} \qquad (15.46)$$

where $(mc_p)_{\text{hex}}$ is the heat capacity of the matrix material, ω the rotational speed (revolutions/s), and $C_c = (\dot{m}c_p)_{\text{cold fluid}}$.

If the heat capacity flow rates of the hot and cold fluids are equal, i.e., $C_c = C_h$ (see Table 15.2), one has

$$\varepsilon = \frac{NTU_0}{1 + NTU_0} \times \left[1 - \frac{1}{9(C_r^*)^{1.93}} \right] \tag{15.47}$$

where

$$NTU_0 = \frac{1}{C_{min}} \left[\frac{1}{(1/\alpha A)_c + (1/\alpha A)_h} \right] \tag{15.48}$$

For heat exchangers of material with high thermal conductivity, a correction due to longitudinal heat conduction must be introduced. One then has

$$\varepsilon = \varepsilon \times \text{correction}_\lambda \tag{15.49}$$

Reference [23] gives when $C_c = C_h$.

$$\text{Correction}_\lambda = 1 - \left[\frac{1}{1 + NTU_0((1 + \lambda\Phi)/(1 + \lambda NTU_0))} - \frac{1}{1 + NTU_0} \right] \tag{15.50}$$

where $\lambda = \lambda_{hex} A_{hc}/LC_{min}$, λ_{hex} is the thermal conductivity of the heat exchanger material, A_{hc} is the heat conducting area in the longitudinal direction, and $\Phi \approx \sqrt{\lambda NTU_0/(1 + \lambda NTU_0)}$.

References

[1] M.N. Özisik, Heat Transfer—A Basic Approach, McGraw-Hill, New York (1985).

[2] J.P. Holman, Heat Transfer, 10th ed., McGraw-Hill, New York (2009).

[3] Tubular Exchanger Manufacturers Association, Standards, TEMA, New York (1959).

[4] R.A. Bowman, A.C. Mueller and W.M. Nagle, Mean temperature difference in design, Trans. ASME, 62, 283–294 (1940).

[5] K.A. Gardner, Variable heat transfer rate correction in multipass exchangers, shell-side film controlling, Trans. ASME, 67, 31–38 (1945).

[6] R.A. Stevens, J. Fernandes and J.R. Woolf, Mean temperature difference in one, two and three-pass cross flow heat exchangers, Trans. ASME, 79, 287–297 (1957).

[7] Heat Exchangers Design Handbook, Hemisphere Publ. Corp., Washington, DC (1983).

[8] W.M. Kays and A.L. London, Compact Heat Exchangers, 3rd ed., McGraw-Hill, New York (1984).

[9] D. Eriksson and B. Sundén, Plate fin-and-tube heat exchangers: A literature survey of heat transfer and friction correlations, in Progress in Engineering Heat Transfer (Eds. B. Grochal, J. Mikielewicz and B. Sundén), pp. 533–540, IFFM Publishers, Gdansk (Poland) (1999).

[10] L. Gray and R.L. Webb, Heat transfer and friction factor for plate finned-tube heat exchangers, Proc. 9th Int. Heat Transfer Conf., 6, 2745–2750 (1986).

[11] A. Zukauskas, High-Performance Single-Phase Heat Exchangers, Hemisphere Publ. Corp., New York (1989).

[12] R.L. Webb and N.H. Kim, Principles of Enhanced Heat Transfer, Taylor and Francis, New York (2005).

[13] R.L. Webb, Enhancement of single-phase heat transfer, in Handbook of Single-Phase Convective Heat Transfer (Eds. S. Kakac, R.K. Shah and W. Aung), John Wiley and Sons, New York (1987).

[14] B. Sundén and I. Karlsson, Enhancement of heat transfer in rotary heat exchangers by streamwise-corrugated flow channels, Experimental Thermal Fluid Sci., 4(3), 305–316 (1991).

[15] B. Sundén and J. Svantesson, Thermal hydraulic performance of new multi-louvered fins, in Heat Transfer 1990 (Ed. G. Hetsroni), vol. 5, pp. 92–96, Hemisphere Publ. Corp., New York (1990).

[16] B. Sundén and J. Svantesson, Heat transfer and pressure drop from louvered surfaces in automotive heat exchangers, Experimental Heat Transfer, 4, 111–125 (1991).

[17] A. Achaichia and T.A. Cowell, Heat transfer and pressure drop characteristics of flat tube and louvered plate fin surfaces, Experimental Thermal Fluid Sci., 1, 147–157 (1988).

[18] T. Tinker, Shell side characteristics of shell and tube heat exchangers, Proc. Gen. Disc. Heat Transfer, Inst. Mech. Engn., London (1951).

[19] A. Zukauskas, Heat transfer from tubes in cross flow, in Advances in Heat Transfer, vol. 18, pp. 87–159, Academic Press, New York (1987).

[20] A. Zukauskas and R. Ulinskas, Heat transfer in tube banks in cross flow, Hemisphere Publ. Corp., New York (1988).

[21] A. Zukauskas, V. Katinas and R. Ulinskas, Fluid dynamics and flow induced vibration of tube banks, Hemisphere Publ. Corp., New York (1988).

[22] S.S. Chen, Flow-induced vibration of circular cylindrical structures, Hemisphere Publ. Corp., New York (1987).

[23] W.M. Rohsenow, J.P. Hartnett and E.N. Ganic, Handbook of Heat Transfer: Applications, 2nd ed., McGraw-Hill, New York (1985).

Further reading

L. Wang, B. Sundén and R.M. Manglik, Plate heat exchangers: Design, applications and performance, WIT Press, UK (2007).

B. Sundén, Heat transfer and heat exchangers, in Kirk–Othmer Encyclopedia in Chemical Technology, John Wiley and Sons, New York, (2007) (on line publication).

R.K. Shah and D.P. Sekulic, Fundamentals of Heat Exchanger Design, John Wiley and Sons, Inc., Hoboken, NJ (2003).

B. Sundén and R.K. Shah, Advances in Compact Heat Exchangers, R.T. Edwards Inc., Philadelphia, PA (2007).

Addendum 1 Derivation regarding unsteady heat conduction for semi-infinite bodies

A1.1 Laplace transform

Let $f(\tau)$ be a real function of the real variable τ for $\tau > 0$. The Laplace transform of $f(\tau)$ is defined according to

$$L\{f(\tau)\} = F(s) = \bar{f} = \int_0^\infty f(\tau) e^{-2\tau}\,d\tau \qquad (A1.1)$$

where s is a complex variable defined as

$$s = \sigma + j\omega$$

where σ and ω are real, and j is the imaginary unit.

Example: Determine the Laplace transform of $f(\tau) = \tau$
   By applying eq. (A1.1), one has

$$F(s) = \int_0^\infty \tau e^{-s\tau}\,d\tau = \left[\tau\frac{e^{-s\tau}}{-s}\right]_0^\infty - \int_0^\infty 1 \cdot \frac{e^{-s\tau}}{-s}\,d\tau = \int_0^\infty \frac{e^{-s\tau}}{s}\,d\tau = \left[-\frac{1}{s^2}e^{-s\tau}\right]_0^\infty$$

Thus, one finds

$$F(s) = \frac{1}{s^2}$$

A1.2 Inverse Laplace transform

As Laplace transform is applied, one needs to come back to the τ-variable at the end, i.e., one needs to use inverse transformation. The inverse Laplace transform is defined by

$$L^{-1}\{F(s)\} = L^{-1}\left\{\bar{f}\right\} = f(\tau) = \frac{1}{2\pi j}\int_{c-j\infty}^{c+j\infty} F(s) e^{s\tau}\,ds \qquad (A1.2)$$

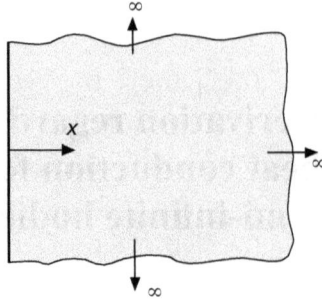

Figure A1.1: A semi-infinite body.

For practical calculations, it is more common to use tables, e.g., Standard Mathematical Tables, instead of the formula (A1.2).

A1.3   Application on unsteady heat conduction

Consider the unsteady heat conduction in the semi-infinite body shown in Fig. A1.1. The governing equation reads

$$\frac{\partial t}{\partial \tau} = a \frac{\partial^2 t}{\partial x^2} \qquad (A1.3)$$

The initial and boundary conditions are given by

$$\tau = 0: t(x, 0) = t_0 \qquad (A1.4a)$$

$$x = 0, (\tau > 0): t(0, \tau) = t_s \qquad (A1.4b)$$

$$x \to \infty, \tau > 0: t(x, \tau) \text{ finite} \qquad (A1.5)$$

Introduce now $\vartheta = t - t_0$ and then eqs. (A1.3)–(A1.5) can be written as

$$\frac{\partial \vartheta}{\partial \tau} = a \frac{\partial^2 \vartheta}{\partial x^2} \qquad (A1.6)$$

The initial and boundary conditions become

$$\tau = 0: \vartheta(x, 0) = 0 \qquad (A1.7)$$

$$x = 0, (\tau > 0): \vartheta(0, \tau) = \vartheta_s \qquad (A1.8)$$

$$x \to \infty, \tau > 0: \vartheta(x, \tau) \text{ finite} \qquad (A1.9)$$

By applying the Laplace transform on eq. (A1.1), one finds

$$L\{\vartheta(x, \tau)\} = \int_0^\infty \vartheta(x, \tau) e^{-s\tau} d\tau = \overline{\vartheta}(x, s) \qquad (A1.10)$$

$$L\left\{\frac{\partial^2\vartheta}{\partial x^2}\right\} = \int_0^\infty \frac{\partial^2\vartheta}{\partial x^2}e^{-s\tau}d\tau = \frac{\partial^2}{\partial x^2}\int_0^\infty \vartheta e^{-s\tau}d\tau = \frac{\partial^2\overline{\vartheta}}{\partial x^2} \qquad (A1.11)$$

$$L\left\{\frac{\partial\vartheta}{\partial\tau}\right\} = \int_0^\infty \frac{\partial\vartheta}{\partial\tau}e^{-s\tau}d\tau = [e^{-s\tau}\vartheta]_0^\infty + s\int_0^\infty \vartheta e^{-s\tau}d\tau = s\overline{\vartheta} \qquad (A1.12)$$

Introducing eqs. (A1.10)–(A1.12) in eq. (A1.6) gives

$$s\overline{\vartheta} = a\frac{\partial^2\overline{\vartheta}}{\partial x^2}$$

or

$$\frac{\partial^2\overline{\vartheta}}{\partial x^2} - \frac{s}{a}\overline{\vartheta} = 0 \qquad (A1.13)$$

The solution to eq. (A1.13) is given by

$$\overline{\vartheta} = C_1(s)e^{\sqrt{s/a}x} + C_2(s)e^{-\sqrt{s/a}x} \qquad (A1.14)$$

The condition (A1.9), i.e., ϑ should be finite as $x \to \infty$, implies that $C_1(s)$ must be chosen as zero. "The constant" $C_2(s)$ is determined by the condition $\vartheta = \vartheta_s$ at $x = 0$. (Note that index s in ϑ_s means $\vartheta(x = 0, \tau)$.)

$$L\{\vartheta_s\} = \int_0^\infty \vartheta_s e^{-s\tau}d\tau = \left[\vartheta_s\frac{e^{-s\tau}}{s}\right]_0^\infty = \frac{\vartheta_s}{s}$$

One then finds

$$\frac{\vartheta_s}{s} = C_2 \qquad (A1.15)$$

The solution (A1.14) can be written as

$$\overline{\vartheta}(x, s) = \frac{\vartheta_s}{s}e^{-\sqrt{s/a}x} \qquad (A1.16)$$

To find the solution $\vartheta(x, \tau)$, eq. (A1.16) must be inverted. This is most conveniently carried out by using tables of Laplace transforms; see, for example, Ref. [1]. The result is

$$L^{-1}\left\{\frac{1}{s}e^{-\sqrt{s/a}x}\right\} = 1 - \text{erf}\left(\frac{x}{2\sqrt{a\tau}}\right)$$

where erf means the Gaussian error function (see eq. (4.50))

$$\text{erf}\left(\frac{x}{2\sqrt{a\tau}}\right) = \frac{2}{\sqrt{\pi}}\int_0^{x/2\sqrt{a\tau}} e^{-\eta^2}d\eta$$

The solution of eq. (A1.3) with the conditions (A1.4) and (A1.5) can now be written as

$$\vartheta(x, \tau) = t - t_0 = \vartheta_s \left(1 - \mathrm{erf} \left(\frac{x}{2\sqrt{a\tau}} \right) \right) = (t_s - t_0) \left(1 - \mathrm{erf} \left(\frac{x}{2\sqrt{a\tau}} \right) \right)$$

or

$$\frac{t(x, \tau) - t_s}{t_0 - t_s} = \mathrm{erf} \left(\frac{x}{2\sqrt{a\tau}} \right)$$

Eq. (A1.17) is identical to eq. (4.53) of this book.

The case with a uniform surface heat flux at $x = 0$ and the case with convective heating or cooling at $x = 0$ can be derived by using the Laplace transform in a similar way as for the case above.

It should be noted that other mathematical methods than the Laplace transform can be used in the derivation of the temperature field solution, see, for example, Carslaw and Jaeger [2] as well as Eckert and Drake [3].

References

[1] Standard Mathematical Tables, 17th ed., The Chemical Rubber Company, Cleveland, OH (1969).
[2] H.S. Carslaw and J.C. Jaeger, Conduction of Heat in Solids, 2nd ed., Oxford University Press, New York (1959).
[3] E.R.G. Eckert and R.M. Drake Jr., Analysis of Heat and Mass Transfer, McGraw-Hill, New York (1972).

Addendum 2　Derivation of the complete temperature field equation

Consider the momentum and energy exchange for the volume element $dx\,dy\,dz$ in Fig. A2.1. According to the mass conservation equation (eq. (6.4)), the mass inside the element and the incoming and outgoing masses form a thermodynamic system (total mass constant). For this mass, the first law of thermodynamics is applicable, i.e.,

$$\Delta\dot{E} = \dot{Q} - \dot{W} \tag{A2.1}$$

A2.1　Determination of $\Delta\dot{E}$

The total energy change $\Delta\dot{E}$ can be written as

$$\Delta\dot{E} = \Delta\dot{E}_{\text{in-out}} + \Delta\dot{E}_{dx\,dy\,dz} \tag{A2.2}$$

For $\Delta\dot{E}_{dx\,dy\,dz}$ one has

$$\Delta\dot{E}_{dx\,dy\,dz} = \frac{\partial}{\partial\tau}(\rho e^0)dx\,dy\,dz \tag{A2.3}$$

where e^0 is the energy per mass unit.
$\Delta\dot{E}_{\text{in-out}}$ can be written as

$$\Delta\dot{E}_{\text{in-out}} = \frac{\partial}{\partial x}(\rho u e^0)dx\,dy\,dz + \frac{\partial}{\partial y}(\rho v e^0)dx\,dy\,dz + \frac{\partial}{\partial z}(\rho w e^0)dx\,dy\,dz \tag{A2.4}$$

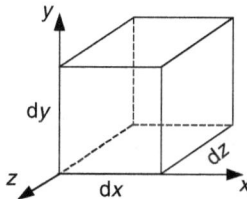

Figure A2.1:　Volume element $dx\,dy\,dz$.

where the first term expresses the net transport in the x-direction, the second and third terms the corresponding net transports in the y- and z-direction, respectively.

The energy e^0 is the sum of the internal energy e and the kinetic energy (per mass unit) $V^2/2 = (u^2 + v^2 + w^2)/2$.

Thus, one finds

$$e^0 = e + \frac{V^2}{2} \tag{A2.5}$$

The energy change $\Delta\dot{E}$ can now be expressed as

$$\Delta\dot{E} = \left[\frac{\partial}{\partial\tau}\left(\rho\left(e + \frac{V^2}{2}\right)\right) + \frac{\partial}{\partial x}\left(\rho u\left(\frac{e + V^2}{2}\right)\right) \right.$$
$$\left. + \frac{\partial}{\partial y}\left(\rho v\left(\frac{e + V^2}{2}\right)\right) + \frac{\partial}{\partial z}\left(\rho w\left(\frac{e + V^2}{2}\right)\right) \right] dx\, dy\, dz \tag{A2.6}$$

A2.1.1 Determination of the heat transfer rate $\dot{Q}$

Similar to the procedure in Chapters 1 and 6, $\dot{Q}$ is found to be

$$\dot{Q} = (\Delta\dot{Q}_x + \Delta\dot{Q}_y + \Delta\dot{Q}_z)$$
$$= \left(\frac{\partial}{\partial x}\left(\lambda\frac{\partial t}{\partial x}\right) + \frac{\partial}{\partial y}\left(\lambda\frac{\partial t}{\partial y}\right) + \frac{\partial}{\partial z}\left(\lambda\frac{\partial t}{\partial z}\right) \right) dx\, dy\, dz \tag{A2.7}$$

A2.2 Determination of the work rate $\dot{W}$

Consider at first the element $dx\, dy$ in the xy-plane according to Fig. A2.2.

In the x-direction, the stresses σ_{xx}, σ_{yx}, and σ_{zx} are acting. These forces perform a work on the fluid within the element. For the normal stress σ_{xx}, one has as indicated in Fig. A2.2:

$$\rightarrow (\sigma_{xx}u + \frac{\partial}{\partial x}(\sigma_{xx}u)dx)dy\, dz$$

$$\leftarrow \sigma_{xx}u\, dy\, dz$$

The net contribution from σ_{xx} is: $\dfrac{\partial}{\partial x}(\sigma_{xx}u)dx\, dy\, dz$

In a similar way, the shear stresses σ_{yx} and σ_{zx} in the x-direction perform works according to

Net contribution from σ_{yx}: $\dfrac{\partial}{\partial y}(\sigma_{yx}u)dx\, dy\, dz$

Net contribution from σ_{zx}: $\dfrac{\partial}{\partial z}(\sigma_{zx}u)dx\, dy\, dz$

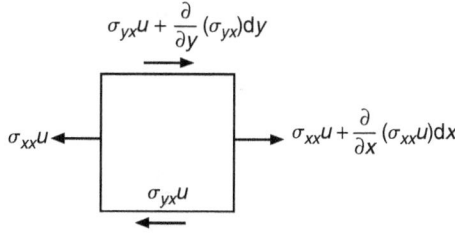

Figure A2.2: The element $dx\,dy$ in the xy-plane, work by forces in the x-direction.

In eq. (A2.1), $\dot{W}$ is positive if the system performs work. In this case, the surrounding is performing work on the system and thus the work is negative. One then has

$$\dot{W}_x = -\left(\frac{\partial}{\partial x}(\sigma_{xx}u) + \frac{\partial}{\partial y}(\sigma_{yx}u) + \frac{\partial}{\partial z}(\sigma_{zx}u)\right) dx\,dy\,dz \qquad (A2.8)$$

$$\dot{W}_y = -\left(\frac{\partial}{\partial x}(\sigma_{xy}v) + \frac{\partial}{\partial y}(\sigma_{yy}v) + \frac{\partial}{\partial z}(\sigma_{zy}v)\right) dx\,dy\,dz \qquad (A2.9)$$

$$\dot{W}_z = -\left(\frac{\partial}{\partial x}(\sigma_{xz}w) + \frac{\partial}{\partial y}(\sigma_{yz}w) + \frac{\partial}{\partial z}(\sigma_{zz}w)\right) dx\,dy\,dz \qquad (A2.10)$$

The work by the surface forces becomes

$$\dot{W}_{\text{surface}} = \dot{W}_x + \dot{W}_y + \dot{W}_z$$

If tensorial formulation is used, this work can be written as

$$\dot{W}_{\text{surface}} = -\frac{\partial}{\partial x_j}(\sigma_{ji}u_i)dx\,dy\,dz \qquad (A2.11a)$$

If, in addition, a volume force $F_i = (F_x, F_y, F_z)$ is acting on the volume element, its contribution to the work can be written as

$$\dot{W}_{\text{volume}} = -(F_x u + F_y v + F_z w)\rho\,dx\,dy\,dz = -F_i u_i \rho\,dx\,dy\,dz \qquad (A2.11b)$$

The total work is governed by

$$\dot{W} = \dot{W}_{\text{surface}} + \dot{W}_{\text{volume}}$$

A2.3   The energy equation in its primary form

Equations (A2.6), (A2.7), (A2.11a), and (A2.11b) are now inserted in eq. (A2.1). If tensorial notation is used, then

$$\frac{\partial}{\partial \tau}\left(\rho\left(e+\frac{V^2}{2}\right)\right) + \frac{\partial}{\partial x_i}\left(\rho u_i\left(e+\frac{V^2}{2}\right)\right) = \frac{\partial}{\partial x_i}\left(\lambda\frac{\partial t}{\partial x_i}\right) + \frac{\partial}{\partial x_j}(\sigma_{ji}u_i) + \rho F_i u_i$$

(A2.12)

Here eq. (A2.12) is called the primary form of the energy equation.

A2.3.1   Rewriting the energy equation

Equation (A2.12) will now be transferred to a more appropriate form. The left-hand side in eq. (A2.12) can be written as

$$\frac{\partial}{\partial \tau}(\rho e) + \frac{\partial}{\partial x_i}(\rho u_i e) + \frac{\partial}{\partial \tau}\left(\rho\frac{V^2}{2}\right) + \frac{\partial}{\partial x_i}\left(\rho u_i\frac{V^2}{2}\right)$$

(A2.13)

By applying the mass conservation eq. (6.4), i.e.,

$$\frac{\partial \rho}{\partial \tau} + \frac{\partial}{\partial x_i}(\rho u_i) = 0$$

Equation (A2.13) can be written as

$$\rho\frac{\partial e}{\partial \tau} + \rho u_i\frac{\partial e}{\partial x_i} + \rho\frac{\partial}{\partial \tau}\left(\frac{V^2}{2}\right) + \rho u_i\frac{\partial}{\partial x_i}\left(\frac{V^2}{2}\right)$$

(A2.14)

The Navier–Stokes' equations (eqs. (6.7)–(6.9)) can by tensorial notation be written as

$$\rho\frac{\partial u_i}{\partial \tau} + \rho u_j\frac{\partial u_i}{\partial x_j} = \rho F_i + \frac{\partial}{\partial x_j}(\sigma_{ji})$$

(A2.15)

If eq. (A2.15) is multiplied by u_i one finds

$$\rho\frac{\partial}{\partial \tau}\left(\frac{u_i u_i}{2}\right) + \rho u_j\frac{\partial}{\partial x_j}\left(\frac{u_i u_i}{2}\right) = \rho F_i u_i + u_i\frac{\partial}{\partial x_j}(\sigma_{ji})$$

However, $u_i u_i/2 = (u^2 + v^2 + w^2)/2$, and hence it is possible to write the above-mentioned equation as

$$\rho\frac{\partial}{\partial \tau}\left(\frac{V^2}{2}\right) + \rho u_j\frac{\partial}{\partial x_j}\left(\frac{V^2}{2}\right) = \rho F_i u_i + u_i\frac{\partial}{\partial x_j}(\sigma_{ji})$$

(A2.16)

Equation (A2.16) inserted into eq. (A2.14) gives

$$\rho\frac{\partial e}{\partial \tau} + \rho u_i\frac{\partial e}{\partial x_i} + \rho F_i u_i + u_i\frac{\partial}{\partial x_j}(\sigma_{ji})$$

(A2.17)

Now, if eq. (A2.17) is introduced into eq. (A2.12), the energy equation can be written as

$$\rho\frac{\partial e}{\partial \tau} + \rho u_i \frac{\partial e}{\partial x_i} = \frac{\partial}{\partial x_i}\left(\lambda\frac{\partial t}{\partial x_i}\right) + \sigma_{ji}\frac{\partial u_i}{\partial x_j} \tag{A2.18}$$

Equation (A2.18) is a very general form of the energy equation. The last term on the right-hand side of eq. (A2.18) will now be rewritten. According to Chapter 6, eqs. (6.12), (6.4), and (6.15), σ_{ji} can be written as

$$\sigma_{ji} = -p\delta_{ji} + \mu\left(\frac{\partial u_j}{\partial x_i} + \frac{\partial u_i}{\partial x_j} - \frac{2}{3}\Delta\delta_{ij}\right) \tag{A2.19}$$

The term $\sigma_{ji}(\partial u_i/\partial x_j)$ can now be written as

$$\sigma_{ji}\frac{\partial u_i}{\partial x_j} = -p\frac{\partial u_i}{\partial x_i} + \mu\frac{\partial u_i}{\partial x_j}\left(\frac{\partial u_j}{\partial x_i} + \frac{\partial u_i}{\partial x_j}\right) - \frac{2}{3}\mu\Delta\frac{\partial u_i}{\partial x_i} \tag{A2.20}$$

According to Chapter 6, one has $\Delta = \partial u_i/\partial x_i$.
    If eq. (A2.20) is inserted in eq. (A2.18), one finds

$$\rho\frac{\partial e}{\partial \tau} + \rho u_i\frac{\partial e}{\partial x_i} = \frac{\partial}{\partial x_i}\left(\lambda\frac{\partial t}{\partial x_i}\right) - p\Delta + \mu\frac{\partial u_i}{\partial x_j}\left(\frac{\partial u_j}{\partial x_i} + \frac{\partial u_i}{\partial x_j}\right) - \frac{2}{3}\mu\Delta^2 \tag{A2.21}$$

Equation (A2.21) is one of many possible forms of the energy equation.
    However, additional rearrangement is necessary to have it more handy. The internal energy e can be expressed by introducing the enthalpy h according to $e = h - p/\rho$. The left-hand side in eq. (A2.21) can therefore be written as

$$\rho\frac{\partial e}{\partial \tau} + \rho u_i\frac{\partial e}{\partial x_i} = \rho\frac{\partial h}{\partial t} + \rho u_i\frac{\partial h}{\partial x_i} - \frac{\partial p}{\partial t} + \frac{p}{\rho}\frac{\partial \rho}{\partial t} - u_i\frac{\partial p}{\partial x_i} + \frac{p}{\rho}u_i\frac{\partial \rho}{\partial x_i}$$

$$= \rho\left(\frac{\partial h}{\partial \tau} + u_i\frac{\partial h}{\partial x_i}\right) - \frac{\partial p}{\partial t} - u_i\frac{\partial p}{\partial x_i} + \frac{p}{\rho}\left(\frac{\partial \rho}{\partial t} + u_i\frac{\partial \rho}{\partial x_i}\right) \tag{A2.22}$$

By considering the mass conservation equation, the last term in eq. (A2.22) can be written as

$$\frac{\partial \rho}{\partial t} + u_i\frac{\partial \rho}{\partial x_i} = -\rho\frac{\partial u_i}{\partial x_i} = -\rho\Delta$$

Finally, one then has

$$\rho\frac{\partial e}{\partial \tau} + \rho u_i\frac{\partial e}{\partial x_i} = \rho\left(\frac{\partial h}{\partial \tau} + u_i\frac{\partial h}{\partial x_i}\right) - \left(\frac{\partial p}{\partial t} + u_i\frac{\partial p}{\partial x_i}\right) - p\Delta \tag{A2.23}$$

Equation (A2.23) in eq. (A2.21) gives

$$\rho \left(\frac{\partial h}{\partial \tau} + u_i \frac{\partial h}{\partial x_i} \right) = \frac{\partial}{\partial x_i} \left(\lambda \frac{\partial t}{\partial x_i} \right) + \frac{\partial p}{\partial \tau} + u_i \frac{\partial p}{\partial x_i} + \mu \frac{\partial u_i}{\partial x_j} \left(\frac{\partial u_j}{\partial x_i} + \frac{\partial u_i}{\partial x_j} \right) - \frac{2}{3} \mu \Delta^2$$

$$(A2.24)$$

Equation (A2.24) is the *general form of the energy equation* being suitable for continued simplifications.

If the enthalpy $h = c_p t$ is introduced, the so-called *general form of the temperature field equation* is obtained, i.e.,

$$\rho c_p \left(\frac{\partial t}{\partial \tau} + u_i \frac{\partial t}{\partial x_i} \right) = \frac{\partial}{\partial x_i} \left(\lambda \frac{\partial t}{\partial x_i} \right) + \frac{\partial p}{\partial \tau} + u_i \frac{\partial p}{\partial x_i} + \mu \frac{\partial u_i}{\partial x_j} \left(\frac{\partial u_j}{\partial x_i} + \frac{\partial u_i}{\partial x_j} \right) - \frac{2}{3} \mu \Delta^2$$

$$(A2.25)$$

A2.4 The boundary layer form of the temperature field equation

Without any derivation, the temperature field equation for a two-dimensional boundary layer is given. This equation is

$$\rho c_p \left(\frac{\partial t}{\partial \tau} + u \frac{\partial t}{\partial x} + v \frac{\partial t}{\partial y} \right) = \frac{\partial}{\partial y} \left(\lambda \frac{\partial t}{\partial y} \right) + \frac{\partial p}{\partial \tau} + u \frac{\partial p}{\partial x} + \mu \left(\frac{\partial u}{\partial y} \right)^2 \quad (A2.26)$$

The three last terms in the right-hand side are commonly neglected, but the last one of these is most important particularly at high flow velocities (frictional heating).

The final form is then given by

$$\rho c_p \left(\frac{\partial t}{\partial \tau} + u \frac{\partial t}{\partial x} + v \frac{\partial t}{\partial y} \right) = \frac{\partial}{\partial y} \left(\lambda \frac{\partial t}{\partial y} \right) + \mu \left(\frac{\partial u}{\partial y} \right)^2 \quad (A2.27)$$

Addendum 3   Heat transfer at high velocities

A3.1   Introduction

In Chapters 6–10, it was assumed that the influence on the energy balance by the work of the viscous forces was negligible. As the flow velocity becomes sufficiently high, this assumption will not be valid but instead the so-called frictional or viscous heating needs to be considered. The temperature gradients then becomes so large that the temperature dependence of the thermophysical properties has to be considered.

This phenomenon is usually called aerodynamic heating and is of relevance and importance for aircraft, space vehicles, and missiles at supersonic velocity. The analysis for this type of problem is complicated. However, for many engineering cases involving high flow velocities along a flat plate and constant wall temperature, the wall heat flux can be calculated by the heat transfer coefficient α for the low velocity case multiplied by the temperature difference $t_w - t_{aw}$. t_w is the wall temperature, and t_{aw} is the so-called adiabatic wall temperature. The theory and background will be summarized here.

A3.2   High velocity flow along a flat plate

Consider the high speed motion of an incompressible fluid along a flat plate. The fluid is assumed to have constant thermophysical properties. According to Chapter 6 and Addendum 2, the boundary layer equations for the laminar case read as follows:

Mass

$$\frac{\partial u}{\partial x} + \frac{\partial v}{\partial y} = 0 \tag{A3.1}$$

Momentum

$$u\frac{\partial u}{\partial x} + v\frac{\partial u}{\partial y} = v\frac{\partial^2 u}{\partial y^2} \tag{A3.2}$$

Energy

$$u\frac{\partial t}{\partial x} + v\frac{\partial t}{\partial y} = a\frac{\partial^2 t}{\partial y^2} + \frac{\mu}{\rho c_p}\left(\frac{\partial u}{\partial y}\right)^2 \tag{A3.3}$$

The last term in eq. (A3.3) represents the work by the viscous forces or the so-called frictional or viscous heating.

The boundary conditions of eq. (A3.1)–(A3.3) read as

$$y = 0: u = v = 0, \quad t = t_w \tag{A3.4}$$

$$y \to \infty: u \to U_\infty, \quad t \to t_\infty \tag{A3.5}$$

The flow field is not changed compared to the low speed case. The solution was given in Chapter 7 for $m = 0$.

The solution of the temperature field is split up in two parts, the homogeneous part and a particular solution. The homogeneous solution is obtained for the case when the frictional heating term in eq. (A3.3) is neglected and this was in principal given in Chapter 7 for $m = 0$ and $\gamma = 0$. The homogeneous solution of (A3.3) can be written as

$$t^{(h)} = t_\infty + (t_w - t_\infty)\theta(\eta) \tag{A3.6}$$

where $\eta = (\sqrt{U_\infty/2vx})y$.

As a particular solution of eq. (A3.3), one looks for the solution being valid for an adiabatic plate, i.e., $q_w = 0$. With η defined earlier and the stream function ψ according to Chapter 7 (see below) and for $m = 0$

$$\psi = \sqrt{2vU_\infty x}f(\eta)$$

Equation (A3.3) can be written as

$$\frac{d^2t}{d\eta^2} + \Pr f\frac{dt}{d\eta} + 2\frac{U_\infty^2}{2c_p}\Pr f''^2 = 0 \tag{A3.7}$$

The boundary conditions are

$$\eta = 0: \frac{dt}{d\eta} = 0 \quad (q_w = 0)$$

$$\eta \to \infty: t \to t_\infty$$

A dimensionless temperature θ^a defined later (eq. (A3.8)) is introduced

$$\theta^a = \frac{t - t_\infty}{U_\infty^2/2c_p} \tag{A3.8}$$

Equation (A3.7) can then be written as

$$\frac{d^2\theta^a}{d\eta^2} + \Pr f\frac{d\theta^a}{d\eta} + 2\Pr f''^2 = 0 \tag{A3.9}$$

With the boundary conditions

$$\eta = 0: \frac{d\theta^a}{d\eta} = 0 \qquad\qquad\qquad (A3.10)$$

$$\eta \to \infty: \theta^a \to 0 \qquad\qquad\qquad (A3.11)$$

Equation (A3.9) with the boundary conditions (A3.10) and (A3.11) was solved by Pohlhausen [1]. It was shown that the difference between the adiabatic wall temperature (the plate temperature for $q_w = 0$) and the fluid temperature could be written as

$$t_{aw} - t_\infty = r \cdot \frac{U_\infty^2}{2c_p} = \theta^a(0)\frac{U_\infty^2}{2c_p} \qquad\qquad (A3.12)$$

where r is called the recovery factor, which depends on the Pr number of the fluid. For moderate Pr numbers (e.g., gases, water), the recovery factor becomes

$$r \cong \sqrt{\mathrm{Pr}} \quad \text{for } 0.6 < Pr < 15 \qquad\qquad (A3.13)$$

The solution of eq. (A3.3) with the conditions (A3.4) and (A3.5) is obtained by combining the homogeneous and the particular solutions, i.e.,

$$t = t^{(h)} + t^{(p)}$$

or

$$t = C_1 t_\infty + C_2(t_w - t_\infty)\theta(\eta) + \theta^a \frac{U_\infty^2}{2c_p} + t_\infty$$

The boundary condition (A3.5) gives $C_1 = 0$, while the boundary condition (A3.4) gives

$$t_w = C_2(t_w - t_\infty) \cdot 1 + (t_{aw} - t_\infty) + t_\infty$$

The constant C_2 is found from this expression as

$$C_2 = \frac{t_w - t_{aw}}{t_w - t_\infty}$$

The solution of eq. (A3.3) can therefore be written as

$$t - t_\infty = (t_w - t_{aw})\theta(\eta) + \theta^a \frac{U_\infty^2}{2c_p} \qquad\qquad (A3.14)$$

A3.3   Calculation of the heat transfer

The heat flux from the plate is written as usual, i.e.,

$$q_w = -\lambda \left(\frac{\partial t}{\partial y}\right)_w = -\lambda \left(\frac{\partial t}{\partial \eta}\right)_w \sqrt{\frac{U_\infty}{2vx}}$$

From eq. (A3.14) one finds

$$\left(\frac{\partial t}{\partial \eta}\right)_w = (t_w - t_{aw}) \left(\frac{\partial \theta}{\partial \eta}\right)_w + \left(\frac{\partial \theta^a}{\partial \eta}\right)_w \frac{U_\infty^2}{2c_p}$$

But $(\partial \theta^a / \partial \eta)_w = 0$ and $(\partial \theta / \partial \eta)_w$ are the temperature derivatives for the low speed case. From Chapter 7 one has (see eq. (7.34))

$$\left(\frac{\partial \theta}{\partial \eta}\right)_w = -\sqrt{2}\frac{\alpha}{\lambda}\sqrt{\frac{vx}{U_\infty}}$$

The heat flux can now be written as

$$q_w = \alpha(t_w - t_{aw}) \tag{A3.15}$$

Thus a simple method to calculate the heat flux at high flow velocities has been found.

The heat transfer coefficient for the corresponding low speed case is multiplied by the temperature difference $t_w - t_{aw}$. t_{aw} is the so-called adiabatic wall temperature. This temperature is determined by using the recovery factor r (see eq. (A3.12)).

A3.4   Turbulent flow

The above-said analysis is valid for laminar cases. For a turbulent boundary layer, it has been found that the heat flux at high velocities can be calculated in a similar way. The recovery factor is determined by (see Ref. [2])

$$r = Pr^{1/3} \tag{A3.16}$$

The heat transfer coefficient for the low flow velocity case was given in Chapter 9.

A3.5   Influence of the temperature dependence of
the thermophysical properties

In the introduction, it was mentioned that at high flow velocities the thermophysical properties will be affected due to the large temperature gradients. This means that these properties may vary considerably across the boundary layer. A complete analysis is too extensive but Eckert [3] recommends that the thermophysical properties are evaluated at a reference temperature t^* according to

$$t^* = t_\infty + 0.5(t_\mathrm{w} - t_\infty) + 0.22(t_\mathrm{aw} - t_\infty) \qquad \text{(A3.17)}$$

References

[1] E. Pohlhausen, Der Wärmeaustausch zvischen festen Körpern und Flüssigkeiten mit kleiner Reibung und kleiner Wärmeleitung, Z. Angew. Math. Mech., 1, 115–121 (1921).

[2] J. Kaye, Survey of friction coefficients, recovery factors and heat transfer coefficients for supersonic flow, J. Appl. Sci., 21, 117–119 (1954).

[3] E.R.G. Eckert, Engineering relations for heat transfer and friction in high-velocity laminar and turbulent boundary layer flow over surface with constant pressure and temperature, Trans. ASME, 78, 1273–1284 (1956).

Further reading

E.R.G. Eckert, Survey of boundary layer heat transfer at high velocities and high temperatures, WADC Technical Report, 59–624 (1960).

E.R.G. Eckert and R.M. Drake Jr., Analysis of Heat and Mass Transfer, McGraw-Hill, New York (1987).

Addendum 4   Collection of problems in heat transfer

This addendum contains a collection of heat transfer problems. The methods of analysis presented in the various preceding chapters can be applied to solve these problems. The addendum groups the problems according to the list of contents below.

Contents

Heat Conduction, E1–E30

Forced Convection, E31–E62

Free Convection, E63–E72

Radiation, E73–E82

Condensation, E83–E91

Boiling and Two Phase Flow, E92–E107

Heat Exchangers, E108–E122

Answers

NOTATIONS

Note the difference in notation for some symbols in American and European books.

	American	European
Thermal conductivity W/(m K)	k	λ
Thermal diffusivity m²/s	α	a
Heat transfer coefficient (W/m² K)	h	α
Velocity of sound (m/s)	a	c

HEAT CONDUCTION

E1. Calculate the temperature distribution in the cork layer shown below:

$\lambda\,(T = 0°C) = 0.010$ W/(m K)

$\lambda\,(T = 30°C) = 0.040$ W/(m K)

The thermal conductivity is a linear function of the temperature.
   Compare the results with a constant conductivity $\lambda = 0.040$ W/(m K).

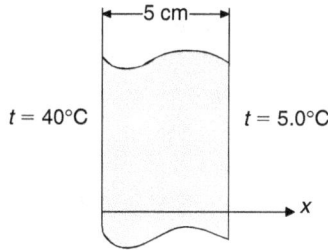

E2. A slab of insulation consists of an inhomogeneous material, see figure below. The thermal conductivity varies in a linear way across the material. Determine an expression for the heat transfer rate across the material.

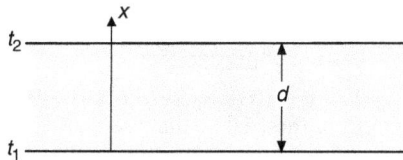

$$\lambda(x) = \lambda_1 + (\lambda_2 - \lambda_1)\frac{x}{d}$$

E3. One face of a 10 cm plate is maintained at 120°C, while the other face is maintained at 15°C, see figure below. The thermal conductivity of the plate is an exponential function of the temperature according to $\lambda = 30 \cdot e^{0.02t}$ W/(m K).

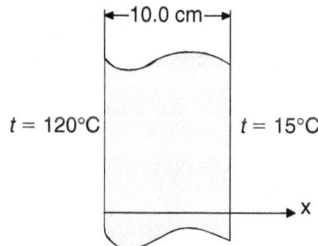

Determine the temperature at the position, $x = 6$ cm.

Compare the result when the thermal conductivity is constant $\lambda = 30$ W/(m K).

E4. A steam pipe with the outer diameter $D_o = 25$ mm is to be insulated with a layer of asbestos, $\lambda = 0.21$ W/(m K). Study the influence of the insulation thickness on the heat flow when the heat transfer coefficient with the surrounding air is 5.7 W/(m^2 K).

E5. A thermometer well is placed in a tube with hot gas. The length of the well is $L = 60$ mm and the wall thickness of the well is $b = 1.6$ mm. The temperature of the tube wall is $100°$C and the thermometer reads $160°$C (see figure below).

Calculate the temperature of the gas if the heat transfer coefficient $\alpha = 80$ W/m^2 K and the conductivity of the well $\lambda_{well} = 50$ W/(m K).

E6. Superheated steam at $315°$C and 100 kPa flows with a velocity of 20 m/s in a $d = 10$ cm tube. In order to measure the temperature of the steam, a thermometer is placed in a well, see figure below. The diameter of the thermometer well is $d = 1.5$ cm with a material thickness of $b = 0.1$ cm.

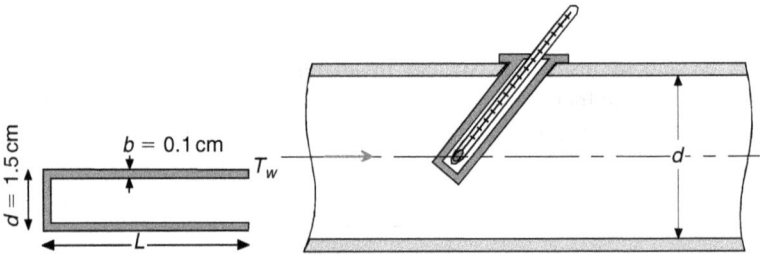

Calculate the length of the well so that the relative error in the temperature measurement does not exceed 0.5%, i.e., $(T_{x=L} - T_{steam})/(T_w - T_{steam}) \le 0.005$. The heat transfer coefficient between the steam and the well is 105 W/(m^2 K) and the thermal conductivity of the well material is $\lambda_{well} = 55$ W/(m K).

E7. Rectangular cooling fins of iron ($\lambda = 55$ W/(m K)) with the thickness 3.0 mm are used to enhance the heat transfer in a heat exchanger.

Determine if it is advantageous to use a finned surface for the following cases:

(a) Convective heat transfer to air: $10 \le \alpha \le 100$ W/(m^2 K).
(b) Convective heat transfer to water: $500 \le \alpha \le 5000$ W/(m^2 K).

E8. For the design of a heat exchanger, rectangular fins are considered. The fins should fulfill the criterion of given weight-maximum heat transfer and the fin effectiveness should be 5 (effectiveness relative to the base area).

(a) Design the fins for $\lambda = 45$ W/(m K) and $\alpha = 50$ W/(m^2 K).

(b) Is the condition for advantageous fins fulfilled?

(c) Determine the fin efficiency.

E9. A triangular fin is designed according to the figure below. The thermal conductivity is $\lambda = 45\,\text{W}/(\text{m K})$ and the heat transfer coefficient is $\alpha = 50\,\text{W}/(\text{m}^2\,\text{K})$.

(a) Is the fin optimal regarding given weight-maximum heat transfer?

(b) Calculate the fin effectiveness (effectiveness relative the base area).

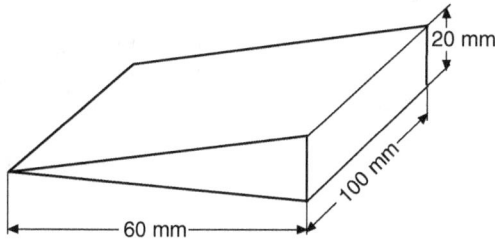

E10. A rectangular aluminum fin which is optimal regarding given weight—maximum heat transfer has an effectiveness (effectiveness relative to the base area) of 6.5.

Determine the effectiveness for a stainless steel fin for the same conditions. The thermal conductivity for aluminum is $\lambda = 225\,\text{W}/(\text{m K})$ and for stainless steel $\lambda = 18\,\text{W}/(\text{m K})$.

E11. A rectangular fin was designed to be optimal regarding given weight-maximum heat transfer. However, by a manufacturing mistake, the fin became 10% longer than the optimal length.

How will this effect the efficiency, effectiveness, and the heat flow from the fin with all other conditions unchanged?

E12. A rectangular fin was designed to be optimal regarding given weight-maximum heat transfer. However, by a manufacturing mistake, the fin became 10% shorter than the optimal length and the thickness 20% higher.

Determine the changes of the effectiveness and efficiency.

E13. A rectangular steel fin ($\lambda = 20\,\text{W}/(\text{m K})$) with the thickness 20 mm are designed for given weight-maximum heat transfer.

Calculate the heat flow per meter fin when the base temperature is 50°C and the fin is cooled by convection ($\alpha = 10\,\text{W}/(\text{m}^2\,\text{K})$) to the surrounding air at 20°C. Also determine the heat flow from a triangular fin with the same base area and length.

E14. For a specific condition for a triangular fin, the following relation applies: $\vartheta_2/\vartheta_1 = (T_{\text{tip}} - T_\infty)/(T_{\text{base}} - T_\infty) = 0.503$. If the heat transfer coefficient is increased by a factor 5, what will the change of the efficiency be?

E15. A heat exchanger tube with the outer diameter 25 mm has a surface temperature of 380 K. Aluminum fins with the length 19 mm ($r_o = 12.5 + 19$ mm) and the thickness 2.5 mm are applied on the tube surface. The tube contains 26 fins per meter length, equally spaced.

How much will the heat transfer increase compared to a smooth tube?

The surrounding air is at 300 K and the heat transfer coefficient is $\alpha = 10$ W/(m^2 K). The thermal conductivity for aluminum is $\lambda = 180$ W/(m K).

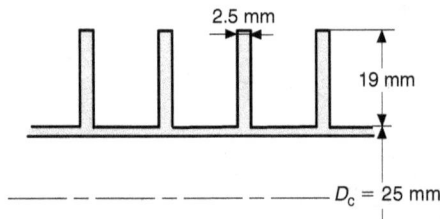

E16. A cylindrical transformer coil made of insulated copper wires has an inner diameter of 16.8 cm and an outer diameter of 24.4 cm. The fraction $\phi = 0.60$ of the total cross section of the coil is copper, and the rest is insulation. The current density in the conductors is $j = 200$ A/cm^2, the resistivity of the copper is $\rho = 200$ Ω cm^2/m. The effective thermal conductivity (across the copper wires) is $\lambda_{coil} = 0.350$ W/(m K) and the heat transfer coefficient on both sides of the coil, which are cooled by air at 20.7°C, is $\alpha = 22$ W/(m^2 K).

Determine the maximum temperature in the coil.

Hint: The coil can be approximated as a plane wall, with the heat generation per unit volume as $Q' = \varphi \cdot j^2 \cdot \rho$ W/m^3.

E17. 8000 W/m^3 heat is generated in a Nickel–Steel sphere with the diameter 10 cm, $\lambda = 10$ W/(m K). The sphere is cooled by convection to the surrounding air at 20°C and the heat transfer coefficient $\alpha = 10$ W/(m^2 K).

Calculate the center temperature at steady conditions.

E18. Hot water 0.5 kg/s flows in a heating steel pipe. The inlet water temperature is 85°C and the pipe is buried at 1.5 m below ground level. The inner diameter is 30 mm and the thickness is 2 mm, and the outer surface has a 15 mm insulation. The pipe length is 200 m and the surface ground temperature is 5°C.

What is the outlet temperature?

$\lambda_{steel} = 45$ W/(m K), $\lambda_{ground} = 0.45$ W/(m K), $\lambda_{insulation} = 0.04$ W/(m K), $\rho_{steel} = 7820$ kg/m^3, $\alpha_{water-steel} = 12,000$ W/(m^2 K), $c_{p,\,water} = 4.2$ kJ/(kg K).

E19. Hot water, 0.6 kg/s, is transported in an uninsulated buried steel pipe with the inner diameter 35 mm and the wall thickness 2 mm. The pipe is 200 m long and the water enters at 75°C. What is the required depth at which the pipe should be buried if the maximum permitted heat loss is 10 kW?

Physical properties: $\lambda_{steel} = 45$ W/(m K), $\lambda_{soil} = 0.45$ W/(m K), $\alpha_{water\text{-}steel} = 12,000$ W/(m^2 K), $c_{p,\,water} = 4.2$ kJ/(kg K), temperature at the ground level is 5°C.

E20. An aluminum sphere with the diameter 10 cm has initially a uniform temperature of 200°C, when it is exposed to a surrounding at 70°C (convective cooling). After 30 s, the temperature at the radial position $r = 3$ cm is 150°C.

Calculate the surface temperature of the sphere.

The thermal diffusivity for aluminum is $a = 8.4 \cdot 10^{-5}$ m^2/s.

E21. A long cylinder with the diameter 0.056 m and the homogeneous temperature 100°C is placed in a fluid with the constant temperature 500°C. Calculate the required time for the surface to reach a temperature of 160°C below the fluid temperature. Also calculate the centerline temperature at this time.

$\alpha = 420$ W/(m^2 K), $\lambda = 35$ W/(m K), $a = \lambda/(\rho \cdot c_p) = 13 \cdot 10^{-6}$ m^2/s.

E22. A rectangular steel block $0.6 \times 0.6 \times 1.2$ m^3 has initially the temperature 20°C when it is placed in an oven with the temperature 400°C. The heat transfer coefficient between the oven gas and the steel block is 565 W/(m^2 K).

How long will it take for the center temperature to reach 370°C?

The following data apply:

$\lambda_{steel} = 43$ W/(m K), $a_{steel} = 0.053$ m^2/h.

E23. A solid body as shown in the figure below has an initial temperature 800°C when it is exposed to cooling airstream at 20°C.

How long will it take for the corner temperature to reach 400°C.

The following data apply:

$\lambda = 3$ W/(m K), $\rho = 4300$ kg/m^3

$c_p = 240$ J/(kg K) and $\alpha = 200$ W/(m^2 K).

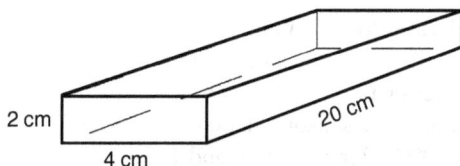

E24. A rectangular steel block $0.4 \times 0.6 \times 1.8$ m^3 has initially the temperature 820°C. The block is cooled by a gas stream at 50°C. The heat transfer coefficient is 400 W/m^2 K. Determine how long will it take for the center temperature to reach 590°C.

The following data apply: $\lambda_{steel} = 43$ W/(m K), $a_{steel} = 1.48 \cdot 10^{-5}$ m^2/s.

A big steel block ($\lambda = 45$ W/(m K), $a_{steel} = 1.48 \cdot 10^{-5}$ m^2/s) has initially the homogenous temperature 35°C. Heat is transferred to the steel block under such conditions that its surface temperature is brought to the constant temperature 250°C.

The steel block can be considered as a semi-infinite body.

(a) Determine the temperature at the depth 2.5 cm after 0.5 min.
(b) If the surface heat flux to the block is $3.2 \cdot 10^5 \, \text{W/m}^2$, what will the corresponding temperature be?

E26. A big aluminum block has initially a homogeneous temperature 200°C, when suddenly the surface temperature is brought to 70°C.

Calculate the total heat removal when the temperature at the depth 4 cm is 120°C. The block can be considered as a semi-infinite body.

The physical properties for the aluminum block are thermal conductivity $\lambda = 215 \, \text{W/(m K)}$ and thermal diffusivity $a = 8.4 \times 10^{-5} \, \text{m}^2/\text{s}$.

E27. During an experiment with a steel body ($\lambda = 40 \, \text{W/(m K)}$), which can be considered as a semi-infinite body, a temperature measurement at the time 10 min revealed the following results:

$$T(x = 3 \, \text{cm}, \ \tau = 10 \ \text{min}) = 59.5°C \text{ and } T(x = 12 \, \text{cm}, \ \tau = 10 \ \text{min}) = 95°C$$

Calculate the total heat flow across the surface $x = 0$ during the time interval $0 \leq \tau \leq 10$ min.

The surface can be assumed to have a constant temperature for all $\tau \geq 0$.
$a = 12 \times 10^{-6} \, \text{m}^2/\text{s}$.

E28. During an experiment with a steel body ($\lambda = 20 \, \text{W/(m K)}$) which can be considered as a semi-infinite body, a temperature measurement at the time 12 min revealed the following results:

$$T(x = 4 \, \text{cm}, \ \tau = 12 \ \text{min}) = 74°C \text{ and } T(x = 18 \, \text{cm}, \ \tau = 12 \ \text{min}) = 110°C.$$

Calculate the total heat flow across the surface $x = 0$ during the time $0 \leq \tau \leq 12$ min.

The surface can be assumed to have a constant temperature for all $\tau \geq 0$.
$a = 12 \times 10^{-6} \, \text{m}^2/\text{s}$.

E29. A steel block ($\lambda = 20 \, \text{W/(m K)}$), which may be considered as a semi-infinite body, has initially a uniform temperature. The surface of the block is suddenly exposed to a temperature which is higher than the initial temperature of the block. A temperature sensor at a depth $x = 6$ cm from the surface shows the temperature 90°C after $\tau = 2$ min and 135°C after $\tau = 4$ min.

Calculate the energy per unit area transferred to the block during the first 10 min.

The thermal diffusivity of steel is $a = 1.4 \times 10^{-5} \, \text{m}^2/\text{s}$.

E30. A steel block ($\lambda = 25 \, \text{W/(m K)}$, $a = 1.4 \times 10^{-5} \, \text{m}^2/\text{s}$) is initially at a uniform temperature of 35°C. The surface of the steel block is suddenly brought in contact with an aluminum block ($\lambda = 170 \, \text{W/(m K)}$, $a = 8 \times 10^{-5} \, \text{m}^2/\text{s}$) at an initial temperature of 180°C.

Calculate the interface temperature between the blocks.
Hint: Each block can be considered as a semi-infinite body.

FORCED CONVECTION

E31. A liquid flows tangentially along a flat plate, $U_\infty = 2$ m/s and $T_\infty = 20°C$. The temperature of the plate is 70°C.

Determine the thickness of the boundary layers of the flow and temperature at the position $x = 0.2$ m (measured from the leading edge of the plate).

E32. The following physical properties apply for the fluid, $Pr = 15$ and $\nu = 15 \cdot 10^{-6}$ m²/s.

The front of a two-dimensional vehicle is formed as a right angle wedge. The surface temperature varies according to $T_w - T_\infty = 5x^4$ with $T_w(x = 0) = 25°C$ and $T_\infty = 25°C$.

Calculate $C_f \sqrt{Re_x}$ and $Nu/\sqrt{Re_x}$ when the Prandtl number is 1.

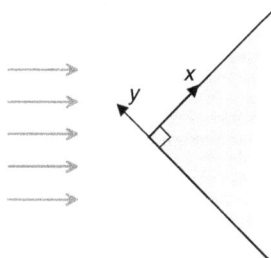

E33. Air flows tangentially along a flat plate, $U_\infty = 3$ m/s and $T_\infty = 20°C$. The temperature of the plate is 40°C.

(a) Determine the local heat transfer coefficient at $x = 0.2$ m (from the leading edge of the plate) and the temperature at the position $y = 2.5$ mm normal to the plate.

(b) Also calculate the total heat flow from the plate in the region $0 \le x \le 0.8$ m.

E34. The front of a two dimensional vehicle is formed as a right angle wedge. The surface of the vehicle is maintained at a constant temperature of 40°C and the ambient air temperature is 20°C.

Calculate

(a) The friction drag per unit width.

(b) Heat loss per unit width.

E35. Air flows, $U_\infty = 1.5$ m/s, along a flat plate. The length of the plate is 2 m and the surface temperature varies as $T_w - T_\infty = 10x^2 + 4x^{4}°C$.

Calculate the heat transfer rate at the position $x = 0.8$ m.

$\nu = 15.2 \cdot 10^{-6}$ m²/s, $\lambda = 0.025$ W/(m K), $Pr = 0.72$.

E36. Air flows, $U_\infty = 5\,\text{m/s}$ and $T_\infty = 20°\text{C}$, along a flat plate. The length of the plate is 4 m and the surface temperature $T_w = 40°\text{C}$. Calculate the heat transfer rate per unit width.

E37. Air flows, $U_\infty = 3.5\,\text{m/s}$, along a flat plate with the length 5.0 m and width 2.2 m. The surface temperature of the plate is given by $T_w - T_\infty = 3 + 1.2x + 0.35x^{3}°\text{C}$.

Calculate the heat flux at $x = 0.8\,\text{m}$ and $x = 4.5\,\text{m}$. The physical properties for air are $\lambda_f = 0.03\,\text{W/(m K)}$, $\text{Pr} = 0.72$, $\nu = 15 \cdot 10^{-6}\,\text{m}^2/\text{s}$.

E38. Air flows, $U_\infty = 2.5\,\text{m/s}$, along a flat plate with the length 6.0 m and width 2.0 m. The surface temperature of the plate is given by $T_w - T_\infty = 4 + 0.85x^2 + 0.25x^{4}°\text{C}$. Calculate the heat transfer rate for the turbulent part of the boundary layer.

The physical properties for air are $\lambda = 0.025\,\text{W/(m K)}$, $\text{Pr} = 0.72$, $\nu = 15 \cdot 10^{-6}\,\text{m}^2/\text{s}$.

E39. Air flows along a flat plate with the length 4.0 m and width 1.0 m. The surface temperature of the plate is given by $T_w - T_\infty = 3 + x^2 + 0.2x^{4}°\text{C}$. The air velocity is $U_\infty = 2.5\,\text{m/s}$. Calculate the heat transfer rate from the plate.

The physical properties for air are $\lambda = 0.025\,\text{W/(m K)}$, $\text{Pr} = 0.72$, $\nu = 15 \cdot 10^{-6}\,\text{m}^2/\text{s}$.

E40. Air at 60°C flows over a porous flat plate with the length 10 m and width 2 m at a velocity of 1.5 m/s. In order to achieve a surface temperature of 20°C, cooling air (20°C) is blown through the plate. The maximum heat flux is limited to 45 W/m² at a position $x = 1\,\text{m}$.

Calculate the amount of cooling air (kg/m² s) at this position.

Assume physical properties for air as $\lambda = 0.026\,\text{W/(m K)}$, $\text{Pr} = 0.7$, $\nu = 17.2 \cdot 10^{-6}\,\text{m}^2/\text{s}$, and $c_p = 1.0\,\text{kJ/(kg K)}$.

E41. Air flows along a 4.7 m long and 1.0 m wide porous flat plate at $U_\infty = 2.0\,\text{m/s}$ and $T_\infty = 60°\text{C}$. Through the plate air is blown at a rate of $v_w\sqrt{\text{Re}_x}/U_\infty = 0.2$. The temperature of the plate and the cooling air through the plate is 20°C.

Determine the effective heat transfer coefficient and mass flow rate of the cooling air at a position where $\text{Re}_x = 10^5$. Compare the result with the case $v_w = 0$.

E42. Air flows along a 8 m long and 2.0 m wide porous flat plate at $U_\infty = 10\,\text{m/s}$ and $T_\infty = 90°\text{C}$. The plate is cooled by blowing cooling air at 20°C (also the temperature of the plate) through the plate. At the position $x = 1.5\,\text{m}$ (from the leading edge of the plate), the maximum permitted heat flux is 780 W/m².

Determine the amount of cooling air (kg/m²/s) at this position.

$\lambda = 0.025\,\text{W/(m K)}$, $\text{Pr} = 0.7$, $\nu = 17 \cdot 10^{-6}\,\text{m}^2/\text{s}$ and $c_p = 1.0\,\text{kJ/(kg K)}$.

E43. The surface temperature on a circular cylinder, $D_o = 50\,\text{mm}$, in a cross flow, varies as $T_w - T_\infty = 3 + 2x^2 - 0.9x^{6}°\text{C}$, where x is measured from the stagnation point.

Determine an expression for the heat flux at $x = 0$ when $Re = U_\infty D/\nu = 75{,}000$, also determine an approximation at $x = 1$ mm.

Use physical properties for air at 20°C.

E44. Oil flows in a 15 mm circular tube, with the constant wall temperature $T_w = 60°C$, at $U = 0.60$ m/s. The oil is to be heated from 30°C to 40°C. Determine the length of the tube.

Physical properties for the oil are $Pr = 275$, $\rho = 900$ kg/m³, $c_p = 1.8$ kJ/(kg K), $\mu_B = 2.3 \cdot 10^{-2}$ N s/m², $\mu_w = 0.75 \cdot 10^{-2}$ N s/m², $\lambda = 0.15$ W/(m K).

E45. A liquid, $\nu = 50 \cdot 10^{-6}$ m²/s and $\lambda = 0.150.2$ W/(m K), flows in a 10 mm tube at $U = 1$ m/s. Determine the wall temperature at the axial position where the centerline temperature is 40°C when the tube is heated with a constant wall heat flux $q_w = 1.3$ kW/m². The flow and temperature fields are assumed to be fully developed.

E46. Liquid sodium should be heated from 120°C to 160°C (bulk temperatures). The sodium flows in a tube with the diameter 50.0 mm and the mass flow rate 0.023 kg/s. The wall of the tube is heated electrically, generating a constant wall heat flux.

Determine the length of the tube when the maximum permitted wall temperature is 200°C.

Physical properties for sodium are (mean values):

$c_p = 1.34$ kJ/(kg K), $\rho = 900$ kg/m³, $\lambda = 80$ W/(m K), $\mu = 0.42 \cdot 10^{-3}$ kg/(ms), $Pr = 0.007$.

The flow and temperature fields can be assumed to be fully developed.

E47. Oil should be heated from 120°C to 150°C (bulk temperatures). The sodium flows in a tube with the diameter 2.0 cm and the bulk velocity 1.75 m/s. The wall of the tube is heated electrically, generating a constant wall heat flux.

Determine the length of the tube when the maximum permitted wall temperature is 250°C.

Physical properties for the oil are (mean values):

$c_p = 1.88$ kJ/(kg K), $\rho = 900$ kg/m³, $\lambda = 14$ W/(m K), $\mu = 0.82 \cdot 10^{-3}$ kg/(ms).

The flow and temperature fields can be assumed to be fully developed.

E48. Oil flows in a quadratic channel 4.0 cm² with a constant wall temperature 100°C, at a velocity 1.75 m/s. Calculate the outlet temperature, when the inlet temperature is 22°C and the length is 10 m.

Physical properties for the oil are (mean values):

$c_p = 1.88$ kJ/(kg K), $\rho = 880$ kg/m³, $\lambda = 14$ W/(m K), $\mu = 0.82 \cdot 10^{-3}$ kg/(ms).

The flow and temperature fields can be assumed to be fully developed.

E49. Water vapor at atmospheric pressure condenses on the outside of a circular copper tube, $d_i = 20$ mm and d_o. Liquid water flows through the tube with the homogeneous inlet temperature 20°C. Determine the outlet temperature if the tube length is 3 m when (a) the mass flow rate is 0.02 kg/s, (b) the mass flow rate is 0.2 kg/s.

E50. Water with the bulk temperature 40°C flows through a 5.0 cm smooth tube at the velocity 3.0 m/s. Determine the heat transfer coefficient by using the Reynolds–Colburn analogy. Compare the result with the Dittus–Boelter equation.

E51. Air flows through a tube with the velocity 30 m/s. The diameter of the tube is 5.0 cm and the length is 1.2 m. Determine the mean heat transfer coefficient considering developing flow. Compare the results with the Dittus–Boelter equation. Physical properties can be taken at 50°C.

E52. Water flows through an 8.0 cm diameter tube at the velocity 4.0 m/s. The wall temperature is 100°C, and the water enters at 40°C and leaves at 85°C. Determine the water temperature at the normal distance 3 mm from the wall at the outlet. The flow and the thermal conditions can be assumed to be fully developed.

E53. Liquid sodium should be heated from 120°C to 160°C (bulk temperature). The sodium flows in a tube with the diameter 2.5 cm and the mass flow rate 2.3 kg/s. The wall of the tube is heated electrically, generating a constant wall heat flux.

Determine the length of the tube when the maximum permitted wall temperature is 200°C.

Physical properties for sodium are (mean values):

$c_p = 1.34$ kJ/(kg K), $\rho = 900$ kg/m^3, $\lambda = 0.80$ W/(m K), $\mu = 0.42 \cdot 10^{-3}$ kg/(ms), Pr $= 0.007$.

The flow and temperature fields can be assumed to be fully developed.

E54. An electrically heated wire, $D = 0.1$ mm, is exposed to an airstream (cross flow) at 20°C. The temperature of the wire is 50°C and the heat transfer rate from the wire to the air is 17.8 W/m. Determine the air velocity.

E55. Atmospheric cooling air (20°C) flows across a circular cylinder, $D = 10$ cm and $L = 2$ m, at the velocity 13 m/s. In order to enhance the heat transfer, the cylinder is equipped with 10 equally spaced circular fins with the outer diameter 18 cm and thickness 3 mm. The thermal conductivity of the fins is 90 W/(m K). Assume that the heat transfer coefficient is unchanged (compared to a bare tube) and calculate the heat transfer rate when the surface temperature of the cylinder is 250°C (base surface temperature).

E56. Air at 100 kPa and 30°C flows across a circular cylinder at the velocity 30 m/s. The diameter of the cylinder is 5.0 cm and its surface temperature is 150°C. Calculate the heat transfer rate per unit length.

E57. $40 \cdot 10^3$ W/m³ heat is generated in a long circular cylinder with the diameter 12 cm. The cylinder material is a steel alloy with the thermal conductivity 10 W/(m K). The cylinder is placed in an air cross flow at 13 m/s and 18°C. Determine the centerline steady-state temperature of the cylinder.

E58. A long circular copper cylinder, $\rho = 8950$ kg/m³, $\lambda = 383$ W/(m K) and $c_p = 386$ J/(kg K), with the diameter 50 mm is to be cooled by a cross flow airstream at 20°C. The cylinder has initially the homogeneous temperature 140°C and should be cooled to 60°C. Determine the air velocity when the maximum cooling time is 15 min.

E59. Pressurized water at high temperature is commonly used for heating. Common heating equipment are tube banks where the water flows through the tubes and air flows on the outside across the tubes. For a specific *staggered* tube bank, the following data apply: $D = 16$ mm, $S_L = 32$ mm, $S_T = 32$ mm, 7 tubes deep (main flow direction) and 8 tubes high (normal to main flow direction).

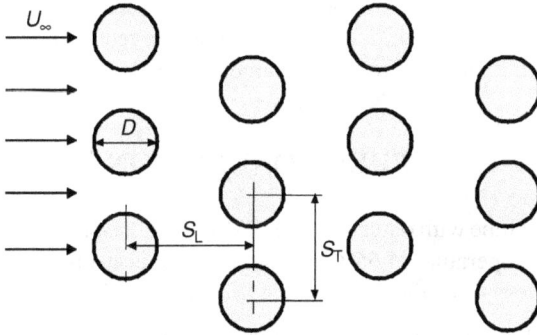

The tube length is 1.5 m. At normal working conditions, the tube surface temperature is 70°C and the conditions for the air flow measured before the tube bank are $U_\infty = 6$ m/s and $T_\infty = 15$°C. Determine the heat transfer coefficient on the air side and the heat transfer rate. Also calculate the pressure drop on the air side.

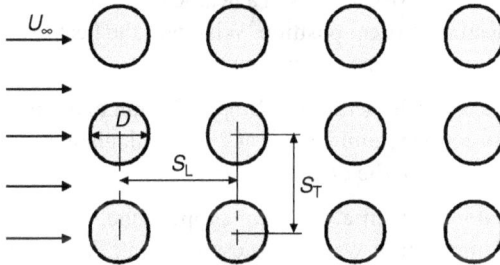

E60. A tube bank consists of a number of 30 mm tubes in an in-line arrangement with $S_L = S_T = 45$ mm and a tube length of 1 m. There are 10 rows in the main flow direction and 7 rows normal to the flow direction. In the tubes steam condenses and the outer tube wall is at 100°C. The free stream velocity for the cooling air (outside the tubes) is $U_\infty = 15$ m/s and the temperature is 27°C. Calculate the air temperature at the outlet and the pressure drop on the air side.

E61. A tube bank with 30 mm diameter tubes in an *in-line* arrangement has equal normal and parallel tube distances $S_L = S_T = 45$ mm and the tube length 1 m. There are 10 rows in the main flow direction and 7 rows normal to the flow direction. Inside the tube steam condenses giving a tube temperature of 100°C. The cooling air velocity is $U_\infty = 15$ m/s with the inlet temperature at 27°C. In order to enhance the performance, the number of tubes in the normal flow direction is increased to $N_T = 9$. How will this effect the heat transfer and pressure drop.

$\nu = 17 \cdot 10^{-6}$ m^2/s, $\lambda = 0.027$ W/(m K), Pr = 0.72, $c_p = 1008$ J/(kg K), $\rho = 1.0$ kg/m^3

E62. A copper sphere, $D = 10$ mm, has the initial and homogeneous temperature 75°C. The sphere is to be cooled to 35°C by an airstream with the velocity of 10 m/s at 100 kPa and 25°C. Calculate the time required for the cooling when the sphere is assumed to be at a homogeneous temperature at each instant.

FREE CONVECTION

E63. A circular tube with outer diameter 75 mm and the length 1.5 m has a constant surface temperature of 65°C. The ambient air is at 1 bar and 25°C. Calculate the heat loss when the tube is placed vertically and horizontally, respectively.

E64. An 8 m long horizontal copper tube with the diameter 120 mm is cooled by natural convection to ambient air at 20°C. The heat loss is measured to 150 W. Calculate the heat loss if the tube is placed vertically, under the same conditions.

E65. A horizontal tube with the outer diameter 0.05 m and the surface temperature 60°C is used for transport of a hot liquid. Cooling by natural convection with the ambient air at 20°C results in a heat loss of 109 W. Due to a redesign, the tube is placed in vertical position. What will the heat loss then be, assuming that the other conditions are unchanged.

E66. A horizontal cylinder with the diameter 75 mm and length 2 m is cooled by natural convection to ambient air at 20°C. Calculate the surface temperature if the heat loss from the cylinder is 75 W.

E67. 0.5 kg/s water flows in a 6 m long copper tube. The inner diameter of the tube is 40 mm with a wall thickness of 1 mm. The outside of the tube is cooled by natural convection to ambient air at 20°C. Calculate the water outlet temperature when it enters at 80°C. The physical properties for the water can be taken at 80°C while the reference temperature for the air is 20°C.

E68. Two horizontal and parallel plates are separated by a 5 cm air space and the temperatures are 15°C and 28°C, respectively. The plates are quadratic with the side length 2 m. Calculate the heat flow when (a) the hot plate is on top, (b) the cold plate is on top.

E69. A bed room window with double glass has the width 1 m and the height 0.8 m. The air space between the glasses is 5 cm. On a specific autumn day, the outer glass has the temperature 5°C while the inner glass has the temperature 18°C. Calculate the heat loss through the window. (Neglect radiative heat transfer.)

E70. On a cold winter day, the temperature on the inner glass of a double-glass window is measured to 13°C and the outdoor temperature is −4°C. The window has the width 1.5 m and the height 0.9 m and the distance between the glass sheets is 5 cm. Calculate the heat loss and the temperature of the outer glass.

E71. Estimate the temperature difference between two parallel and horizontal plates, when the heat transfer between the plates is 8.0 kW/m². The plates are separated by 25 mm by a water layer at the temperature 4°C. The hot plate is placed at the bottom.

E72. A 15 mm air space is used as insulation between two parallel horizontal plates. The lower plate is at 50°C while the upper one is at 15°C. (a) Calculate the heat transfer rate between the plates. (b) What would the answer be if a third plate is placed in between the two plates.

RADIATION

E73. A furnace with the temperature 1000°C has a small opening covered with quartz glass. The radiative properties of the glass are:

$0 \leq \lambda \leq 4\,\mu m; \tau = 0.9; \varepsilon = 0.1; \rho = 0$
$4 < \lambda < \infty; \tau = 0; \varepsilon = 0.8; \rho = 0.2$

The furnace can be considered as a black body. Calculate the transmitted energy to a room with the temperature 30°C.

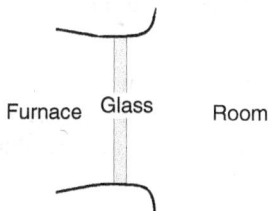

Furnace Glass   Room

E74. Two black body parallel plates 0.5 × 1.0 m is separated by 0.5 m. One of the plates has a surface temperature 1000°C and the other is at 500°C. Calculate the heat transfer rate due to radiation.

E75. Two parallel plates 0.5 × 1.0 m are separated by 0.5 m. One of the plates has a surface temperature 1000°C and the other is at 500°C. The emissivity of the plates is 0.2 and 0.5, respectively. The plates are placed in a big room with the wall temperature 27°C. Heat is exchanged between the plates and from the plates to the room. Only the plate surfaces facing each other need to be considered in the analysis. Calculate the heat transfer rate to each plate and to the room.

E76. The electronic equipment in an optical device generates heat at a rate of 2.5 W. The printed circuit card where the electronics are mounted is enclosed between a cover plate and a mounting plate according to the figure below.

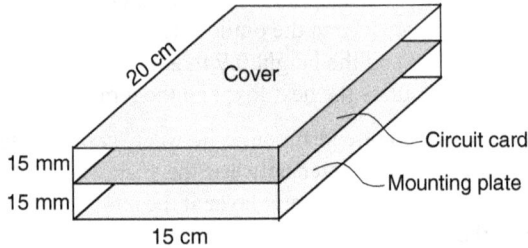

The cover plate and the mounting plate can be considered as black bodies while the emissivity of the circuit card is 0.8. The temperatures of the cover plate and mounting plate are 30°C and 40°C, respectively. The device operates in an evacuated space and only radiative heat transfer occurs. Calculate the temperature of the circuit card.

Only the surfaces facing each other need to be considered in the analysis.

E77. At the center of an evacuated, thin-walled, and opaque cylinder, a filament with the diameter 0.2 mm is placed. The filament has the emissivity, $\varepsilon = 0.9$, and the temperature 900°C and the cylinder has the diameter 50 mm and the emissivity $\varepsilon = 0.8$. The cylinder is placed in a big room at 20°C. Calculate the cylinder wall temperature.

Hint: Between the filament and the cylinder wall, only radiative heat transfer occurs, while the heat transfer between the cylinder wall and the room is due to combined natural convection and thermal radiation. Physical properties for the air can be taken at $T_{\text{ref}} = 30°C$.

E78. A cubic furnace with the side 0.5 m has inner walls that can be considered as black bodies. The gas in the furnace consists of 20% (volume) CO_2 and 80% N_2. The total pressure is 1 atm and the gas temperature is 1500°C. Calculate the heat removal when the walls are maintained at the temperature 300°C.

E79. The space between two infinite parallel plates with the wall temperature 570°C is filled with a gas mixture at 2 atm and 1200°C. The gas consists of 15% (volume) CO_2 and 85% N_2. Calculate the heat removal when the walls are at a distance of 80 cm.

E80. Two parallel plates separated by 0.7 m are maintained at 200°C and 500°C, respectively. The space between the plates is filled with a gas mixture of 20%

CO_2, 15% H_2O–vapor and 65% N_2 (volume percent) at the total pressure 3 atm. The gas temperature is 1000°C. Calculate the heat transfer between the gas and each plate.

E81. A gas mixture at 3 atm and 1400°C consists of 18 volume percent CO_2 and 82% N_2. The gas is confined between two parallel infinite plates, which can be considered as black bodies, at 600°C. Calculate the heat flux (W/m^2) which has to be removed from the plates if the plate distance is 30 cm.

E82. A cylindrical furnace with the diameter 80 cm has the surface emissivity 1.0. The gas in the furnace consists of 10% CO_2, 20% H_2O–vapor, and 70% N_2 (volume percent) and is at 1500 K. The total pressure is 1 atm. Calculate the heat transfer which has to be removed from the furnace walls to keep the walls at 350°C.

CONDENSATION

E83. A vertical plate, 30 × 30 cm, with the temperature 98°C is exposed to saturated steam at atmospheric pressure. Calculate the heat transfer rate and the condensate rate per hour.

E84. Saturated water vapor at $0.15 \cdot 10^5$ Pa condenses on a vertical wall with the width 0.3 m. Calculate the height of the wall if the condensate mass flow rate at the bottom of the wall is $3.5 \cdot 10^{-3}$ kg/s. The temperature of the wall is 46°C. (The liquid properties can be considered as independent of pressure.)

E85. Saturated water vapor at 100°C condenses on the outside of a horizontal tube with the length 2 m and diameter 20 mm. The surface temperature of the tube is 60°C. Determine the mean heat transfer coefficient and the mass flow rate of condensation.

E86. Saturated water vapor at 80°C ($p = 47.4$ kPa) condenses on the outside of a vertical tube with the diameter 20 mm and the wall temperature 40°C. If the condensate mass flow rate is 0.025 kg/s, determine the length of the tube.

E87. Saturated ammonia vapor at -5°C ($p = 0.353$ MPa) condenses on the outside of a vertical tube with the diameter 12.7 mm and the length 0.75 m. The wall temperature of the tube is -15°C. Determine the mean heat transfer coefficient and the condensate mass flow rate.

E88. Saturated water vapor at 60°C ($p = 19.94$ kPa) condenses on the outside of a quadratic tube bank, 10 × 10 rows. The tubes have an outer diameter 25 mm, a length of 2 m, and the surface temperature is 40°C. Determine the mean heat transfer coefficient and the total condensate mass flow rate.

E89. A condenser has 625 tubes with the outer diameter 12.5 mm and the length 3 m. The tubes are in a quadratic arrangement of 25 × 25 rows. Saturated water vapor at 50°C ($p = 12.35$ kPa) condenses on the outside of the tubes which have the wall temperature 30°C. Determine the condensate mass flow rate.

E90. 2.0 kg/s saturated water vapor at 80°C condenses on the outside of the tubes with the outer diameter 12.5 mm and the length 2.3 m. The tubes are placed horizontally in a rectangular in-line-arrangement with 15 parallel tubes in the horizontal plane. Determine the number of tubes if the tube temperature is 20°C.

E91. Saturated water vapor at 50°C ($p = 12.4$ kPa) condenses on the outside of a vertical tube with the diameter 20 mm. Determine the required tube length to attain turbulent film condensation at the bottom of the tube. The tube temperature is 30°C.

BOILING AND TWO-PHASE FLOW

E92. A container with a brass bottom (30×30 cm) contains saturated water (liquid) at 100 kPa. Determine the evaporation rate if the bottom plate has the temperature 117°C.

E93. A pot with a bottom plate of pure copper contains saturated water at atmospheric pressure. Determine the evaporation rate per unit area if the temperature at the bottom plate is 13.9°C higher than the water saturation temperature.

E94. 1.8 kg/h water boils in a copper pot with the diameter 22.5 cm at atmospheric pressure. Determine the temperature at the bottom of the pot.

E95. Consider pool boiling of a R134A at saturated conditions with the following parameters: $T = 26.7$°C, $p = 7.0$ bar, $p_{crit} = 40.7$ bar, $M = 102$ kg/kmol, $\phi = 2.5 \cdot 10^4$ W/m^2, $R_p = 2$ μm. Determine the heat transfer coefficient with Cooper's and Gorenflo's methods.

E96. Determine the critical heat flux (burnout, q_{max}) for Example 94.

E97. In an open pot with a copper bottom, water boils at atmospheric pressure. The temperature at the bottom is 108°C. Determine the bottom temperature if the heat flow rate is increased by 40%.

E98. Saturated water at 100°C is brought to boil by a copper heating element. If the surface heat flux is 400 kW/m^2, what is the surface temperature at the heating element?

E99. Saturated water at atmospheric conditions flows across a cylindrical heating rod at a velocity 2 m/s. The diameter of the rod is 6 mm and the length is 12 cm. Determine the maximum heat flow rate from the rod.

E100. A metallic body formed as a circular cylinder with the surface area 400 cm^2 and the volume 600 cm^3 is cooled by saturated water at atmospheric pressure. Determine the maximum heat flow rate when (a) the body is stationary, (b) the system moves in a way that the relative velocity for the water across the cylinder is 1 m/s.

E101. A very hot circular silver cylinder has the diameter 15 cm and the length 2 m. The cylinder is cooled by saturated water at atmospheric conditions. Determine the maximum heat flow rate when (a) the cylinder is stationary, (b) the system moves in a way that the relative velocity for the water across the cylinder is 4 m/s.

E102. A copper sphere with the volume 700 cm³ is cooled by water at atmospheric conditions. The initial temperature of the sphere is 500°C. If the sphere is stationary, what is the maximum heat flow rate during the cooling process?

E103. A mixture of water and air flows in a 2.5 m long tube having a diameter of 13 cm at a constant temperature 40°C. Determine the pressure drop with the Lockhardt–Martinelli's method when the mass flow rate for the water is 40 and 1.5 kg/s for the air.

E104. Determine the pressure drop due to friction by using the Friedel's model when R123 flows in a horizontal tube. The following data apply:

$D = 10$ mm, $p = 0.38$ bar, $T = 3°C$, $W = 0.11$ kg/s, $x = 0.45$,
$\rho_g = 2.6$ kg/m³, $\rho_f = 1520$ kg/m³, $\sigma = 0.018$ N/m,
$\mu_g = 1.26 \cdot 10^{-5}$ kg/(ms), $\mu_f = 586 \cdot 10^{-6}$ kg/(ms).

E105. 0.6 kg/s H_2O at 200°C flows in a circular tube with the diameter 5 cm. The tube wall is heated in a way that the uniform heat flux becomes $q_w = 184,000$ W/m². Calculate the wall temperature at a position where the vapor mass fraction $x = 0.2$ ($x = m_{vapor}/m_{tot}$).

E106. 1.4 kg/s H_2O at 220°C flows in a circular tube with the diameter 6 cm. The tube wall is heated by a constant uniform heat flux $q_w = 2 \cdot 10^5$ W/m². (a) Determine the wall temperature at a position where $x = 0.4$ ($x = m_{vapor}/m_{tot}$). Assume annular flow. (b) If the flow is isothermal, i.e., $x = 0.4$ for all positions in the tube, determine the pressure drop per unit length with Lockhardt–Martinelli's method.

E107. Water at 1.186 bar is vaporized in a horizontal tube with the inner diameter 12.7 mm. The mass flow rate is 140 kg/h. Determine the heat transfer coefficient with Gungor–Winterton's correlation at a position where $x = 0.3$ and $\phi = 200$ kW/m². The water is saturated at the inlet and the roughness of the tube wall is 2 μm.

HEAT EXCHANGERS

E108. 68 kg/min water is heated from 35°C to 75°C with an oil flow with a specific heat of 1.9 kJ/(kg K). A double-pipe heat exchanger with counterflow is used. The oil enters at 110°C and leaves at 75°C. Determine the heat transfer area when the overall heat transfer coefficient is 320 W/(m² K).

E109. Determine the heat transfer area for example E108, if a shell-and-tube heat exchanger with one shell pass and two tube passes is used. The water flows on the shell side and the oil on the tube side.

E110. 3.8 kg/s water is heated from 38°C to 55°C in a shell-and-tube heat exchanger, where the water flows in the tubes. The heat exchanger is designed with one shell and the medium on the shell side is hot water with the mass flow rate 1.9 kg/s entering at 93°C. To save space, the maximum allowed length of the tubes is 2.5 m. Determine the number of tube passes, the number of tubes, and the tube length. The overall heat transfer coefficient is 1400 W/(m² K) and the flow velocity in the tubes is 0.37 m/s ($d_i = 1.9$ cm, tube material thickness 1 mm).

E111. 5.8 kg/s oil flows on the shell side in a shell-and-tube heat exchanger, with one shell pass. The oil enters at 130°C and leaves at 30°C. The cooling medium is cold water which enters the tubes at 6°C and leaves at 22°C. To save space, the maximum allowed length of the tubes is 2.0 m. Determine the number of tube passes, the number of tubes, and the tube length. The overall heat transfer coefficient is 850 W/(m² K) and the flow velocity in the tubes is 0.4 m/s ($d_i = 18$ mm, tube material thickness 1 mm). $c_{p_{olja}} = 2.2$ kJ/(kg K), $c_{p_{H_2O}} = 4.2$ kJ/(kg K), $\rho_{H_2O} = 10^3$ kg/m³.

E112. 15 kg/s water is heated from 32°C to 48°C in a shell-and-tube heat exchanger, with one shell pass. The flow velocity of the water in the tubes is 0.8 m/s. At the shell side 8.0 kg/s hot oil enters at 99°C. To save space, the maximum allowed length of the tubes is 5.6 m. Determine the number of tube passes, the number of tubes, and the tube length. The flow velocity in the tubes is 0.8 m/s ($d_i = 21$ mm, tube material thickness 1.5 mm). The heat transfer coefficient on the tube side is 2400 W/(m² K) and on the shell side is 1200 W/(m² K). The thermal conductivity of the tube material is $\lambda = 35$ W/(m K). $c_{p,water} = 4.2$ kJ/(kg K) and $c_{p,oil} = 2.3$ kJ/(kg K).

E113. Consider the heat exchanger in Example 109. Determine the outlet temperature of the water and the heat transfer rate if the mass flow rate is changed to 40 kg/min.

E114. A finned tube heat exchanger, see figure below, is used for the heating of 2.36 m³/s air at atmospheric pressure from 16°C to 30°C. Hot water enters the tubes at 82°C. The total heat transfer area is 9.3 m² and the overall heat transfer coefficient is 230 W/(m² K). Determine the outlet temperature for the water and the heat flow rate.

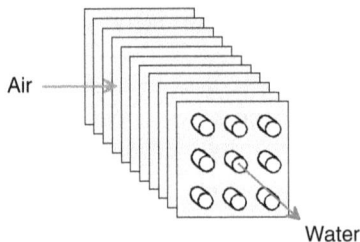

Air

Water

E115. A counterflow double-pipe heat exchanger is used for heating 1.25 kg/s water from 35°C to 80°C by using an oil ($c_p = 2.0$ kJ/(kg K)) which enters at 150°C and leaves at 85°C. The overall heat transfer coefficient is 850 W/(m² K).

An alternate solution for the heating problem is considered, by using two smaller units coupled in series on the water side and parallel on the oil side, see figure below. The oil flow is divided equally between the units and the overall heat transfer coefficient is 850 W/(m² K). If the smaller heat exchangers cost 20% more per unit area, which case is the most economic, one big heat exchanger or two smaller units?

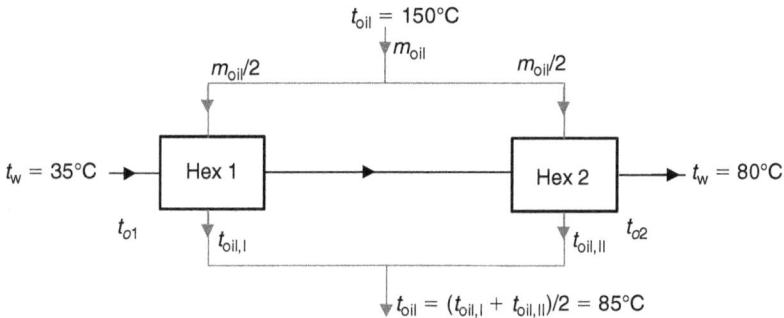

E116. A cross flow heat exchanger with both fluids unmixed is used for the cooling of 0.4 kg/s air from 275°C to 100°C. The cooling medium enters at 50°C and leaves at 75°C. After 8 months' operation, the overall heat transfer coefficient is reduced to 55% (of the original) due to fouling. Determine the outlet temperatures for the air and the water assuming the other data are unchanged.

E117. A shell-and-tube heat exchanger with one shell pass and one tube pass, where water condenses on the shell side at 1 bar, is used for heating of 0.3 m³/s water from 17°C to 40°C. The heat transfer coefficient on the shell side is 10.2 kW/(m² K) and the available pumping power limits the pressure drop on the tube side to 96.5 kPa. Determine the tube length and the number of tubes when steel tubes with the inner diameter 26.6 mm and the outer diameter of 33.4 mm are used.

E118. 10 kg/s benzene (c_p (benzene) = 1.76 kJ/(kg K)) is cooled in a heat exchanger from 70°C to 35°C by the use of 10 kg/s water which enters at 20°C. If the overall heat transfer coefficient is 400 W/(m² K), estimate the required heat transfer area for the following arrangements: (a) cross flow, one pass, both fluids unmixed, (b) shell-and-tube heat exchanger with one shell pass and two tube passes, and (c) cross flow, one pass, water mixed, benzene unmixed.

E119. 5.8 kg/s oil flows on the shell side in a shell-and-tube heat exchanger with two shell passes and four tube passes. The oil enters at 180°C and leaves at

35°C. The oil is cooled by water which flows in the tubes. The water enters at 30°C and leaves at 50°C. Determine the heat exchanger area if the overall heat transfer coefficient $U = 420 \, \text{W/(m}^2 \, \text{K)}$ and $c_{p_{oil}} = 2.3 \, \text{kJ/(kg K)}$.

E120. Water with the inlet temperature 15°C is used for cooling 10 kg/s ammonia-liquid in a shell-and-tube heat exchanger with one shell pass and two tube passes. The ammonia enters at 120°C and leaves at 40°C. The overall heat transfer coefficient $U = 1500 \, \text{W/(m}^2 \, \text{K)}$ and the heat transfer area is 90 m². Determine the water flow rate in the tubes.

E121. Design a shell-and-tube heat exchanger (1 shell pass, 1 tube pass) where saturated water vapor ($\dot{m} = 39.6$ ton/h) at 0.35 bar condenses completely on the shell side. On the tube side, 1000 ton/h water enters at 20°C and leaves at 40°C. The length of the steel tubes is 4 m and the tube inner diameter is 19 mm with a material thickness of 2 mm. The tubes are arranged in a quadratic in-line-arrangement with the pitch $S/D = 1.75$. The thermal resistance in the tube material can be neglected.

E122. 10 kg/s liquid ammonia is cooled from 120°C to 40°C in a shell-and-tube heat exchanger with 2 shell passes and 4 tube passes. The tubes are placed in an in-line-arrangement with the pitches $S_T = S_L = 32$ mm. The tube length is 3 m, the outer diameter is 26.6 mm, and the wall thickness is 2 mm. The maximum flow velocity for the ammonia, which flows in the tubes, is 2 m/s. The cooling medium is water which enters at 15°C. Determine the number of tubes, the water mass flow rate, and the water outlet temperature.

ANSWERS

1. (a) $t^2 + 20 \cdot t = -45.5 \cdot 10^3 \cdot x + 2400$; (b) $t = -700 \cdot x + 40$
2. $q = \dfrac{\lambda_2 - \lambda_1}{\ln(\lambda_2/\lambda_1)} \cdot \dfrac{t_1 - t_2}{d}$
3. (a) 82.5°C; (b) 57°C
4. See Figure 6 in the textbook
5. $t_{gas} \approx 185°C$
6. 13.7 cm
7. (a) advantageous; (b) may be advantageous
8. (a) $b = 5.7$ cm, $L = 22.7$ cm; (b) advantageous; (c) $\varphi = 0.63$
9. (a) nej; (b) $\eta = 5.05$
10. $\eta_{stainless} = 2.1$
11. $\varphi_{actual}/\varphi_{optimal} = 0.94$
    $\eta_{actual}/\eta_{optimal} = 1.03$
    $\dot{Q}_{actual}/\dot{Q}_{optimal} = 1.03$
12. $\eta_a/\eta_{opt} = 0.84$, $\varphi_a/\varphi_{opt} = 1.13$
13. $\dot{Q}/z = 67.6$ W/m
14. $\varphi_2/\varphi_1 = 0.58$
15. $\dot{Q}_{withfins}/\dot{Q}_{withoutfins} = 2.8$
16. 87°C
17. 33.7°C
18. $t_{v_2} = 83°C$
19. $N = 0.45$ m
20. $t \approx 147°C$
21. $\tau \approx= 80$ s., $t_{centre} \approx= 310°C$
22. $\tau \approx 1.5$ h
23. $\tau \approx 10$ s
24. $\tau \approx 1100$ s
25. (a) 118.5°C; (b) 79.3°C
26. $Q/A \approx 21$ MJ/m²
27. $Q/A \approx 23$ MJ/m²
28. $Q/A \approx 10.7$ J/m²
29. $Q/A \approx 41$ M/m²
30. $t_{interface} = 142.3°C$
31. $\delta_{flow} = 6.1$ mm, $\delta_{thermal} = 2.3$ mm
32. $C_F\sqrt{Re_x} = 1.52$ $Nu_x/\sqrt{Re_x} = 0.97$
33. (a) $\alpha = 7.5$ W/(m² K), $t = 26.6°C$; (b) $\dot{Q} = 119$ W

34. $P_D = 1.07 \rho c L \sqrt{vc}$ N/m

    $\dot{Q}/b = 23.4 \cdot L^{2/3} \cdot \lambda_f \sqrt{c/v}$ W/m

35. $q_w = 43$ W/m^2

36. $\dot{Q}/b = 880$ W/m

37. (a) $q_w = 23.8$ W/m^2;   (b) $q_w = 590$ W/m^2

38. $\dot{Q} = 8.4$ kW/m

39. 780 W

40. $\dot{m}/A = 1.8 \cdot 10^{-3}$ kg/m^2 s

41. $\dot{Q}_{v_w} = 122$ W, $\dot{Q}_o = 197$ W, $\dot{m}_w/A = 1.5 \cdot 10^{-3}$ kg/m^2 s

42. $\dot{m}/A = 0.04$ kg/m^2 s

43. $q_w = 407.5 + 437.3 \cdot 10^6 \cdot x^2 - 267.7 \cdot 10^{18} \cdot x^6$; q_w $(x = 1$ mm$) = 577.1$ W/m^2

44. $L \approx 19\text{–}34$ m

45. $t_w = 64.4°C$

46. $L \approx 0.028$ m

47. $L \approx 115$ m

48. $t_{B_2} = 23.1°C$

    Are the assumptions correct?

49. (a) 47.5°C;   (b) 64.5°C

50. Prandtl–Taylor: 12,276–12,782 W/m^2 K; Reynolds–Colburn: 8700–9160 W/m^2 K; Dittus–Boelter: 8710–10,090 W/m^2 K

51. $\alpha_{D/L\text{—correction}} = 127$ W/m^2 K

    $\alpha_{\text{Dittus–Boelter}} = 95$ W/m^2 K

52. $t = 87°C$

53. $L \approx 1.87$ m

54. $U \approx 32.5$ m/s

55. $\Delta\dot{Q} \approx 3.1$ kW

56. $\dot{Q}/L \approx 2.2$ kW/m

57. 46.8°C

58. $U \approx 10$ m/s

59. $\dot{Q} \approx 29$ kW, $\alpha \approx 142$ W/m^2 K, $\Delta p \approx 200$ Pa

60. $t_{\text{air,exit}} = 45°C$, $\Delta p \approx 2.4$ kPa

61. $N_T = 7$; $\dot{Q} \approx 100$ kW, $\Delta p \approx 3.0$ kPa

    $N_T = 9$; $\dot{Q} \approx 110$ kW, $\Delta p \approx 1.9$ kPa

62. 50–80 s

63. (a) $\dot{Q} = 65$ W;   (b) $\dot{Q} = 80$ W

64. About the same

65. 88 W

66. $t \approx 52°C$

67. $t_{B_2} = 79.9°C$

68. (a) $\dot{Q} \approx 26.5\,W$;   (b) $\dot{Q} \approx 122\,W$

69. $\dot{Q} \approx 15.7\,W$

70. $\dot{Q} \approx 21\,W$, $t_{outer\ glass} = 3.7°C$

71. $\Delta t \approx 19°C$

72. (a) $q \approx 140\,W/m^2\,K$;   (b) $q \approx 61\,W/m^2$

73. $q_{net} \approx 113\,kW/m^2$

74. $\dot{Q} = 18.3\,kW$

75. $\dot{Q}_1 = 14.4\,kW$, $\dot{Q}_2 = 2.6\,kW$, $\dot{Q}_3 = -17\,kW$

76. $t \approx 44°C$

77. $T_w \approx 326\,K$

78. $\dot{Q} = 49\,kW$

79. $29\,kW/m^2$

80. $\dot{Q}_1/A \approx 69.3\,kW/m^2$, $\dot{Q}_2/A \approx 45.7\,kW/m^2$

81. $\dot{Q}/A \approx 61\,kW/m^2$

82. $\dot{Q} \approx 110\,kW$

83. $\dot{Q} \approx 2.8\,kW$, $\dot{m} = 1.25 \cdot 10^{-3}\,kg/s$

84. $L \approx 0.48\,m$

85. $\alpha_m = 9018\,W/m^2\,K$, $\dot{m} = 20.1 \cdot 10^{-3}\,kg/s$

86. $L \approx 4.6 - 5.3\,m$

87. $\alpha_m = 5240\,W/m^2\,K$, $\dot{m} = 1.23 \cdot 10^{-3}\,kg/s$

88. $\alpha_m = 5032\,W/m^2\,K$, $\dot{m} = 0.7\,kg/s$

89. $\dot{m} = 2.72\,kg/s$

90. $N = 15$

91. $L \approx 11\text{--}13\,m$

92. $\dot{m} = 0.31\,kg/s$

93. $\dot{m} = 0.174\,kg/s$

94. $t_w = 105.9°C$

95. $\alpha \approx 7.7\,kW/m^2\,K$, $\alpha \approx 9.1\,kW/m^2\,K$

96. $q_{max} = 1.12\,MW/m^2$

97. $109°C$

98. $t_w = 114°C$

99. $q_{max} = 4.2\,MW/m^2$

100. (a) $q_{max} = 1.0\,MW/m^2$;   (b) $q_{max} = 1.7\,MW/m^2$

101. (a) $Q_{max} = 940\,kW$;   (b) $Q_{max} = 3610\,kW$

102. $q_{max} = 940\,kW/m^2$

103. $\Delta p \approx 27\,kPa$

104. $\Delta p \approx 227\,kPa/m$

105. $t_w = 210.5°C$
106. (a) 227°C;   (b) 1660 Pa/m
107. 37,100 W/m² K
108. $A \approx 16\,m^2$
109. $A \approx 20\,m^2$
110. Two tube passes, 37 tubes/pass, $L = 1.6\,m$
111. $n \approx 187$, 2 tube passes, $L = 1.25\,m$
112. $n = 55$, 4 tube passes, $L = 5.5\,m$
113. $\dot{Q} = 148\,kW$
     $t_{c_{out}} = 88°C$
114. $\dot{Q} = 41\,kW$
     $t_{v_{out}} = 21.2°C$
115. (hex price 2) $= 1.3 \times$ (hex price 1)
116. $t_{h_{out}} = 142°C$
     $t_{c_{out}} = 69°C$
117. $n = 80$, $L = 11.3\,m$
118. (a) 76.5 m²;   (b) 90 m²;   (c) 85.5 m²
119. $A = 133.4\,m^2$
120. 3.2 kg/s
121. $n = 1100$, $D_{shell} > 1.36\,m$
122. $n = 102$, $\dot{m}_v = 2.1$ kg/s, $t_{v_{out}} = 65°C$

Index

WITPRESS ...for scientists by scientists

Advanced Computational Methods in Heat Transfer XII

Edited by: **B. SUNDÉN**, *Lund University, Sweden;* **C.A. BREBBIA**, *Wessex Institute of Technology, UK and* **D. POLJAK**, *University of Split, Croatia*

Containing papers presented at the twelfth in a series of biennial international conferences on Advanced Computational Methods and Experiments in Heat Transfer, this book covers the latest developments in this important field. Heat Transfer plays a major role in emerging application fields such as sustainable development and the reduction of greenhouse gases, as well as micro- and nano-scale structures and bio-engineering. Typical applications include heat exchangers, gas turbine cooling, turbulent combustion and fires, electronics cooling, melting and solidification.

The nature of heat transfer problems is complex, involving many different simultaneously occurring mechanisms (e.g., heat conduction, convection, turbulence, thermal radiation. phase change). Their complexity makes it imperative that we develop reliable and accurate computational methods to replace or complement expensive and time-consuming experimental trial and error work. Tremendous advances have been achieved during recent years due to improved numerical solutions of non-linear partial differential equations and more powerful computers capable of performing efficient and rapid calculations. Nevertheless, to further progress, it will also be necessary to develop theoretical and predictive computational procedures – both basic and innovative – and in applied research. Accurate experimental investigations are needed to validate the numerical calculations.

The book includes such topics as: Heat Exchangers; Advances in Computational Methods; Natural and Forced Convection and Radiation; Multiphase Flow Heat Transfer; Modelling and Experiments; Heat Transfer in Energy Producing Devices; Nanotechnologies; Bioheat Transfer; Heat Recovery; Advanced Thermal Materials; Heat and Mass Transfer Problems; Radiofrequency and Thermal Dosimetry; Heat Transfer in Therapeutic Applications.

WIT Transactions on Engineering Sciences, Vol 75
ISBN: 978-1-84564-602-8   eISBN: 978-1-84564-603-5
Forthcoming 2012 / apx 350pp / apx £150.00

WITPRESS *...for scientists by scientists*

Solar Thermal and Biomass Energy

G. LORENZINI and *C. BISERNI*, *University of Bologna, Italy and* **G. FLACCO**, *Bologna, Italy*

This book covers renewable energy sources. In particular, it reviews the state of the art in thermal solar techniques and biomasses. The first topic treated is solar radiation. The authors discuss the possible utilisation of solar energy through thermal solar techniques, starting with low-temperature techniques and moving on successively to medium- and high-temperature ones. This discussion includes, of course, explanations of those devices and materials employed in using those technologies.

The last part of the book covers biomasses. Starting with the definition of this wide-ranging field, the authors successively treat how to obtain energy from biomasses and all the environmental aspects linked to this topic. Very few books treat so broadly the state of the art in both thermal solar and biomass energy, especially the latter, a topic about which there is considerable confusion due to the very wide range of related technologies.

ISBN: 978-1-84564-147-4   eISBN: 978-1-84564-386-7
Published 2010 / 224pp / £85.00

Thermal Engineering in Power Systems

Edited by: **R.S. AMANO**, *University of Wisconsin-Milwaukee, USA and* **B. SUNDÉN**, *Lund University, Sweden*

Research and development in thermal engineering for power systems are of significant importance to many scientists who work in power-related industries and laboratories. This book focuses on a variety of research areas including Components of Compressors and Turbines that are used for both electric power systems and aero engines, Fuel Cells, Energy Conversion, and Energy Reuse and Recycling Systems.

To be competitive in today's market, power systems need to reduce operating costs, increase capacity and deal with many other tough issues. Heat Transfer and fluid flow issues are of great significance to power systems. Design and R&D engineers in the power industry will therefore find this state-of-the-art book on those issues very useful in their efforts to develop sustainable energy systems.

Series: Developments in Heat Transfer, Vol 22
ISBN: 978-1-84564-062-0   eISBN: 978-1-84564-323-2
Published 2008 / 416pp / £137.00

WITPRESS *...for scientists by scientists*

Computational Fluid Dynamics and Heat Transfer

Emerging Topics

Edited by: **R.S. AMANO**, *University of Wisconsin-Milwaukee, USA and* **B. SUNDÉN**, *Lund University, Sweden*

Heat transfer and fluid flow issues are of great significance. This state-of-the-art edited book with reference to new and innovative numerical methods is an important reference for researchers in both academia and research organizations, as well as industrial scientists and college students.

The book provides comprehensive chapters on research and developments in emerging topics in computational methods, e.g., the finite volume method, finite element method, and turbulent flow computational methods. Fundamentals of the numerical methods, comparison of various higher-order schemes for convection-diffusion terms, turbulence modelling, the pressure-velocity coupling, mesh generation and the handling of arbitrary geometries are presented. Results from engineering applications are provided. Chapters have been co-authored by eminent researchers.

Series: Development in Heat Transfer, Vol 23
ISBN: 978-1-84564-144-3   e-ISBN: 978-1-84564-402-4
Published 2011 / 512pp / £195.00

Heat Transfer in Gas Turbines

Edited by: **B. SUNDÉN**, *Lund University, Sweden and* **M. FAGHRI**, *University of Rhode Island, USA*

Containing invited contributions from some of the most prominent specialists working in this field, this unique title reflects recent active research and covers a broad spectrum of heat transfer phenomena in gas turbines. All of the chapters follow a unified outline and presentation to aid accessibility and the book provides invaluable information for both graduate researchers and R&D engineers in industry and consultancy.

Partial Contents: Heat Transfer Issues in Gas Turbine Systems; Combustion Chamber Wall Cooling – The Example of Multihole Devices; Conjugate Heat Transfer – An Advanced Computational Method for the Cooling Design of Modern Gas Turbine Blades and Vanes; Enhanced Internal Cooling of Gas Turbine Airfoils; Computations of Internal and Film Cooling; Heat Transfer Predictions of Stator/Rotor Blades; Recuperators and Regenerators in Gas Turbine Systems.

Series: Developments in Heat Transfer, Vol 8
ISBN: 978-1-85312-666-6   ISBN-10: 1-85312-666-7
Published 2001 / 536pp / £175.00

WIT*PRESS* ...*for scientists by scientists*

Heat Transfer in Food Processing

Recent Developments and Applications

Edited by: **S. YANNIOTIS**, *Agricultural University of Athens, Greece and* **B. SUNDÉN**, *Lund University, Sweden*

Heat transfer is one of the most important and most common engineering disciplines in food processing. There are many unit operations in the food industry where steady-state or unsteady-state heat transfer is taking place. These operations are of primary importance and affect the design of equipment as well as safety, nutritional and sensory aspects of the product.

The chapters in this book deal mainly with: heat transfer applications; methods that have considerable physical property variations with temperature; methods not yet widespread in the food industry; or methods that are less developed in the food engineering literature. The application of numerical methods has received special attention with a separate chapter, as well as emphasis in almost every chapter. A chapter on artificial neural networks (ANN) has also been included since ANN is a promising alternative to conventional methods for modelling, optimisation, etc in cases where a clear relationship between the variables is not known, or the system is too complex to be modelled with conventional mathematical methods.

Series: Developments in Heat Transfer, Vol 21
ISBN: 978-1-85312-932-2   eISBN: 978-1-84564-292-1
Published 2007 / 288pp / £95.00

WIT*Press*
Ashurst Lodge, Ashurst, Southampton,
SO40 7AA, UK.
Tel: 44 (0) 238 029 3223
Fax: 44 (0) 238 029 2853
E-Mail: witpress@witpress.com

www.ingramcontent.com/pod-product-compliance
Lightning Source LLC
Chambersburg PA
CBHW060800220326
41598CB00022B/2501